Global Biodiversity

Volume 2

Selected Countries in Europe

Global Biodiversity

Volume 2

Selected Countries in Europe

Edited By
T. Pullaiah, PhD

Apple Academic Press Inc.
3333 Mistwell Crescent
Oakville, ON L6L 0A2
Canada

Apple Academic Press Inc.
9 Spinnaker Way
Waretown, NJ 08758
USA

© 2019 by Apple Academic Press, Inc.

First issued in paperback 2021

Exclusive worldwide distribution by CRC Press, a member of Taylor & Francis Group

No claim to original U.S. Government works

Global Biodiversity, Volume 2: Selected Countries in Europe
ISBN 13: 978-1-77463-132-4 (pbk)
ISBN 13: 978-1-77188-717-5 (hbk)

Global Biodiversity, 4-volume set

ISBN 13: 978-1-77188-751-9 (hbk)

Library and Archives Canada Cataloguing in Publication

Global biodiversity (Oakville, Ont.) Global biodiversity / edited by T. Pullaiah, PhD.

Includes bibliographical references and indexes.
Contents: Volume 2. Selected countries in Europe.
Issued in print and electronic formats.
ISBN 978-1-77188-717-5 (v. 2: hardcover).--ISBN 978-0-42948-775-0 (v. 2 : PDF)

1. Biodiversity--Asia. 2. Biodiversity--Europe. 3. Biodiversity--Africa. I. Pullaiah, T., editor II. Title.

QH541.15.B56G66 2018   578.7   C2018-905091-8   C2018-905092-6

CIP data on file with US Library of Congress

Apple Academic Press also publishes its books in a variety of electronic formats. Some content that appears in print may not be available in electronic format. For information about Apple Academic Press products, visit our website at **www.appleacademicpress.com** and the CRC Press website at **www.crcpress.com**

Contents

Contents

About the Editor

T. Pullaiah, PhD
Former Professor, Department of Botany,
Sri Krishnadevaraya University, Anantapur, Andhra Pradesh, India,
E-mail: pullaiah.thammineni@gmail.com

T. Pullaiah, PhD, is a former Professor at the Department of Botany at Sri Krishnadevaraya University in Andhra Pradesh, India, where he has taught for more than 35 years. He has held several positions at the university, including Dean, Faculty of Biosciences, Head of the Department of Botany, Head of the Department of Biotechnology, and Member of Academic Senate. He was President of Indian Botanical Society (2014), President of the Indian Association for Angiosperm Taxonomy (2013) and Fellow of Andhra Pradesh Akademi of Sciences. He was awarded the Panchanan Maheshwari Gold Medal, the Dr. G. Panigrahi Memorial Lecture award of the Indian Botanical Society and Prof. Y.D. Tyagi Gold Medal of the Indian Association for Angiosperm Taxonomy, and a Best Teacher Award from Government of Andhra Pradesh. Under his guidance 54 students obtained their doctoral degrees. He has authored 46 books, edited 17 books, and published over 330 research papers, including reviews and book chapters. His books include *Ethnobotany of India* (5 volumes published by Apple Academic Press), *Flora of Andhra Pradesh* (5 volumes), *Flora of Eastern Ghats* (4 volumes), *Flora of Telangana* (3 volumes), *Encyclopaedia of World Medicinal Plants* (5 volumes, 2nd edition), and *Encyclopaedia of Herbal Antioxidants* (3 volumes). He was also a member of Species Survival Commission of the International Union for Conservation of Nature (IUCN). Professor Pullaiah received his PhD from Andhra University, India, attended Moscow State University, Russia, and worked as a Postdoctoral Fellow during 1976–78.

Contributors

Guram Aleksidze
Georgian Academy of Agricultural Sciences, I. Javakhishvili str. #51, 0102, Tbilisi, Georgia,
E-mail: guram_aleksidze@yahoo.com

Senka Barudanovic
Center for Ecology and Natural Resources Academician Sulejman Redzic, Faculty of Science,
University of Sarajevo, 71 000 Sarajevo, Bosnia-Herzegovina, E-mail: sebarudanovic@gmail.com

Kjetil Bevanger
Scientific Advisor and Senior Research Scientist, The Norwegian Institute for Nature Research (NINA),
Pb. 5685 Sluppen, NO–7485 Trondheim, Norway, E-mail: Kjetil.bevanger@nina.no

Theophanis Constantinidis
University of Athens, Faculty of Biology, Department of Ecology and Systematics,
157 84 Athens, Greece, E-mail: constgr@biol.uoa.gr

Gianniantonio Domina
Department of Agricultural, Food and Forest Sciences, University of Palermo,
viale delle Scienze bldg.4, 90138, Palermo, Italy, E-mail: gianniantonio.domina@unipa.it

Ali A. Dönmez
Department of Biology, Faculty of Science, Hacettepe University, 06800 Beytepe, Ankara, Turkey,
Email: donmez@hacettepe.edu.tr

Ulf Gärdenfors
Professor of conservation biology and Deputy Director of the Swedish Species Information Centre,
University of Agricultural Sciences, Uppsala, Sweden, E-mail: Ulf.Gardenfors@slu.se

Anatoli Giorgadze
Georgian Academy of Agricultural Sciences, I. Javakhishvili str. #51, 0102, Tbilisi, Georgia,
E-mail: anatoli5@mail.ru

Philippe Grandcolas
Director, Institut de Systématique, Evolution, Biodiversité, Muséum national d'Histoire naturelle,
CNRS, UPMC, EPHE, France, E-mail: pg@mnhn.fr

Anna Guttová
Plant Science and Biodiversity Centre, Slovak Academy of Sciences, Dúbravská cesta 9, 845 23
Bratislava, Slovakia, E-mail: anna.guttova@savba.sk

Alica Hindáková
Plant Science and Biodiversity Centre, Slovak Academy of Sciences, Dúbravská cesta 9, 845 23
Bratislava, Slovakia, E-mail: alica.hindakova@savba.sk

Givi Japaridze
Georgian Academy of Agricultural Sciences, I. Javakhishvili str. #51, 0102, Tbilisi, Georgia,
E-mail: japaridze.givi@yahoo.com

Tamar Kacharava
Georgian Technical University, Kostava 77, 0175, Tbilisi, Georgia, E-mail: thamkach@mail.ru

Jaromír Kučera
Plant Science and Biodiversity Centre, Slovak Academy of Sciences, Dúbravská cesta 9, 845 23
Bratislava, Slovakia, E-mail: jaromir.kucera@savba.sk

Viktor Kučera
Plant Science and Biodiversity Centre, Slovak Academy of Sciences, Dúbravská cesta 9, 845 23 Bratislava, Slovakia, E-mail: viktor.kucera@savba.sk

Anastasios Legakis
University of Athens, Faculty of Biology, Zoological Museum, 157 84 Athens, Greece, E-mail: alegakis@biol.uoa.gr

Marta Mútňanová
State Nature Conservancy of Slovak Republic, Tajovského 28B, 974 01 Banská Bystrica, Slovakia, E-mail: marta.mutnanov@sopsr.sk

Biljana Panjković
Institute for Nature Conservation of Vojvodina Province, Serbia, E-mail: biljana.panjkovic@pzzp.rs

Panos V. Petrakis
Hellenic Agricultural Organization Demeter, Institute for Mediterranean Forest Research, Laboratory of Entomology, Terma Alkmanos, 115 28 Athens, Greece, E-mail: pvpetrakis@fria.gr

Ranee Om Prakash
Curator-Flowering Plants, Department of Life Sciences, Natural History Museum, South Kensington, SW7 5BD, United Kingdom, E-mail: r.prakash@nhm.ac.uk

Slobodan Puzović
Institute for Nature Conservation of Vojvodina Province, Serbia, E-mail: slobodan.puzovic@pzzp.rs

Fred Rumsey
Senior Curator-in Charge, British, European and Historic Herbaria, Department of Life Sciences, Natural History Museum, South Kensington, SW7 5BD, United Kingdom

Jozef Šibík
Plant Science and Biodiversity Centre, Slovak Academy of Sciences, Dúbravská cesta 9, 845 23 Bratislava, Slovakia, E-mail: jozef.sibik@savba.sk

Libor Ulrych
State Nature Conservancy of Slovak Republic, Tajovského 28B, 974 01 Banská Bystrica, Slovakia, E-mail: libor.ulrych@sopsr.sk

Peter Urban
Matej Bel University, Faculty of Natural Sciences, Department of Biology and Ecology, Tajovského 40, 974 01 Banská Bystrica, Slovakia

Zoltán Varga
Department of Evolutionary Zoology, University of Debrecen, Hungary, E-mail: varga.zoltan@science.unideb.hu

Sedat V. Yerli
Department of Biology, Faculty of Science, Hacettepe University, 06800 Beytepe, Ankara, Turkey

Marzio Zapparoli
Department for Innovation in Biological, Agrofood and Forest systems, University of Tuscia, Via San Camillo de Lellis s.n.c. – 01100 Viterbo, Italy, E-mail: zapparol@unitus.it

Abbreviations

ALERC	Association of Local Environmental Records Centers
APV	Autonomous Province of Vojvodina
ARC	Amphibian and Reptile Conservation
BBS	British Bryological Society
BCT	Bat Conservation Trust
BDS	British Dragonfly Society
BPS	British Pteridological Society
BRC	Biological Records Centre
BSBI	Botanical Society of Britain and Ireland
BTO	British Trust for Ornithology
CEDaR	Centre for Environmental Data and Recording
CEE	Central Eastern European
CEH	Centre for Ecology & Hydrology
CET	Central England Temperature
CIEEM	Chartered Institute of Ecology and Environmental Management
CR	critically endangered
CWR	crop wild relatives
DD	deficient data
EU	European Union
EX	extinct taxa
GBIF	Global Biodiversity Information Portal
HBMS	Hungarian Biodiversity Monitoring System
IAS	invasive alien species
IBA	important bird area
INCS	Institute for Nature Conservation of Serbia
IPA	important plant areas
JNCC	Joint Nature Conservation Committee
MBA	Marine Biological Association
NBN	National Biodiversity Network
NCI	Natural Capital Index
NHM	Natural History Museum
NNSS	Non-Native Species Secretariat
NPP	net primary production
NT	near threatened
NVC	National Vegetation Classification

OTs	overseas territories
PBA	prime butterfly areas
PEEN	Pan-European Ecological Network
PTES	People's Trust for Endangered Species
RCAT	Research Centre for Agrobotany at Tápiószele
RIPP	Research Institute of Plant Production
RLI	red list index
RSPB	Royal Society for the Protection of Birds
SAC	special areas of conservation
SAHFOS	Sir Alister Hardy Foundation for Ocean Science
SPAs	special protection areas
SRS	special responsibility species
STI	Swedish Taxonomy Initiative
UKBAP	UK Biodiversity Action Plan
WDC	Whale and Dolphin Conservation
WWF	World Wildlife Fund
WWT	Wildfowl & Wetlands Trust
ZSL	Zoological Society of London

Preface

The term 'biodiversity' came into common usage in the conservation community after the 1986 National Forum on BioDiversity, held in Washington, DC, and publication of selected papers from that event, titled *Biodiversity*, edited by Wilson (1988). Wilson credits Walter G. Rosen for coining the term. Biodiversity and conservation came into prominence after the Earth Summit, held at Rio de Janeiro in 1992. Most of the nations passed biodiversity and conservation acts in their countries. Biodiversity is now the buzzword of everyone from parliamentarians to laymen, professors, and scientists to amateurs. There is a need to take stock on biodiversity of each nation. The present attempt is in this direction.

The main aim of the book is to provide data on biodiversity of each nation. It summarizes all the available data on plants, animals, cultivated plants, domesticated animals, their wild relatives, and microbes of different nations. Another aim of the book series is to educate people about the wealth of biodiversity of different countries. It also aims to project the gaps in knowledge and conservation. The ultimate aim of the book is for the conservation of biodiversity and its sustainable utilization.

The present series of the four edited volumes is a humble attempt to summarize the biodiversity of different nations. Volume 1 covers Biodiversity of Selected Countries in Asia, Volume 2 presents Biodiversity in Europe, Volume 3 looks at Biodiversity in Africa, and Volume 4 contains Biodiversity in the Americas and Australia. In these four volumes, each chapter discusses the biodiversity of one country. Competent authors have been selected to summarize information on the various aspects of biodiversity. This includes brief details of the country, ecosystem diversity/ vegetation/biomes, and species diversity, which include plants, animals and microbes. The chapters give statistical data on plants, animals, and microbes of that country, and supported by relevant tables and figures. They also give accounts on genetic diversity with emphasis on crop plants or cultivated plants, domesticated animals, and their wild relatives. Also mentioned are the endangered plants and animals and their protected areas. The book is profusely illustrated. We hope it will be a desktop reference book for years to come.

Biodiversity of some countries could not be presented in this book. This needs explanation. I tried to contact as many specialists as possible from these countries but was unable to convince these experts to write chapter on biodiversity of their country.

The book will be useful to professors, biology teachers, researchers, scientists, students of biology, foresters, agricultural scientists, wild life managers, botanical gardens, zoos, and aquaria. Outside the scientific field it will be useful for lawmakers (parliamentarians), local administrators, nature lovers, trekkers, economists, and even sociologists.

Since it is a voluminous subject, we might have not covered the entire gamut; however, we tried to put together as much information as possible. Readers are requested to give their suggestions for improvement for future editions.

I would like to express my grateful thanks to all the authors who contributed on the biodiversity of their countries. I thank them for their cooperation and erudition.

I wish to express my appreciation and help rendered by Ms. Sandra Sickels, Rakesh Kumar, and the staff of Apple Academic Press. Their patience and perseverance has made this book a reality and is greatly appreciated.

—T. Pullaiah, PhD

CHAPTER 1

Biodiversity in Bosnia-Herzegovina

SENKA BARUDANOVIC

Center for Ecology and Natural Resources Academician Sulejman Redzic, Faculty of Science, University of Sarajevo, 71 000 Sarajevo, Bosnia-Herzegovina, E-mail: sebarudanovic@gmail.com

1.1 Geographical Information

Bosnia and Herzegovina is situated in Southeastern Europe. With a total land area of 51,129 km², it lies in the western part of the Balkan Peninsula. It has common frontiers with Republic of Croatia (931 km²), Serbia (375 km²) and Montenegro (249 km²). The northern border to Croatia follows the Sava river. The biggest part of eastern border follows the Drina river and western the Una river. In the south, Bosnia-Herzegovina has access to the Adriatic Sea.

The land is mainly hilly to mountainous with an average altitude of 500 meters. Of the total land area, 5% are lowlands, 24% hills, 42% mountains and 29% Karst area. Most of the Dinaric Alps are situated here, descending gradually in the northern direction and suddenly in the southern direction.

The climate of Bosnia-Herzegovina is influenced by the Atlantic Ocean in the west, the Mediterranean in the south, the continental masses of Europe in the north, and of Asia in the northeast and east. An average temperature in January, for example, on Bjelašnica Mountain amounts –7.2°C, while in Neum city on the Adriatic coast, it is 6.5°C. Annual precipitation in Bosnia and Herzegovina is unevenly distributed, whereas it increases from the South towards Dinaric massifs, and declines again towards peri-Pannonian margin. Snow occurs regularly in winter, covering mountain peaks over 6 months a year. The climate of Bosnia and Herzegovina is comfortable for human life and health, being therefore an important natural wealth of our country.

Forests and forest soils cover 2,709,769 ha (53% of the territory). Woods cover 2,209,732 ha (about 43%), and barren land covers 500,037 ha (about 10%). The total agricultural land covers 2.5 million hectares or 0.7 ha per capita.

There are six large river basins (Una, Vrbas, Bosna, Drina, Sava, and Neretva) of which the first five belong to the Black Sea catchment and the last one to the Adriatic Sea catchment. All of the mentioned watercourses emerge under the Dinaric Alps. The underground water collection, which country has got plenty of, occurs in lose depositions, around large river beds, karst fissures, trenches, and caves. Thermal and mineral springs occur around ingenious bedrocks and tectonic disruptions. The hottest thermal spring (with the water temperature of +58°C) is Ilidža's spa, and other well-known mineral springs are Kiseljak, Kakanj, Srednje, Busovača, Srebrenica, Žepa, Tešanj, Maglaj, Žepče, etc. (Redžić et al., 2008).

1.2   Ecosystem Diversity

Numerous factors have caused a high degree of biodiversity in Bosnia and Herzegovina. Amongst them are different climate effects, complex orography, complex geological substrate, and the mosaic of soils. Very diverse natural and semi-natural ecosystems are widespread in our country (Barudanović et al., 2015).

Only those types of ecosystems, which reflect a high degree of species diversity, are described in this text; that includes the presence of numerous endemic and rare species on the Dinaric Alps and the Balkan Peninsula.

1.2.1   Ecosystems on Permanent Snow Patches

Ecosystems on permanent snow patches are fragmentarily developed in Bosnia and Herzegovina. They prevail at the highest mountain peaks, in habitats sheltered from Sunrays, usually on the slopes directed towards the North. These habitats are small in size and are usually surrounded by ecosystems of screes and rocks. The vegetative period is quite short and lasts three months at most. The ecoclimate is humid, cold, and nival with definitive temperature extremes in the winter period. Mean annual temperature is between 0 and 2°C. The geological substrate is made of middle- and upper-Triassic limestone, dolomite or rocks with a silicate composition. The soil is of the regosol type, and more rarely of very shallow calcomelanosol or rankers (Lakušić et al., 1979, 1982, 1987).

The ecosystems on permanent snow patches form the class of *Salicetea herbaceae,* which is further divided into:

1   group on a carbonate substrate (order *Salicetalia retusae-serpillifo-liae,* alliance *Salicion retusae),* with these characteristic species: *Salix retusa, Soldanella alpina, Silene pusilla, Viola zoysii, Galium aniso-phyllum, Tortella tortuosa, Veronica aphylla, Homogyne discolor, Plantago atrata, Poa alpina, Leontodon helveticus,* and others;

2 group on a silicate substrate (order *Salicetalia herbaceae*, alliance *Ranunculion crenati*), with these characteristic species: *Ranunculus crenatus, Festuca halleri, Sieversia montana, Gnaphalium supinum, Sedum alpestre, Soldanella alpina, Phleum gerardi, Veronica alpina, Saxifraga androsacea, Taraxacum alpinum, Myosotis alpestris, Chrysanthemum alpinum*, and others.

1.2.2 Ecosystems of Rock Crevices

Ecosystems of rock crevices represent a significant part of a specific biodiversity of Bosnia and Herzegovina. On the vertical profile, they are distributed from high mountain peaks, through canyons where the most of rivers in Bosnia and Herzegovina flows, to crevices of seaside Dinarides on the south of the country.

Ecosystems of rock crevices are developed in habitats with definitive slopes (40–90°), which are exposed to explicit variations of seasonal and daily temperatures as well as gusts of wind. The types of rock and soil, as well as conditions of humidity have a significant impact on the survival of certain plant species in these habitats. The geological substrate is made of carbonate and silicate rocks of different age and origin, and the soils are shallow and poorly developed (Barudanović et al., 2015).

A high degree of endemism is the reason behind a great scientific interest in the communities of these kinds, which has resulted in a great number of described vegetative units.

The order *Potentilletalia caulescentis* has a southeastern border of its distribution in Bosnia and Herzegovina. Communities of this order appear on the northern slopes of the sub-Alpine and alpine belt.

The communities of the endemic order *Amphoricarpetalia* are common in our land (Lakušić et al., 1989; Lakušić and Redzic, 1989b). The highest numbers in the mentioned communities are counted among these species: *Achillea clavennae, Amphoricarpus autariatus, Asperula hercegovina, Aster bellidiastrum, Athamanta haynaldi, Avenastrum blavii, Campanula balcanica, Dianthus kitaibelii, Hieracium bifidum*, and others (Lakušić et al., 1969).

The vegetation of the order *Moltkeetalia petraeae* is developed in the crevices of carbonate rocks of Supra-Mediterranean and sub-Mediterranean belt, as well as in the canyons in the continental part of Bosnia and Herzegovina (Lakušić et al., 1991). The most often species found there are: *Allium saxatile, Asperula longiflora, Athamantha haynaldi, Delphinium fissum,*

Euphorbia myrsinites, Euphrasia illyrica, Iris bosniaca, Jurinea mollis, and others (Redžić et al., 2007a, 2007b, 2011).

The communities of the order *Asplenietalia septentrionalis* are developed in the crevices of silicate rocks. Center of their distribution is the mountain Vranica (Lakušić et al., 1979). By their floristic composition, they are characterized by low species diversity. The most common ones among them are: *Asplenium septentrionale, Rhodiola rosea, Sedum alpestre, Sempervivum schlechani, Thymus alpestris*, and others.

Through the investigation of canyon communities, it has been found that the exposition and slope play a small role in the differentiation of these communities. A larger role is played by the elevation and the change of climate factors. As is confirmed by a number of vegetation and floristic studies in the region of the Dinarides (Lakušić and Redžić, 1991; Redžić et al., 2008), the canyon habitats represent refugia for relict life forms, both plants and animals. Canyon communities have an ascertainment of the presence of a large number of endemic and refugial-relict species of plants.

1.2.3 Ecosystems of Screes

Ecosystems of screes are also of an azonal character. They develop anywhere between the Mediterranean belt towards the very mountain peaks. The alpine and sub-Alpine screes developed with the help of glaciers and rock tearing beneath the mountain peaks, while those in the river canyons, sub-Mediterranean and Mediterranean belt developed under the influence of the constant heating and cooling of rocks (Redžić et al., 2008). The habitat is characterized by the constant moving of the geological substrate, and the soils are quite shallow. The other ecological conditions vary in a wide range, which particularly refers to the temperature, relative air humidity, and attainable water.

The ecosystems of screes are particularly developed in the alpine and sub-Alpine belt of Bosnia and Herzegovina. Communities are here of unique floristic composition, visage, and origin. The characteristic group of species on the screes of alpine belt in the region of Maglić mountain is composed of: *Valeriana bertiscea, Euphorbia capitulata, Bunium alpinum, Cardamine glauca, Myosotis suaveolens, Edraianthus serpyllifolius* subsp. *pilosulus, Viola zoysii, Armeria canescens, Sedum megallense,* and a small number of other species (Lakušić et al., 1987).

In the sub-Alpine belt of these same mountains, the characteristic scree species are: *Saxifraga glabella, Papaver kerneri, Crepis aurea* ssp. *bosniaca, Ranunculus sartorianus, Edraianthus serpyllifolius, Veronica aphylla, Taraxacum alpinum*, and others.

The upland belt is dominated by somewhat milder climate conditions, where the screes commonly contain these species: *Drypis spinosa* ssp. *linneana, Silene marginata, Scrophularia tristis, Laserpitium marginatum, Myosotis suaveolens, Festuca bosniaca, Achillea abrotanoides, Alyssum bosniacum, Scabiosa leucophylla, Heracleum orsinii* var. *balcanicum, Hladnikia golaka, Arabis flavescens, Seseli malyi, Euphrasia liburnica,* and others.

In the zone of thermophilic oak forests and underbrushes with hornbeam, the scree habitats are often found containing *Geranium macrorrhizum, Corydalis ochroleuca, Moehringia muscosa, Veronica urticifolia,* and *Convallaria majalis* (Barudanović et al., 2015).

In the supra-Mediterranean screes, the role of edificators is played by *Stipa calamagrostis,* and is accompanied by the species from the rockery vegetations. The warmest screes of the Mediterranean belt are populated by *Peltaria alliacea.*

In the range of the class *Thlaspietea rotundifolii,* a larger number of vegetative units are described. The class is composed of five vegetative orders: *Polygonetalia alpini, Arabidetalia flavescentis, Dripetalia spinosae, Achnantheretalia calamagrostis,* and *Epilobietalia fleischeri* (Redzic et al., 2011).

1.2.4 The Mountain Meadows of Bosnia and Herzegovina

The mountain meadows of Bosnia and Herzegovina, the largest part of the Dinaric Alps, are situated in Bosnia and Herzegovina. Moving from the northwest towards the southeast, it is composed of the mountains *Plješivica, Šator, Dinara, Cincar, Čabulja, Čvrsnica, Prenj, Vranica, Zec, Majevica, Konjuh, Ozren, Romanija, Treskavica, Bjelašnica, Igman, Zelengora, Maglić, Lebršnik, Volujak, Velež, Orjen,* and a number of other smaller mountain ranges.

In the highest parts of these mountains, above the forest line, ecosystems of mountain meadows are developed. They are composed of a large number of endemic species, glacial relicts, and rare and/or endangered species.

The ecological conditions, which are prevalent in the habitats of mountain meadows, are the following: low mean annual temperature, a long period of physiological and physical droughts, strong air currents, short vegetative seasons, and large daily temperature differences (Redžić et al., 2007a; Mišić and Lakušić, 1990; Lakušić, 1976).

The first factor for the diversity of mountain meadows in our country is the diversity of rocks (Lakušić et al., 1982), which causes different pH value of the soil on their habitats. By this factor, the ecosystems of mountain meadows split into two classes: the class Juncetea trifidi on the silicate, and the class Elyno-Seslerietea on the carbonate geological substrate. The second factor for the differentiation of mountain meadows is their position on the vertical profile, what divides them into the alpine and the sub-Alpine meadows. The third factor for the diversity is the biogeographical affiliation within a certain group of Dinaric Alps in Bosnia-Herzegovina on the horizontal profile.

The mean annual temperatures in the habitats of mountain meadows vary between 0 and 2°C, and the relative mean annual air humidity is between 60 and 70%. The floristic composition of alpine meadows on acid soils is usually composed of the following species: *Sesleria comosa, Carex curvula, Potentilla aurea, Homogyne alpina, Pulsatilla alba, Luzula campestris, Juncus trifidus, Antennaria dioica, Deschampsia flexuosa, Nardus stricta,* etc. and the ones most commonly found on the carbonate soils are: *Carex laevis, Agrostis alpina, Potentilla crantzii, Thymus balcanus, Hypochoeris illyrica, Alyssum bosniacum, Sesleria tenuifolia* var. *juncifolia, Anthyllis jacquinii, Helianthemum alpestre, Linum capitatum, Gentiana dinarica, G. crispata, G. verna, Arctostaphylos uva-ursi, Dryas octopetala* (Lakušić et al., 1969).

The sub-Alpine meadows are located on the milder slopes and deeper soils than alpine. The average yearly temperature in these habitats varies between 5 and 2°C, and the relative mean annual air humidity is between 65 and 75%. The special floristic markings in these ecosystems on acid soils give us the following species: *Lilium bosniacum, Gentianella crispata, Crepis conyzifolia, Anemone narcissiflora, Potentilla aurea* ssp. *piperorum, Potentilla heptaphylla, Achillea Igulata, Poa violacea, Dianthus cruentus, Jasione orbiculata, Gentiana acaulis, Hieracium pilosella* (Bjelčić, 1966; Mišić, 1984; Muratovic et al., 2005; Šoljan et al., 2014). On the carbonate soils, the composition of sub-Alpine meadows is made of a number of endemic and tertiary-relict species, such as: *Festuca bosniaca, Stachys serotina, S. recta* ssp. *subcrenata, Dianthus tristis, Bromus erectus* var. *dissolutus, Achillea abrotanoides, Thymus balcanus, Scabiosa leucophylla, Myosotis suaveolens, Acinos alpinus* ssp. *dinaricus, Verbascum durmitoreum,* etc. (Lakušić, 1990).

1.2.5 The Ecosystems of Mountain Heaths

The ecosystems of mountain heaths have a special meaning for the people in Bosnia and Herzegovina. A large number of blueberries, cranberries, and medicinal and vitaminous species from these ecosystems represent a significant source of income for the people from the mountainous regions (Lakušić, 1969, 1979, 1982). From the biogeographical aspect, the ecosystems of heaths in our land have an extrazonal character. By their physiognomy, the ecosystems of mountain heaths represent an equivalent of a tundra biome. They are developed on the northern, northeastern or southwestern exposure in the sub-Alpine belt of Vranica, Bjelašnica, Hranisava, and other mountains. Their floristic composition is made of the following species: *Vaccinium myrtillus, V. vitis-idaea, Arctostaphylos uva-ursi, Sieversia montana, Festuca nigrescens, Antennaria dioica, Homogyne alpina, Luzula campestris, Cetraria islandica, Scorzonera rosea,* and many others.

1.2.6 The Ecosystems with Mountain Pine

The ecosystems with mountain pine in Bosnia and Herzegovina have an explicitly zonal character. Today, they are represented on the slopes of Volujak, Maglić, Zelengora, Lelija, Videš, the rocks of Vilenjak, Bijelo, and Crno lakes. In their habitats, the snow is around for the larger part of the year. The geological substrate is made of different rocks at various parts of Bosnia and Herzegovina (acid vulcanic, sedimentary carbonate, dolomite, or metamorphic rocks). The soils are shallow, and rich with humus. On Maglić mountain, the ecosystem with mountain pine (*Pinus mugo*) on acidic soils is composed of: *Vaccinium myrtillus, V. vitis-idaea, Homogyne alpina, Oxalis acetosella, Luzula sylvatica, Deschampsia flexuosa, Cetraria islandica, Gentiana punctata,* etc. (Lakušić et al., 1987).

On Čvrsnica dolomite rocks, this ecosystem is composed of: *Erica carnea, Dryas octopetala, Gentiana dinarica, Thymus balcanus, Hieracium murorum, Acinos alpinus* ssp. *dinaricus,* and others.

1.2.7 Spruce Forests

Spruce forests are prevalent in the upland and sub-Alpine belt of Bosnia and Herzegovina's mountains. The mean annual temperatures in these habitats vary between 4 and 8°C and the relative mean annual air humidity varies between 70 and 90%. The rocks are of different compositions, and the soils are deep and acidic or acidified (Redžić et al., 2008). The herbaceous layer most often consists of: *Vaccinium myrtillus, V. vitis-idaea, Luzula sylvatica, L. luzulina, Hieracium murorum, Dryopteris filix-mas, Pyrola secunda, Oxalis acetosella, Melampyrum sylvaticum, Homogyne alpina,* as well as moss from the genus *Rhytidiadelphus, Hylocomium, Bryum, Mnium, Polytrichum,* and others (Barudanović et al., 2015).

1.2.8   Forests of Picea omorika

Since Pančić's discovery of spruce in the year of 1876, to this day, *Picea omorika* remains one of the most interesting subjects of investigation both in our country and entire Europe (Dizdarević et al., 1985; Lakušić, 1987). The area of Pančić's spruce in our country is split into two parts, and both are tied to the catchment of Drina river. Research shows that communities containing *Picea omorika* throughout the history of these areas have been widely distributed. It is presumed that they have glacial origins, and that their area has been expanding and contracting during dynamic periods of glaciations and interglaciations (Dizdarević et al., 1985).

1.2.9   Tall Herbs

Tall herbs are specific communities of herbaceous, usually broad-leaved species. They develop inside of broad-leaved – deciduous and coniferous forests of the mountain belt. They represent azonal vegetation, present in the conditions of high air and soil humidity, in small ranges of variation of other ecological factors. They are situated in small depressions, usually near mountain creeks.

The floristic composition of these communities is made of: *Petasites albus, P. doerfleri, Adenostyles alliariae, Doronicum grandiflorum, Cicerbita pancicii, Campanula latifolia, Impatiens noli-tangere, Cirsium waldsteinii, Aconitum variegatum, Telekia speciosa, Epilobium montanum, Angelica sylvestris, Hypericum montanum*, and many others.

1.2.10   Beech Forests

In the middle and continental Dinaric Alps in Bosnia and Herzegovina, the belt of beech forests is spread between (300) 700 and 1,500 meters a.s.l. The mean annual air temperatures in habitats vary between 12 and 4 (3)°C, and relative mean annual air humidity varies between (65) 70 and 80 (85)%. The geological substrate in these habitats is heterogeneous.

Beech forests in Bosnia and Herzegovina are differentiated into the following groups: sub-Alpine beech forests, beech-fir forests, upland beech forests, beech and autumn moor grass forests, and peri-Pannonian beech forests.

The most productive communities in Bosnia and Herzegovina are beech-fir forests, and spruce (*Picea abies*) has a significant participation in these communities as well (Barudanović, 2003). They find their optimum in northern, western, and northwestern slopes of Bosnia and Herzegovina's mountains. The mean annual temperature in these habitats varies between 6 and 7°C. The community of beech-fir forests is described with the *Abieti-Fagetum sylvaticae* association. In Bosnia and Herzegovina, numerous new subassociations have been described, among which are: *A.-F.m. festucetosum drymeiae, A.-F.m. aceretosum pseudoplatani,* and *A.-F.m. piceetosum.*

Today's syntaxonomy shows that in Bosnia and Herzegovina, communities of beech which belong to at least two different alliances, namely to Illyrian beech forests and Central European beech forests. The basis to the differentiation of these communities are: orographic, geological and pedological conditions, the effects of the Oceanic climate in the Illyrian province (northern and western area), and the Continental in the Moesian province (central and eastern part of Bosnia and Herzegovina), etc. All these effects have birthed specific floristic compositions and communities of beech on Bosnia and Herzegovina's part of Dinaric Alps, which resulted in very high scientific interest and a great number of described associations and alliances of beech forests in our country (Lakušić, 1989).

1.2.11 Oak Forests

Forest communities and underbrushes in which different species of the genus *Quercus* play the role of a modifier have an essential significance in the substantiation of ecosystem services. In Bosnia and Herzegovina, since the olden times, these ecosystems have provided the people with the basic needs (Figure 1.1), be it firewood, construction wood, or food and the care for other people and livestock.

The significance of the species of *Quercus* genus is clearly revealed in the biogeographic differentiation of ecosystems on the horizontal profile of Bosnia and Herzegovina. From the southernmost parts of Bosnia and Herzegovina, where ecosystems of underbrushes are widespread (*Q. ilex, Q. coccifera*), to the warmer regions of Herzegovina (*Q. pubescens, Q. trojana, Q. conferta*), and its karst fields (*Q. robur*), to the Supra-Mediterranean landscapes (*Q. conferta, Q. cerris, Q. pubescens*), to the hills and wide valleys of Posavina (*Q. robur, Q. petraea*), the species of this genus give a stamp to the image of our country (Barudanović et al., 2015).

1 Sessile Oak Forests (*Quercus petraea*). Oak–hornbeam forests, developed as a climatogenic phytocoenosis of the hilly belt of Bosnia, are of a mesophilic character. They are located between 300 and 800 m

Figure 1.1   Rural areas in the hilly belt; Breza, Bosnia-Herzegovina.

of elevation, with the mean annual temperatures varying between 12 and 8°C, and mean annual air humidity between 70 and 80%. The geological substrate is diverse: from limestone and dolomite, marl and shale, serpentine, peridotite, diabase, etc., on which most common are brown limestone soils, distric and eutric cambisol, and other deeper soils.

The floristic composition is made of *Quercus petraea, Carpinus betulus, Acer campestre, Prunus avium, Acer pseudoplatanus, Fraxinus excelsior, Sorbus torminalis, Acer tataricum, Euonymus europaeus, Asarum europaeum, Stellaria holostea, Hedera helix*, and others.

In the lower part of the hilly belt, on the sunny and dry habitats of the central and Pannonian Bosnia, thermophilic communities of sessile oak with black pea (*Lathyro nigrae – Quercetum petraea*) are developed. On the peridotite, xerophilous communities of this oak with heath (*Erica carnea*) are widespread.

2   Pedunculate Oak Forests (*Quercus robur*). Towards the north of Bosnia and Herzegovina, especially in the Pannonian areas, the commu-

nities of pedunculate oak are found more and more often. They are fragmentarily developed in the heartland, on the rims of karst fields (Lakušić et al., 1991; Stefanović, 1990). In the region of northern Bosnia, the communities of pedunculate oak have in history spread from south of Sava through the valleys of Una, Sana, Vrbas, Bosna, Usora, Ukrina, and Spreča, as well as throughout basins of smaller influents of Sava, between Brčko and Bijeljina, and today, their habitats are heavily fragmented. In pedunculate oak forests, the following trees are the most common: *Quercus robur, Carpinus betulus, Fraxinus angustifolia, Ulmus minor, Acer campestre, Alnus glutinosa, Populus alba, Populus nigra,* and others.

3   Turkey Oak forests (*Quercus cerris*) – Turkey Oak forests are widespread on high plateaus on the rims of karst fields (Gatačko, Nevesinjsko, Popovo, Duvanjsko) as well as on the plateaus of western Bosnia (Stefanović, 1991). The communities of turkey oak belong to the alliance *Quercion petraea – cerris,* and develop optimally in mean annual temperatures between 9 and 13°C, and mean annual relative air humidity between 60 and 70%. The communities of turkey oak develop in drier, but more or less acidic soils. Along with turkey oak, the composition of such forests most often consists of *Fraxinus ornus, Sorbus torminalis, Acer campestre, Acer obtusatum, Cornus sanguinea, Cotinus coggygria, Crataegus monogyna, Rosa canina, Coryllus avellana,* and others.

4   Downy Oak forests (*Quercus pubescens*). In the sub-Mediterranean landscapes, on the verticale profile from 300 to 800 meters elevation, communities of downy oak forests are widespread. They develop in continental regions as well, in the canyons of rivers Una, Sana, Vrbas, Drina, Neretva and gorges of Bosnia. They are characterized by a high degree of biodiversity, and plenty endemic and tertiary relicts find their necessary conditions in their habitats too, as well as a number of medicinal, edible and vitaminous species (Redžić et al., 2008).

5   Hungarian Oak forests (*Quercus farnetto*), the largest areas of these forests, are located in Herzegovina, in Bosnia; however, they are developed along the river Drina. In the understory, species such as *Prunus spinosa, Pyrus pyraster, Ulmus minor, Sorbus torminalis, Carpinus orientalis, Acer obtusatum, Acer tataricum, Cotinus coggygria, Viburnum lantana, Tamus communis, Ligustrum vulgare, Crataegus monogyna, Cornus sanguinea, Clematis vitalba, Euvoni-*

mus europeus, and others are numerous (Fukarek, 1964). They differentiate into a couple associations. The communities of turkey oak and Hungarian oak (*Quercum farnetto-cerris*) is a climatogenic community in the lower parts of the hilly belt on the east and southeast of Bosnia and Herzegovina.

6   Macedonian Oak forests (*Quercus trojana*). Recent forests of Macedonian Oak (today in the stage of degradation) inhabit regions of the sub-Mediterranean and Mediterranean belt of Bosnia and Herzegovina (Redžić et al., 2008). The habitats of Macedonian Oak are located on steep rockeries of limestone, karst terrain, degraded surfaces subject to erosion, coves, etc. Mean annual temperatures in these habitats vary between 12 and 15°C, and the mean annual relative air humidity is between 50 and 70%.

7   Evergreen Oak forests (*Quercus ilex*). In Bosnia and Herzegovina, ecosystems of evergreen oak take up the narrow Eu-Mediterranean belt in Bosnia and Herzegovina. In this area, today, degradation stages are present under the name *Maquis shrubland*. The habitats of these communities spread between sea level, up to 100 meters of elevation. The climate is hot or warm (in the summer period), and as for the humidity, it is semiarid or arid. In the zone of the climatogenic vegetation of the evergreen oak, ecosystems of coastal cliffs are present (*Plantagini – Staticetum cancellatae*), ecosystems of seas and oceans (*Zosteretum marinae adriaticum*), ecosystems of brackish waters with sea lettuce (*Ulvaetum lactucae*), and ecosystems of coastal wetlands (*Juncetum maritimo – acuti*) (Redžić et al., 2008).

1.2.12   European Hop-Hornbeam and Oriental Hornbeam Forests

Xerothermal low-pitched forests and underbrushes are widespread on the coastal Dinarides between 300 and 1,200 meters elevation. They are also present in the form of extrazonal vegetation in the gorges and canyons of the central parts of the country. In the underbrushes of the lower part of the sub-Mediterranean belt, oriental hornbeam (*Carpinus orientalis*) is the most common, while in the forests and underbrushes of the upper part, European Hop-Hornbeam (*Ostrya carpinifolia*) is the most common. Xerothermal forests and underbrushes are today widespread in Bosnia and Herzegovina in what used to be downy, Macedonian, Turkey, Hungarian, and even the thermophilic forests of sessile oak habitats (Lakušić et al., 1982). The mean annual temperature varies between 10 and 15°C, and the mean annual relative air humidity is between 50 and 65%.

The communities with the European Hop-Hornbeam (alliance *Seslerio – Ostryon*) have an optimum on the steep slopes in the canyons of Bosnia and Herzegovina's rivers (Lakušić et al., 1982). Their composition is made of species such as: *Ostrya carpinifolia, Tilia platyphyllos, Acer obtusatum, Sesleria autumnalis, Mercurialis ovata, Teucrium chamaedrys, Cyclamen purpurascens, Galium schultesii, Hepatica nobilis, Brachypodium pinnatum*, and many others.

The composition of the communities in the *Carpinion orientalis* alliance is composed of the following species: *Carpinus orientalis, Fraxinus ornus, Acer monspessulanum, Quercus pubescens, Ostrya carpinifolia, Coronilla emerus* subsp. *emeroides, Colutea arborescens, Vincetoxicum hirundinaria, Melittis melissophyllum, Mercurialis ovata, Carex humilis, Dioscorea balcanica, Petteria ramentacea*, etc. Along the mentioned species, communities with oriental hornbeam in the Dinaric Alps are characterized by numerous endemic species such as *Hyacinthella dalmatica, Helleborus hercegovinus, Crocus tommasinianus, Lilium cathaniae, Adenophora lilifolia*, and many others (Redzic and Šoljan, 1988b).

1.2.13 Pine Forests

Pine forests belong to the type of temperate conifer forests and are present on the profile of Dinaric Alps from the coast of the Adriatic Sea upwards of 2,200 meters of elevation. Alongside zonal forests in Bosnia and Herzegovina, different species of the genus *Pinus* today play a modifier role in ecosystems of relict-refugial landscapes. These are the habitats, which have suffered the smallest degree of change from the preglacial to the postglacial period. The position of tertiary relict ecosystems in Bosnia and Herzegovina is tied first and foremost to canyons, gorges and steep slopes of the mountains in the basins of rivers Una, Vrbas, Bosna, Drina, and Neretva. In the biogeographic sense, these ecosystems belong to a special province of relict pine forests, whose area includes forests of Bosnian pine (*Pinus heldreichii*), Illyrian black pine (*Pinus nigra* ssp. *austriaca*) and Dalmatian black pine (*Pinus dalmatica*) (Redžić et al., 2008).

1 Illyrian black pine forests. Black pine inhabits steep, southernmost terrains, on which warm winds blow. The average yearly temperatures in ecosystems of Illyrian black pine vary between 8 and 12°C. The soil is most commonly calcomelanosol, rendzina or ranker.

Black pine forests on dolomite are very specific in their floristic composition. Many tertiary endemic and relict species live inside of them. Some are explicitly tied to dolomite rocks, such as *Alyssum moelendorfianum, Acinos orontius, Thymus aureopunctatus* (Redžić et al., 2011; Riter-Studnička, 1963). The forests of black pine take up large areas in the ophiolitic region of Bosnia and Herzegovina, where the geological substrate is dominated by serpentinite and peridotite. The black pine is commonly joined by the Scots pine *Pinus sylvestris* (region of mountain Kozara, Uzlomac, Borja, Ozren, Majevica, the basin area of river Bosna, mountain Konjuh, Sjemeč) and the valley of the river Lim (Riter-Studnička, 1963).

2 Bosnian pine forests (*Pinus heldreichii*). The area of Bosnian pine in Bosnia and Herzegovina takes up the region of the mountains Orjen, Velež, Prenj, Čvrsnica, Čabulja, Vrana, and Hranisava. The forests of the Bosnian pine are developed as a zonal vegetation of the upper part of the Supra-Mediterranean and sub-Alpine belt on the coastal and central Dinaric Alps (Redžić et al., 2011).

On the mountains around Neretva, Bosnian pine is present in the alpine and the sub-Alpine belt. On the highest peaks, Bosnian pine forms a special type of ecosystem of rock crevices.

1.2.14 *Ecosystems on Karst Fields*

Karst fields are one of nature's phenomena in Bosnia-Herzegovina. The relief is a specific with mainly underground water circulation within soluble rocks (limestone, dolomite, tuff). Karst fields of Bosnia and Herzegovina are differentiated in several groups, following the direction of Dinaric Alps. Many of them are endemic centers of flora and fauna. The largest field (400 sq. km) is Livanjsko polje, where specific conditions brought together eco-logically different ecosystem types. In the field's northwestern part post-glacial processes of alkaline bogs formation are still ongoing. The unique type of hydromorphous soil, which occurs here, planohystosol, is vitally important for the survival of wilderness in swamps. Some endemic and/or rare plant species on karst fields are: *Helleborus hercegovinus, Ranunculus croaticus, Corydalis leiosperma, Hesperis dinarica, Rhamnus intermedius, Bupleurum karglii, Athamantha haynaldii, Scrophularia bosniaca, Onosma visianii,* and others. The Red book of Federation of BiH marked the next species as vulnerable to anthropogenic pressures: *Salvia bertolonii, Utric-ularia vulgaris, Scilla litardierei, Narcissus radiiflorus,* and *Iris illyrica.*

The karst fields in Bosnia-Herzegovina are an integral part of migratory bird's routes. Some of globally threatened bird species stop here, as *Anthya nyroca, Aquila pomarina, Falco neumanni, Crex crex,* and others (Redžić et al., 2008).

1.2.15 Wetland Ecosystems

Wetland ecosystems are of local character in Bosnia and Herzegovina, where specific orographic conditions appear. There is a diversity of wetland ecosystem types in our country. One occurs along large watercourses (Una, Vrbas, Bosna, Drina, Neretva) on whose shores are riparian forests of willow, alder, purple and marsh willow developed. Alkaline blanket bogs occur at lower altitude, and raised bogs with a domination of bog-mosses occur in the zone of dark coniferous forests. In the sub-Alpine belt, in the small depressions and around springs develop a special form of boreo-relict blanket bogs. Mountain lakes of Bosnia and Herzegovina's Dinaric Alps [Šatorsko, Kukavičko, Rastićevsko and Turjača on plateau of Kupres, Prokoško lake on Vranica Mountain (Figure 1.2), Blatačko lake on

Figure 1.2 Prokosko lake; sub-Alpine belt of Vranica Mountain.

Bjelašnica Mountain, Idovačko lake on Raduša Mountain, Blidinje lake in Dugo Polje between Čvrsnica Mountain and Vran Muntain, Uloško lake on Crvanj Mountain, Boračko lake beneath Prenj Mountain, Veliko, Blatno, Crno and Bijelo lake on Treskavica Mountain, Kotlaničko, Orlovačko, Crno, Bijelo, Štirinsko, Kladopoljsko, Donje Bare and Gornje Bare on Zelengora Mountain] are surrounded by specific types of wetland vegetation (Redžić et al., 2008).

1.3 Species Diversity

Flora, Fauna, and Fungi of Bosnia and Herzegovina are considered to be among the most diverse in Europe as a whole, especially in term of a high level of endemic and relict species (Redžić et al., 2008). At this point, it is necessary to emphasize that Bosnia-Herzegovina is a young country, which still does not have an inventory of plant and animal species, as well as an inventory of fungi. However, according to a report of a group of 30 Bosnian scientists assessment species diversity showed great richness in a relatively small territory of our country.

1.3.1 Plant Species Diversity

High level of floristic diversity is based upon diversity of cyanophytes, algae, mosses and vascular plants. It is identified to date 1,859 species from 217 genera within a group of cyanophytes and algae. Table 1.1 shows diversity within different taxa.

Table 1.1 Taxonomic Diversity of Cyanobacteria and Algae in Bosnia-Herzegovina

Taxon	Genera	Species
Cyanobacteria	36	303
Rodophyta	7	20
Charophyceae	33	319
Chlorophyceae	65	242
Euglenophyta	4	21
Dinophyta	5	20
Bacillariophyceae	57	881
Xanthophyceae	4	21
Chrysophyceae	12	32

(Adapted from Redžić et al., 2008.)

Floristic richness of Bosnia and Herzegovina is shown in Table 1.2. Aside from the species, whose number is listed in Table 1.2, in Bosnia-Herzegovina so far 1,086 subspecies have been identified. Within Spermatophyta, families with the highest species diversity are: Asteraceae, Fabaceae, Poaceae, Rosaceae, Brassicaceae, parsley family, Lamiaceae, Cyperaceae, figworts, pinks, lilies, and buttercups. There is also a great number of families with only one genus and species (approximately 30% of total number). The most specific characteristic of our flora is a great stake of paleo and neoendemic species, tertiary and glacial relicts, maintained by refugial habitats (such as cliffs, canyons, and mountain cirques). At the current state of knowledge, it is estimated that a number of 450 endemic plant taxa in Bosnia-Herzegovina. Newly undertaken investigations indicate that this number is much bigger, especially as far as poorly investigated genera are concerned (e.g., *Alchemilla, Potentilla, Rosa, Rubus, Hieracium, Centaurea, Carex, Festuca*).

Stenoendemic plant species of Bosnia and Herzegovina have always provoked the greatest interest of domestic and European botanists. Among these species are the following: *Acinos orontius* (K. Maly) Šilić, *Alyssum moellendorfianum* Aschers. ex G. Beck, *Asperula hercegovina* Degen, *Barbarea bosniaca* Murb., *Campanula hercegovina* Degen & Fiala, *Centaurea bosniaca* (Murb.) Hayek, *Dianthus freynii* Vandas, *Edraianthus niveus* G.Beck, *Minuartia handelli* Mattf., *Oxytropis prenja* G. Beck, *Symphyandra hofmanni* Pantocsek, etc. (Figure 1.3)

1.3.2 Animal Species Diversity

Thanks to the diversity of certain animal groups, Bosnia and Herzegovina also belong to the top of European biodiversity. The most unique is a fauna of karst sources, mountain torrents, and canyons. The fauna of Bosnia and Herzegovina is far from being fully investigated. For the regnum of Protozoa, there are missing data on species and phyla, while phyla of Metazoa, such as Plathelminthes, Nemertina, Nematoda, Rotatoria, Pogonophora happen to be the least, or not at all investigated in Bosnia Herzegovina. Within invertebrates, special attention of scientist is paid to different groups

Table 1.2 Taxonomic Diversity of Plants

Taxon	Families	Genera	Species
Bryophyta	52	187	565
Pteridophyta	14	26	61
Spermatophyta	161	858	3,256

(Adapted from Redžić et al., 2008.)

Figure 1.3 Plant species diversity of mountain meadows (*Crocus neapolitanus* (Ker Gawl.) Loisel., Vranica mountain).

of Arthropoda. Within this group, known species richness is as follows: Acarina (208), Amphipoda (31), Decapoda (5), Chilopoda (9), Diplopoda (55), Pauropoda (23), Symphila (12), Colembola (224), Ephemeroptera (58), Plecoptera (74), Trichoptera (215), Protura (18), Diplura (15), Zygentoma (2), Mantodea (4), Blattodea (17), Heteroptera (705), Adephaga (701), Polyfaga: *Lymexylidae* (1) *Buprestidae* (129), *Hydrophilidae* (47), *Sphaeridiidae* (30), *Sphaeritiidae* (1), *Dascillidae* (1), *Trogidae* (3), *Geotrupidae* (9), *Scarabaeidae* (159), *Lucanidae* (7), *Chrysomelidae* (322), *Cerambicidae* (218), *Scolitidae* (55); Lepidoptera (1622), Caelifera (70), Ensifera (85), Hymenoptera (353) (Redžić et al., 2008).

Due to the diversity of aquatic habitats, and the occurrence of different kind of watercourses, limnofauna of invertebrates of Bosnia and Herzegovina ought to be very diverse (50 species of annelids that belong to 19 genera; 8 species of leaches belonging to 7 genera, etc.). River crustaceans encompass 31 species, of which 16 are endemic. The most diverse is a fauna of aquatic insects, with a high level of endemism. Thus, fauna of mayflies comprises 58 species belonging to 20 genera, of which 5 are Dinaric, Balkan or Dinaric-alpine endemic species. Water moths ought to be numerous groups with 215 detected species. About 50 of these species are endemic in Balkan, and 24 of them are endemic having the Dinaric areal. The most interesting is genus *Drusus* (Redžić et al., 2008).

Herzegovina's caves are inhabited by interesting and endemic forms of life, as *Eremulus simplex* (Willmann, 1940), *Autognata willmanni* (Willmann, 1941), *Chamobates petrinjensis* (Willmann, 1940), and *Carabodes bosniae* (Frank, 1965).

Fauna of vertebrates in Bosnia and Herzegovina is represented as shown in Table 1.3.

Fish fauna in Bosnia and Herzegovina is relatively well investigated. The highest diversity is recognized within the family *Cyprinidae* (26 genera and 51 species) and *Salmonidae* (5/8).

Among tail-less amphibians the most abundant is genus *Rana* with 7 species and among caudate amphibians this is genus *Triturus* with 5 species.

Reptiles inhabit freshwater, ponds, marshes and almost all terrestrial ecosystems (especially extreme habitats, such as rocky grassland) belonging to the 38 species (45 subspecies) from 12 families. The highest reptile diversity is evident in the Mediterranean region and Supra-Mediterranean belt. However, some species are spread up to the highest mountain peaks (Bosnian and Orsiny's viper live in mountain swards, then on screes and rock crevices all around Bosnia and Herzegovina).

Birds diversity recorded 326 species (60 families). Most of them are stationary (nesting), while migratory spend some time in ecosystems of Bosnia and Herzegovina only by seasons (wetlands: Buško Blato, Hutovo Blato, Bardača, a lower flow of Drina river, etc.).

The fauna of mammals encompasses 85 species (19 families). Most of the species live in terrestrial habitats, while a small number of them inhabits aquatic ecosystems (Redžić et al., 2008).

Table 1.3 The Diversity of Vertebrates in Bosnia and Herzegovina

Animal group	Number of species	Endemic species	Threatened species in BIH
Fish	119	12	36
Amphibians	20	6	3
Reptiles	38	12	11
Birds	326	–	97
Mammals	85 (+2?)	9	24

(Adapted from Redžić et al., 2008.)

1.3.3 Fungi Species Diversity

Fungi inhabit both terrestrial and aquatic environment. It is being estimated that in Bosnia and Herzegovina live a high number of fungi species, but to date, only 552 species were identified. There are many macromycetes in Bosnia and Herzegovina, which has high economic potential and value being, therefore, an important income source for a local community. The most important among them are: morels (*Morchella sp.*), bolete (*Boletus sp.*), chanterelle (*Cantarelus cybarius*), parasol (*Macrolepiota sp.*), and saffron milk cap (*Lactarius deliciosus*).

1.4 Genetic Diversity

High diversity of genetic resources in Bosnia and Herzegovina is contained in a great number of original animal breeds and plant sorts. Through the existence of different civilizations on the territory of Bosnia and Herzegovina, many animal breeds were domesticated. Majority evolved as distinct ecotypes, but there are only sporadic scientific data on identified sorts and breeds. Between them are horses (Bosanski brdski), cattle (Buša i Gatačko), sheeps (Pramenka), goats (Balkanska rogata), pigs (Šiška), dogs (Bosanski tornjak), and pigeons.

The diversity of ecoclimate has supported high and still well-preserved diversity of genetic resources contained in fruits. This is reflected in great number of sorst of cherries (*Prunus avium*: alice, ašlame, hašlamuše, hrušćovi, crnice, bjelice); plums (*Prunus domesticus*: bijele, prskulje, mrkulje, savke); pears (*Pyrus* sp.: ječmenke, krivočke, mednjače, takiše, bijeli karamut, crni karamut, krupnjače, jeribasme); apples (*Malus* sp.: petrovače, golubače, šarenike, zelenike, senabije, šahmanuše, krompiruše, crvenike, etc.), as well as sour cherries, apricots, peaches, almonds, raspberries, blackberries, strawberries, and currants (Redžić et al., 2008).

Among gardening genetic resources, diversity of forms and special ecotypes characterizes pumpkins from genus *Cucurbita*, bean (*Phaseolus vulgaris*: čućo, bubnjo, trešnjo, kućićar, mesni), cabbage from genus *Brassica*, paprika (*Capsicum annuum*), widely known okra (*Hibiscus esculentus*), watermelon called semberka (*Citrullus colocynthis*), melon (*Cucumis melo*), and spectrum of potato's sorts (*Solanum tuberosum*: romanijski, kupreški, fojnički, glamočki, etc.).

A rich cultural tradition of Bosnia and Herzegovina resulted in many traditional products. This is seen in bread making, milk-production, brewery, viniculture and especially cheese production. So far 15 sorts of indigenous cheese have been identified, but for sure many other unknown biotechnological formulas still exist, hidden in the mountain cottages of Dinaric Alps.

Today, the highest yield is achieved in Pannonian and peri-Pannonian belt by different crops (wheat, maize, barley, oat and Johnson-grass), then cultivated vegetables (watermelon, gombo, sunflower, paprika, tomato, aubergine, different sorts of cabbage), herbal genetic resources (plums called "požegače," wallnuts, pears, apples, grapes) and a lot of horticultural species.

It must be pointed out the fact that in the last 25 years happened a large displacement of people from mountain rural areas in Bosnia and Herzegovina, primarily because of the war (1992–1995). Parallel to this process, a numerous traditional knowledge has been lost. The loss caught the most of knowledge on practices to use the rich genetic diversity of plants and animals. Unfortunately, a legislation that would regulate the issue of inventory and protection of this knowledge still does not exist (Barudanović et al., 2015).

Keywords

- *Crocus vernus*
- genetic diversity

- ecosystem diversity
- rural areas

References

Barudanović, S., (2003). Ecological differentiation of broad-leaved forests on Vranica mountain. *PhD Dissertation, Faculty of Science*, University of Sarajevo.

Barudanović, S., Macanović, A., Topalić-Trivunović, L., & Cero, M., (2015). *Ecosystems of Bosnia and Herzegovina in Function of Sustainable Development*, Fojnica, D. D., Sarajevo.

Bjelčić, Ž., (1966). Vegetation of sub-Alpine belt of Jahorina Mountain, Herald of the National Museum of Bosnia and Herzegovina. *Natural Science*, 5, 31–103.

Dizdarević, M., Lakušić, R., Grgić, P., Kutleša, L., Pavlović, B., & Jonlija, R., (1985). Ecological basics for an understanding of the relictness Pancic's spruce. *Bull. Soc. Ekol. SR BiH, Series A, Ecological Monographs, 2*, 7–28.

Fukarek, P., (1964). Northwest Frontier of today's distribution of Hungarian oak (*Quercus conferta* Kit.). Šumarski List, *88*(3–4), 109–123.

Lakušić, R., & Redzic, S., (1989). The flora and the vegetation of vascular plants in Refugial-Relict Ecosystems in the Canyon of the Drina river and its tributaries. *Bull. Dept. Nat. Sci. CANU, 7*, 107–205.

Lakušić, R., & Redžić, S., (1991). Vegetation of refugial-relict ecosystems of Una River Basin. *Bull. Soc. Ecol. SR BiH, Series B., 6*, 25–73.

Lakušić, R., (1976). The natural system of geo-biocenoses on the Dinaric Alps. *Annal of the Institute of Biology of the University of Sarajevo, 28*, 175–191.

Lakušić, R., (1987). Chorologic-morphological and organic-phylogenetic differentiation endemic plant species the Dinarids. *ANUBiH Papers*, *83*(14), 159–166.

Lakušić, R., (1989). Ecological differentiation of the Bosnian-Herzegovinian space, Herald of the National Museum of Bosnia and Herzegovina, *Natural Science*, *28*, 97–102.

Lakušić, R., (1990). *Mountain Plants*, Svjetlost, Institute for textbooks and teaching materials, Sarajevo.

Lakušić, R., Bjelčić, Ž., Šilić, Č., Kutleša, L., Mišić, L., & Grgić, P., (1969). The mountain vegetation of Maglic, Volujak, and Zelengora. *ANUBiH Papers, Department of Natural Sciences and Mathematics, Special Edition, 11*(3), 171–187.

Lakušić, R., Dizdarević, M., Grgić, P., Pavlovic, B., & Redžić, S., (1991). Ecological differentiation of Una catchment area and its value. *Bull. Soc. Ecol. SR BiH, Series B., 6*, 155–159.

Lakušić, R., Dizdarević, M., Grgić, P., Pavlovic, B., Redžić, S., (1989). Flora and vegetation of higher plants and fauna Symphyla, Pauropoda and Mollusca in refuge-relict ecosystems canyon of the river Tara, Piva, Komarnica, Lim and Drina. *Bull. Dept. Nat. Sci. CANU*, 7, 93–105.

Lakušić, R., Mišić, L., Kutleša, L., Muratspahić, D., Redžić, S., & Omerović, S., (1987). Overview of nonforest ecosystems of the National Park "Sutjeska." *Bull. Soc. Ecol. SR BiH, Sarajevo, Series A., Ecological Monographs, 4*, 29–51.

Lakušić, R., Pavlovic, D., & Redžić, S., (1982). Chorologic-ecological and floristical differentiation of forests and bushes with hornbeam (*Carpinus orientalis* Mill.) and European Hop-Hornbeam (*Ostrya carpinifolia* Scop.) in Yugoslavia. *Herald of the Republic Institute for Nature Protection of the Natural History Museum*, Titograd, *15*, 103–116.

Lakušić, R., Pavlović, D., Abadžić, S., Kutleša, L., & Mišić, L., (1982). Ecosystems of Vlašić mountain. *Bull. Soc. Ecol. SR BiH, Sarajevo, Series A., Ecological Monographs, 1*, 1–131.

Lakušić, R., Pavlović, D., Abadžić, S., Kutleša, L., Mišić, L., Redžić, S., Maljević, D., & Bratović, S., (1979). The structure and dynamics of ecosystems on the Vranica Mountain in Bosnia. *Proceedings of the II Congress of Ecologists of Yugoslavia, 1*, 605–714.

Lakušić, R., Redžić, S., Muratspahić, D., & Omerović, S., (1987). The structure and dynamics of ecosystems on the permanent sample plots of the National Park "Sutjeska." *Bull. Soc. Ekol. SR BiH, Series A., Ecological Monographs, 4*, 53–105.

Mišić, L., & Lakušić, R., (1990). *Meadow Plants*. Svjetlost, Institute for Textbook, Sarajevo, Institute for Textbooks and Teaching Materials, Belgrade.

Mišić, L., (1984). Vegetation of Pastures on the Treskavice Mountain. *PhD Dissertation. Faculty of Science*, University of Sarajevo.

Muratovic, E., Bogunic, F., Šoljan, D., & Šiljak-Yakovlev, S., (2005). Does *Lilium bosniacum* merit species rank? A classical and molecular-cytogenetic analysis. *Plant Syst. Evol., 252*, 97–109.

Redzic, S., & Šoljan, D., (1988). *Adenophora lilifolia* (L.) Ledeb. ex A. DC. A rare plant in the flora of Bosnia and Herzegovina, *Herald of the National Museum of Bosnia and Herzegovina, Natural Science*, *27*, 79–84.

Redžić, S., Barudanović, S., & Radević, M., (2008). *Bosnia-Herzegovina – Land of Diversity*. The first report BiH to the CBD. Federal ministry of environment and tourism, Bemust. Sarajevo.

Redžić, S., Barudanović, S., Lelo, S., Lepirica, A., Kotrošen, D., Trakić, S., Hadžiahmetović, A., & Kulijer, D., (2007). *Biodiversity of Endemic Development Centers in Herzegovina as a Contribution to the Aspirations of Targets 2010*, BA-FDCP-CQ-SA-CS-06-TF052697-CS8–10, Sarajevo.

Redžić, S., Barudanović, S., Lelo, S., Lepirica, A., Kotrošen, D., Trakić, S., Hadžiahmetović, A., & Kulijer, D., (2007). *Evaluation of Biodiversity Ecosystems of Karst Fields on the Territory of the Federation of Bosnia and Herzegovina as a Contribution to the Thematic Programs of the Convention on Biological Diversity in Accordance With the Targets 2010*. BA-FDCP-CQ-SA-CS-06-TF052697-CS8–10, Sarajevo.

Redžić, S., Barudanovic, S., Trakic, S., & Kulijer, D., (2011). Vascular plant biodiversity richness and endemo-relictness of the Karst Mountains Prenj, Čvrsnica and Čabulja in Bosnia and Herzegovina (Balkan, W.), *Acta Carsologica, 40*(3), 527–555.

Riter-Studnička, H., (1956). Flora and vegetation on the dolomites of Bosnia and Herzegovina. *Annals of the Institute of Biology of the University of Sarajevo, 1–2,* 73–122.

Riter-Studnička, H., (1963). The vegetation on serpentines in Bosnia. *Annal of the Institute of Biology of the University of Sarajevo, 16,* 91–204.

Šoljan, D., Muratović, E., & Abadžić, S., (2014). *Orchids of Mountains Around Sarajevo*, The Good Book doo, Sarajevo.

Stefanović, V., (1990). Coenological range of the English oak (*Quercus robur*) in Bosnia and Herzegovina. *Annal of the Institute of Biology of the University of Sarajevo, 42,* 73–84.

Stefanović, V., (1991). Phytocoenological relations within Turkey oak forests in the area of upper Pounje. *Bull. Soc. Ekol. SR BiH, Series B., 6,* 75–80.

CHAPTER 2

Biodiversity in France

PHILIPPE GRANDCOLAS

Director, Institut de Systématique, Evolution, Biodiversité, Muséum national d'Histoire naturelle, CNRS, UPMC, EPHE, France, E-mail: pg@mnhn.fr

2.1 Introduction

As a European country, France is at the crossroad of several biogeographic areas and pathways. Its terrestrial territory is limited by the North Sea, the Atlantic Ocean in the West and the Mediterranean Sea in the South, being therefore exposed to different marine influences. Mountains as well are part of this natural delimitation, with Vosges in the Northeast, the Alps forming a mountainous belt in the Southeast and the Pyrenean chain in the Southwest separating France from the Iberian Peninsula. The resulting natural landscape is a very diverse mix of different influences, oceanic on the western side, continental in the East, alpine in Southeast and Mediterranean in the South (Metzger et al., 2005). This landscape is located around 46°00'N latitude and 2°00'E longitude and the altitude peaks at more than 4,000 meters in the Alps and Pyrénées while more than 70% of the territory is below 500 m. Actually, among the biogeographic influences found in Europe, only the Macaronesian one is to be excluded and the boreal one is present in several remote and relict places of small size. The territory has been profoundly marked by past climatic events and the presence of human populations as well.

First, a short account of biodiversity will be presented, providing numbers for large groups of organisms, signaling lacks or gaps in the knowledge as well and indicate resources to have access to primary data. Second, the different ecosystems comprising these species are detailed and are provided a few examples of organisms belonging to different groups and exemplifying different biogeographic trends. Third, emphasis will be put on the dynamics of the biota since the last quaternary climatic events, showing how they evolved and how they could be modified in the future according to climatic change.

2.2 Biodiversity Numbers

The flora and fauna of France are remarkably rich for a temperate and human-modified landscape, because of the combination of different climatic and biogeographic areas. The total number of living species in France is more than 85,000 of which more than 2,000 are endemic and about 2,500 are introduced (Gargominy et al., 2016). There are about 7,000 species of Plants (Tracheophyta), which represent 40% of the European species, therefore less than a few other European Mediterranean countries (Tison and Foucault, 2004). If introduced and cultivated species are considered, the number of plants in France must be set up around 10,000. The largest groups in this flora belong to the Palearctic realm in the families Asteraceae (1,165 species), Rosaceae (580 species), Poaceae (518 species), Fabaceae (436 species), Brassicaceae (275 species), Apiaceae (262 species), and Caryophyllaceae (225 species). Other nonvascular plants (Bryophyta, Hepatica, Algae) totalize around 3,000 species. Fungi comprise more than 3,000 species.

Concerning animals, there are about 1,500 vertebrate species (half of them are marine), among which 284 bird species nesting for 500 species observed in France (Issa and Muller, 2015), 43 amphibians (Lescure and Massary, 2012), 135 mammals. Mammals belong to families Leporidae (4 species), Castoridae (1 species), Cricetidae (14 species), Muridae (5 species), Gliridae (3 species), Sciuridae (2 species), Erinaceidae (1 species), Soricidae (10 species), Talpidae (4 species), Bovidae (4 species), Cervidae (2 species), Suidae (1 species), orders Chiroptera (37 species), Carnivora (16 species) and Cetacea some of which are distributed and not erratically present (Savouré-Soubelet et al., 2016). It must also be mentioned that part of this biodiversity migrates in or from France (e.g., birds) and that conservation or population issues must be considered at a supra-national level.

Insects comprise more than 40,000 species and the other large groups of arthropods are the Acari with several thousands of species and the Spiders with about 1,500 species. Other groups like harvestmen (120 species), Pseudoscorpions (125 species) or Scorpions (5 species) are much less rich.

These counts may be very accurate for vascular plants or vertebrates but they are still very incomplete for arthropods or other small animals. A general picture can be found on GBIF website (GBIF, 2017) or on INPN website, including the distribution of archeological organisms' remains (INPN, 2017). Even if the list of species is quite complete now for many groups or big organisms, their accurate distribution has still to be documented. A selection of all data on France biodiversity on the GBIF portal actually shows many sampling gaps that have been discussed in a recent report available online (Witté and Tourroult, 2017).

2.3 Regions and Ecosystems

The ecosystems mainly of Palearctic temperate origin are distributed in four different areas. The Atlantic area with an oceanic climate is distributed along the west coast but includes also many lowland regions inside the country such as Ile de France (around Paris), with a low elevation landscape originally harboring either oak or hornbeam forests or open vegetation like moorland.

- *Uromenus rugosicollis* (Audinet-Serville, 1838)—an example of Atlantic distribution: This species of katydid (Rough Saddle Bush-cricket) is distributed along the Atlantic coast; it goes north until the Cotentin and southwest until Marseille on the Mediterranean coast (INPN, 2017). It is part of a group of katydids that shows a Mediterranean distribution (Orthoptera Ensifera, subfamily Ephippigerinae).

The continental area is at the opposite on the eastern border from the North to the first and lower parts of the southeast mountains. As indicated by its name, the climate is continental, therefore colder and with a stronger seasonality than the previous one, more or less degraded in an oceanic one. The typical vegetation is oak and beech forest, with some elm stands in wet places.

- *Orobanche alsatica* Kirschl., 1836—an example of continental distribution: This is a parasitic species of the family Orobanchaceae usually developing on the plant *Peucedanum cervaria* (L.) Lapeyr. (family Apiaceae). It occurs at the border of woodlands within high nonligneous vegetation (INPN, 2017; Muller, 2006). Its distribution is mainly around central Europe and its occurrence in France is limited to the central eastern part of the country.

The mountainous area is obviously related to mountains, either in the Pyrénées or in the Alps, and at a lesser degree in Massif Central. Colder and with a shorter summer season, this area is divided into several successive altitudinal zones. From the lower to the higher, can be found the following levels: "collinéen" (<1,300 m) mostly forested with oaks (*Quercus*), "montagnard" (between 1,300 m and 1,800 m) mostly forested with abies firs (*Abies*), beeches (*Fagus*) and some maples (*Acer*), with "subalpin" (1,800–2,400 m) today occupied mainly by anthropogenic grasslands but originally with spruce (*Picea* sp.) and larch (*Larix*) and "alpin" levels (2,400–3,000 m) with alpine grasslands and small bushes (*Juniperus*). Elevation numbers are given for gross information but vary according to the mountain' side and to the latitude.

- *Soldanella alpina* L.—an example of mountainous distribution: According to genetic analysis, this group of plants (family Primulaceae) originates in Central Asia and exemplifies the diversity of plants occurring in mountains of France. *S. alpina* L. is a short plant growing close to the soil, found at high elevation (>2,000 m) in Combes where snow is occurring until late in the season.

The Mediterranean area is limited to the Southern part of the country along the Mediterranean coast from Spain to Italy with the typical original corresponding climate, combining soft winter, rainy spring and hot summers. This area harbored olive tree (*Olea*) and oak (*Q. suber* or *Q. ilex*) forests, shrubby evergreen vegetation on acid or calcareous soils (maquis, garrigue).

- *Quercus suber* L.—an example of Mediterranean distribution: The cork oak (Family Fagaceae) is distributed in the western part of the Mediterranean area in Spain, North Africa, southern France, and Italy. This tree is typical of the Mediterranean area.

2.4 Evolution of the Biota Since Quaternary Era and in the Future

The landscape of France has been deeply modified during the last glaciation events. The ice belt has been moving far to the South and ecosystems have strongly shifted as well. Moving back to the North from more southern areas after glaciation events has been classically documented for many organisms with some presumptive roads of recolonization and diversification (Hewitt, 1999; Taberlet et al., 1998; Schmitt, 2007). Even 6,000 years ago, the distribution of the main ecosystems was still quite different (Prentice et al., 1996). These changes have left some traces in term of genetic structure, endemism and more basically in term of ecosystem richness. A simple comparison with other areas in Europe or in other temperate areas in the world less directly affected by glaciations showed lower species numbers in ecosystems of France for many groups or organisms than in temperate areas less affected by glaciations. This situation has been especially mentioned for trees but is true for many groups.

Since ever after these dramatic events, Man has had a very strong influence on ecosystems and profoundly modified them. The first consequence in the historical term, beginning thousands of years ago (Late Iron Age) and continuing until the eighteenth century, was the decrease in forested areas that have been replaced by cultivated areas and pastures, both in the lowlands and in the mountains. In addition to this big change in the ecosystems, many species have spread in a way that often has to be understood with archeological studies. For example, an animal as common as the rabbit [*Oryctolagus cuniculus* (Linnaeus, 1758)] was distributed mainly in the Southeast and has spread to the North with the practice of breeding and hunting in the Middle Ages (Callou, 2003). Today, this species is distributed all over the landscape and is usually and misleadingly perceived as a common species that has always taken part of the natural communities. The same way, sheep and cowherds (*Ovis aries* and *Bos taurus*) have deeply modified the biota of the Alps mountainous area by grazing since the Late Iron Age and the Roman period (Giguet–Covex et al., 2014).

Today, industrial agricultural and forestry practices significantly decrease the diversity of natural biota, by impacting natural communities and decreasing the species numbers. Primary functions of ecosystems are at risk of failure in several regions (especially pollination and water quality). Paradoxical situations begin to be observed with part of biodiversity higher in semiurban areas than in intensive agricultural landscape (Deguines et al., 2012).

The effect of climate change is well documented, especially for some well-studied groups like birds or butterflies (DeVictor et al., 2012) and there is a shift for southern populations to go to the North.

2.5 Biodiversity in Overseas Territory

France includes as well many tropical overseas territories including French Guiana (Amazonia), les Antilles Françaises in the Caribbean region, New Caledonia in the Pacific, La Réunion and Mayotte in the Indian Ocean, and several other smaller or biologically less rich locations (Gargominy, 2003). Taking into account such places with very rich, endemic and diverse biota pushes up the estimates of species richness to twice the number cited higher (around 160000 species). The inventory of biodiversity in all these regions is still in progress (e.g., Guilbert et al., 2014) which means that the total number of species is actually much higher, including for endemic ones.

2.6 Natural Parks and Reserves

There are 10 national parks and 51 regional parks in France, which represent around 17% of the area of the national territory. Several ones are located in overseas regions. These parks are tolerant to some agricultural, breeding and touristic activities in specific conditions. They welcome more than 5 millions of visitors per year.

Keywords

- biodiversity numbers
- national parks
- overseas territory
- quaternary era

References

Callou, C., (2003). From the warren to the hutch. Archaeozoological study of rabbits in Western Europe. *Mémoires du Muséum National d'Histoire Naturelle, 189*, 1–356.

Deguines, N., Julliard, R., De Flores, M., & Fontaine, C., (2012). The whereabouts of flower visitors: Contrasting land-use preferences revealed by a country-wide survey based on Citizen Science. *PLoS ONE, 7*(9), 45822.

DeVictor, V., Van Swaay, C., Brereton, T., Brotons, L., Chamberlain, D., Heliola, J., et al., (2012). Differences in the climatic debts of birds and butterflies at a continental scale. *Nature Clim. Change, 2*, 121–124.

Gargominy, O., (2003). Biodiversity conservation in overseas French districts. *Collection Planète Nature. Paris: Comité Français pour l'UICN.*

Gargominy, O., Tercerie, S., Régnier, C., Ramage, T., Schoelinck, C., Dupont, P., et al., (2016). *TAXREF v10. 0, Taxonomic Repository for France: Methodology, implemetation and dissemination,*. Muséum national d'Histoirenaturelle, Paris. Rapport SPN–101.

GBIF., (2017). Global Biodiversity Information Facility. Systèmemondiald' informationsur la Biodiversité. http://www.gbif.fr/.

Giguet-Covex, C., Pansu, J., Arnaud, F., Rey, P. J., Griggo, C., Gielly, L., et al., (2014). Long livestock farming history and human landscape shaping revealed by lake sediment DNA. *Nature Commun., 5*, 3211. doi: 10.1038/ncomms4211.

Guilbert, E., Robillard, T., Jourdan, H., & Grandcolas, P., (2014). Zoologia Neocaledonica 8. Biodiversity Studies in New Caledonia. *Mémoires du Muséum National d'Histoire Naturelle, 206*, 1–315.

Hewitt, G. M., (1999). Post-glacial recolonization of Europeanbiota. *Biol. J. Linn. Soc., 68*, 87–112.

INPN., (2017). National Inventory of the Natural Patrimony. https://inpn.mnhn.fr/.

Issa, N., & Muller, Y., (2015). *Atlas of Birds of France. Nesting and Winter presence* Delachaux & Niestlé.

Lescure, J., & De Massary, J. C., (2012). *Atlas of Amphibians and Reptiles of France.* Biotope & Muséum national d'histoire naturelle.

Metzger, M. J., Bunce, R. G. H., Jongman, R. H. G., Mucher, C. A., & Watkins, J. W., (2005). A climatic stratification of the environment of Europe. *Global Ecology and Biogeography*, DOI:10.1111/j.1466-822X.00190.

Muller, S., (2006). *Protected plants of Lorraine: Distribution, Ecology and Conservation.* Biotope.

Prentice, C., Guiot, J., Huntley, B., Jolly, D., & Cheddadi, R., (1996). Reconstructing biomes from palaeoecological data: A general method and its application to European pollen data at 0 and 6 ka. *Climate Dynamics, 12*, 185–194.

Savouré-Soubelet, A., Aulagnier, S., Haffner, P., Moutou, F., Can Canneyt, O., Charrassin, J. B., & Ridoux, V., (2016). *Atlas of feral mammals of France: Marine mammals*. Muséum national d'Histoire naturelle, Paris & IRD, Marseille, vol. 1.

Schmitt, T., (2007). Molecular biogeography of Europe: Pleistocene cycles and postglacial trends. *Frontiers in Zoology, 4*, 11.

Taberlet, P., Fumagalli, L., Wust-Saucy, A. G., & Cosson, J. F., (1998). Comparative phylogeography and post glacial colonization routes in Europe. *Mol. Ecol., 7*, 453–464.

Tison, J. M., & De Foucault, B., (2014). *Flora Gallica: Flora of France*. Biotope.

Witté, I., & Touroult, J., (2017). Identification and mapping of poorly known areas for biodiversity at the national scale estimated from publicly shared data. Rapport SPN 2017–6. MNHN. Paris.

CHAPTER 3

Biodiversity in Georgia

GURAM ALEKSIDZE,[1] GIVI JAPARIDZE,[1] ANATOLI GIORGADZE,[1] and TAMAR KACHARAVA[2]

[1]Georgian Academy of Agricultural Sciences, I. Javakhishvili str. #51, 0102, Tbilisi, Georgia, E-mail: guram_aleksidze@yahoo.com; japaridze.givi@yahoo.com; anatoli5@mail.ru

[2]Georgian Technical University, Kostava 77, 0175, Tbilisi, Georgia, E-mail: thamkach@mail.ru

3.1 Introduction

3.1.1 Geographical Location

Georgia is located between 41°07'–43°35' of north latitude and 40°05'–46°44' of east latitude and occupies 69,700 square kilometers of central and western Transcaucasia. It is a mountainous country, with 53.6% of total territory occupied by mountains, 33.4% by foothills, and 13% by lowland. Vertically, the territory rises from the Black Sea level up to the 5,070-meters Mt. Shkhara. The Likhi mountain-ridge plays an important role in the development of contrast between the landscapes of western and eastern Georgia. In the west, a humid subtropical climate dominates, while in the east, the climate varies from humid subtropical to continental. Complex and diverse geographical features extend both in the horizontal and vertical directions, forming a distinct character for each part of the territory. The average variance of annual precipitation for different areas is rather high. Although the arid eastern part does not exceed 500 mm/year, the western seacoast receives 4500 mm/year.

3.1.2 Climate Condition

The climate in Georgia is quite versatile. Almost all types—from glaciers and eternal snow to subtropical—are present. The most warm and humid is the Black Sea coast and the Rioni river lowland (average annual air temperature reaches 15°C). As height increases by 100 m, temperature falls by approximately 0.3–0.5°C. On average, January is the coldest month of the year, though the lowest absolute temperature was registered in February. The east Georgian lowland and the northern part of the Abkhazia seacoast are notable for the highest average monthly temperature (24–25°C in July). In both these regions, sunny hours reach 2,500 hours annually. The annual

sum of temperatures above 10°C is the highest in the Gardabani and Mar-
neuli regions in the east (over 4,100°C), and in the Sukhumi region on the
western coast (4,632°C). The frostless period exceeds 300 days at the Black
Sea coast; at the territory 2,000 m a.s.l., it decreases to 120 days and above
3,000 m, there are no frostless days. (Didebulidze, 1997).

3.1.3 Soil Condition

Among the soil types, red podzol dominates in the subtropical zone of western Georgia; brown, chestnut and black lands in eastern Georgia; mountain black land in the Meskheti-Javakheti volcanic upland; mountain wood
brown, brown wood and humus carbonate soils in the western and eastern
mountain forests, and mountain-field in the high mountainous zone. In the
inner regions' marshes, salted and alutaceous aluviar soils dominate. Among
the types of soils, mountain-forest brown and mountain-field are most common, followed by yellow and red, alutaceous, and mountain brown (Sabashvili, 1995) (Figure 3.1).

3.2 Global Significance of Georgian Biodiversity

There are 34 hotspots identified at present on the earth, among which two
are Caucasus (partially) and Iran-Anatolia, which cover Caucasus region
and Georgia as well. Caucasus is among 200 global ecoregions which are
selected by the World Wildlife Fund (WWF) considering such criteria as:
plant and animal species diversity, level of endemism, taxonomic uniqueness, evolutionary processes and the flora and fauna historical development
characteristics, vegetation types and variety of biomes rarity at the global
scale.

From the floral point of view Georgia is one of the richest among the
countries of moderate climate. About 21% – 900 species of higher plants –
are listed in regional Red Books of Rare and Endangered Species, including
600 Caucasus and 300 Georgian endemics. Georgia represents one of the
centers of cultural plants origin and diversity. It has developed the vine, cereals, fruit and many wonderful species.

More than 6,350 species of vascular plants are found in the Caucasus. A
quarter of these plants are found nowhere else on Earth – the highest level
of endemism in the temperate world. Thus, the unique phytogenetic pool is
a live monument of natural-cultural heritage; therefore, its study, protection
and restoration are very important.

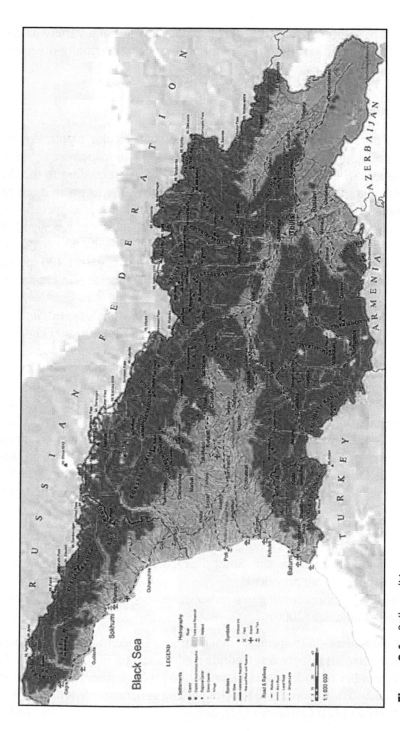

Figure 3.1 Soil conditions.

In spite of Georgia's relatively small area, as a result of a variety of geographical and climatic zones the country possesses an unusually diverse of flora and fauna.

3.3   Diversity of Flora

Georgia's flora comprises about 4,100 species of vascular plants. Among them: Pteridophyta – 74; Gymnospermae – 17; Angiospermae – 4,009 (Dicotyledoneae – 3,254; Monocotyledoneae – 755). According to the floristic – ethnographic districts of Georgia, the number of species amounts to: 1,978 species – in Abkhazeti; 1,100 species – in Svaneti; 1,200 species – in Racha-Lechkhumi; 900 species – in Imereti; 1347 species – in Khevi; 1,000 species – in Tusheti-Khevsureti; 1,125 species – in Shida Kartli (Ksan-Likhvi); 1,400 species – in Kakheti; 1,650 species – in Trialeti; 1,400 species – in Meskheti; 1,500 species – in Javakheti and 200 species – in Gardabani (Gagnidze, 2000).

High endemic level of Georgia's flora shows its richness. In Georgia, approximately 21% of flora is endemic and consists of about 900 endemic species. Among them about 600 species are Caucasian endemic species and about 300 are Georgian endemic species. The number of species in the Caucasus amounts to 6,350; while the number of endemic species to 1,500.

Generic endemism of Georgia's flora is high enough. In the Caucasus and Georgia's flora 16–17 genera are represented. Georgia's endemic (and subendemic) genera are oligotypic and monotypic. The Western Caucasus flora comprises following endemic genera: *Alboviodoxa, Woronowia,* and *Chymsydia*. Greater Caucasus endemic genera are: *Trigoncaryum, Symphyoloma, Pseudobetckea, Charesia, Mandenovia,* and *Scredynskya*. The whole Caucasus endemic genera are: *Grossheimia, Cladochaeta, Pseudovesicaria, Sobolewskya, Gadellia, Agasyllis, Paederotella,* and *Kemulariella*.

Systematic structure of Georgia's flora is characterized by diversity; the structure is Mediterranean and South-European, i.e., Sub-Mediterranean families and genera are prevalent.

In Georgia's flora according to number of species, 10 leading families are singled out (Gagnidze, 2000):

1   ***Compositae*** – 538 species. Total number of endemics – 131. Numerical correlation between Georgia and Caucasus is 51/80 (where 51 is number of endemics in Georgia and 80 is number of endemics in Caucasus). According to species leading genera are: *Hieracium* (47), *Cirsium* (43), *Centaurea* (32), *Senecio* (27), *Pyrethrum* (20), *Anthe-*

mis (19), *Tragopogon* (15), *Alchillea* (14), *Inula* (13), and *Psephellus* (13). Family consists of 4 endemic genera – *Cladochaeta, Alboviodoxa, Kemulariella,* and *Grossheimia.*

2 **Gramineae** – 332 species. Family is poor with endemics. It consists of about 15 Caucasian endemics; they are distinctly differentiated. Leading genera are: *Festuca* (20), *Poa* (15), *Stipa* (9), *Elytrigia* (9), *Bromus* (8), *Bromopsis* (7), *Agrostis* (7), *Aegilops* (7), and *Trisetum* (7).

3 **Leguminosae** – 322 species. About 89 species are endemic; and 34/45 are leading genera. They are: *Astragalus* (72), *Trifolium* (40), *Vicia* (33), *Medicago* (22), *Onobrychis* (19), *Lathyrus* (14), *Trigonella* (10), *Genista* (10), *Hedysarum* (7), and *Orobus* (6). Caucasian-Front Asian genus – *Vavilovia* is included in Georgia's ecotypes.

4 **Rosaceae** – 238 species. About 118 species are endemics. 63/58 are leading genera: *Alchemilia* (61), *Rubus* (36), *Potentilla* (31), *Rosa* (30), *Sorbus* (12), *Pyrus* (11), *Cretaegus* (8), *Cerasus* (4), and *Geum* (3). In Georgia's flora the family is represented by one endemic genus – *Woronowia.*

5 **Cruciferae** – 183 species. About 34 species are endemic. 11/23 are leading genera: *Draba* (16), *Erysiumum* (13), *Arabis* (9), *Isatis* (8), *Thlaspi* (7), *Alyssum* (7), and *Dentaria* (4). The family in Georgia's flora is represented by one endemic ultra-oreophylic lytofite genus – *Pseudovesicaria.* Out of rare genera there are found Caucasus-Front Asian – *Coluteocarpus* and Mediterranean – *Alyssoides.*

6 **Scrophulariaceae** – 179 species. 52 species are endemic. 14/38 are leading genera *Veronica* (45), *Verbascum* (29), *Scrophularia* (24), *Pedicularis* (14), *Euphrasia* (13), *Melapmyrum* (8), *Rhinanthus* (7), and *Digitalis* (4).

7 **Umbelliferae** – 177 species. 58 species are endemic. 21/37 are leading genera: *Heracleum* (22), *Bupleurum* (13), *Chaerophyllum* (11), *Anthriscus* (8), *Pimpinella* (7), *Peucedanum* (6), *Carum* (5), *Pastinaca* (5), and *Ligusticum* (4). In Georgia's flora 4 endemic genera are found – *Chymsydia, Symphyoloma, Mandenovia,* and *Agasyllis.*

8 **Labiatae** – 149 species. 26 species are endemic. 9/17 genera are leading ones: *Stachys* (18), *Salvia* (13), *Nepeta* (13), *Scutellaria* (13), *Thymus* (11), *Teucrium* (7), *Ziziphora* (6), *Betonica* (6), *Lamium* (5), and *Dracocephalum* (4).

9 **Caryophyllaceae** – 135 species. 47 species are endemic. 10/37 genera are leading ones: *Silene* (34), *Cerastum* (25), *Dianthus* (22), *Minu-*

arttia (21), *Gypsophylla* (9), *Stellaria* (7), *Arenaria* (7), *Melandrium* (6), *Saponaria* (4), and *Spergularia* (4). Endemic lytophylic genus – *Charesia* – is found in Greater Caucasian flora.

10. **Liliaceae**-(S.L) – 129 species. 34 species are endemic. 10/24 are leading genera: *Allium* (34), *Gagea* (18), *Muscari* (12), *Scilla* (9), *Ornithogalum* (7), *Lilium* (6), *Asparagus* (6), *Fritillaria* (5), *Polygonatum* (5), and *Colchicum* (4).

Florocoenotic complexes according to their species, ecology and botanical – geographical factors are diverse; especially, tall-herbaceous, limestone ecotope, xerophilous, high-mountain, Colkhian, and coastal wetland complexes.

Out of rare and endemic complexes of wetland, water and swamp complexes with endemic species (*Nymphaea colchica, Trapa colchica,* and *Nuphar lutea*) are most noteworthy. *Kosteletzkya pentacarpa* complexes in grass-spagnum swamps and *Hibiscus ponicum* complexes in Alder thickets are unique.

Limestone ecotopes of Colkhian botanical-geographical provinces are characterized by diverse biotopes. Excess of calciphilous species, high level of endemism and biodiversity are determined by diverse ecological conditions: warm and temperate climate, mountain layers with different chemical composition and abundance of limestone.

In limestone ecotopes dispersion of endemism is observed. The representatives of several genera are found there. A great number of species are characterized by local spreading. Age of limestone ecotopes' flora complexes species are distinct and young and relict species' groups are well separated from each other.

Colkhian refuge is distinguished by its diverse complexes. Among them peculiar phenomenon has to be noted for Colkhida – evergreen broadleaf bushes complexes, that is called Colkhian evergreen understories. Species of following genera: *Rhododendron, Epigaea, Ruscus, Ilex, Daphne, Hedera, Laurocerasus* are found in this complex.

Natural habitats of evergreen dendroflora species are diverse too. Some species have Colkhic-Lazistan nature habitat (*Rhododendron ungernii, R. smirnrnowii, Epigaea gaultheriodes, Ruscus colkhicus,* etc); species – (*Rhododendron ponticum, Laurocerasus officinalis*) between the Caucasus, the Mediterranean and Western Europe are characterized by disjuncted spreading. Outside of Colkhida these species are in a condition of "dying relicts," when in the Colkhian ecosystems they are in a condition of progressive relict. This is the peculiarity of mezophyle evergreen dendro-flora (Gagnidze, 2000).

3.4 Basic Biomes of Georgia

Georgia is distinguished by a great diversity of biomes. It is explained by the diversity of the physicogeographic and climate conditions of the country and in addition by its location at the junction of the phyto-landscapes of different origin. In Georgia, on a relatively small territory, there are multiform ecosystems starting with the near-mountain semideserts of East Georgia and the dense forests of almost subtropical damp climate of Colkhida of the same zone, and finishing with the peculiar biomes of the severe climate of high mountains. The complex relief (especially in the central and eastern parts of the Caucasian mountain range) and complicated configuration of the mountain ranges resulted in the geographical and ecological (environmental) isolation of the species and ecosystems. This could be explained by high level of local endemism (Nakhutsvrishvili, 2000).

The difference between the climate conditions of eastern and western Georgia, the difference of their ecosystems is also revealed in the vertical structure of the zone. In eastern Georgia, the forest zone of semiarid and arid vegetation does not exist at all; forested valleys and foothill slopes start from the very seashore. Therefore, there are only five main zones: the forest zone (to 1900 m), the sub-Alpine zone (1900–2500 m), the Alpine zone (2500–3000m.), the subnival zone (3000–3600 m), and the nival zone (>3600 m). In eastern Georgia, the vertical zone system is more complicated, 6 main zones are found here: the semidesert zone, the dry field and arid, the light forest zone (150–600 m), the forest zone (600–1900 m), the sub-Alpine (1900–2500 m), the Alpine (2500–3000 m), the subnival (3000–3700 m), and nival (>3700 m) zone. In the forest and sub-Alpine zone of the highland of Southern Georgia there are occasional forestall formations of the semi-arid ecosystems with the predominance of mountain grassland vegetation (Nakhutsvrishvili, 2000).

3.4.1 The Biome of the Desert and Semi-Desert

The low-lying marshy lowlands of eastern Georgia are occupied by the semi-desert biomes with intermittent fragments of the solicited desert with the participation of *Salsola ericoides, S. dendroides, Jamanthus pilosus, Suaeda microphylla, Petrosimonia brachiata, Kalidium caspicum*. For this type of the desert vegetation the presence of ephemeral organisms such as *Poa bulbosa, Colpodium humile, Bromus japonicas, Eremopyron orientale, Alissum desertorum*, and others are characteristic. One of the fragments of the desert biome is represented by the communities of *Nitraria schoberi*, which are spread in Shida Kartli (Inner Kartli), Kartli, and Meskheti. One of the variations of the erosive desert is the vegetation covering the washed-away soil of the river Iori, where *Festuca sulcata, Stipa szovitsiana, Artemisia fragrans* and other species can be found. It is the place where *Tulipa eichleri*, a very rare endemic plant grows.

One of the main dominants of the semidesert biome is *Artemisia fragrans* (some botanists assign it to desert vegetation), which is widely spread in eastern Georgia, particularly on the plateau of the river Iori and in Kvemo Kartli (Lower Kartli). The semidesert ecosystem populated by wormwood is *Artemisieto-Salsoletum dendrites*; it grows on the area rich in loam and black soil. Its flora is poor and consists of only 26 species.

3.4.2 Biome of the Steppe

The steppe vegetation is spread in eastern Georgia, a little higher altitude than the semidesert (300–700 m). Due to the anthropogenic influence the steppe biome is interspersed by the elements of the forest, a dry, light forest and bush vegetation. The soil is black, in some places it changes into the chestnut soil type. Within the steppe biome, the climate is dry and subtropical with some features of the continental climate; the winter is rather dry and the summer is hot. The snow cover is insignificant and unsteady. One of the most characteristic ecosystems of the steppe biome is a steppe where *Bothriochloa ischaemum* dominates. Some botanists consider such steppes to be semisteppes, others think that they are semi-Savannah's or savannoids. To prove it they refer to the presence of the Savannah components *Imperata cylindrica* and *Erianthus purpurascens* in the steppes together with *Bothriochloa ischaemum*. The scientists (Nakhutsvrishvili, 2000) share the opinion of those botanists who consider such steppes one of the variations of the steppe. The main component of these steppes is *Bothriochloa ischaemum*, partially *Festuca sulcata*, and following them is *Stipa capillata, S. lessingiana, S. pulcherrima, Koelera macrantha, Cleistogens bulgarica, Glycyrrhiza glabra, Onobrychis kachetica*, and others.

3.4.3 The Biome of the Light Forest and the
Semi-Xerophilus Shrubbery

In the zone of semideserts and steppes of eastern Georgia the vegetation of arid light forest is widespread. This biome consists of xerophilous plants of the Forest and the grass cover, which is quite drought-resistant. This biome is most widespread between the plateaus of the Alazani and Iori, rivers and in Vashlovani reserve; it occupies 5000 hectares. Its basic ecosystems are: *Pistaceeta mutica,* gunipereta *(Guniperus fetidissima, G. policarpos),* and *Pyreto-Calteeta (Pyrus salicifolia, Celtis caucasica).*

3.4.4 The Forest Biome

The forest in Georgia occupies a relatively larger area than other vegetation types. It covers 36.7% of the country's territory. Different dominants of the forest cover various territories of the country. For example, *Fagus orientalis* occupies 51% of the area, *Abies nordmanniana* – 10%, *Quercus iberica* and various species of oaks – 33%, *Picea orientalis* – 6.3%, *Pinus kolchiana* – 3.6%, *Alnus barbata* – 3%, *Castanea sativa* – 2.1%, *Betula litwinowii* and other species of *Betula* – 2%. Georgia's remaining forest area is covered by *Carpinus caucasica, Lilia caucasica, Acer platanoides, A. trautvetteri, Fraxinus excelsior,* and others. Only the Javakheti plateau is covered with forests. A very small territory of Khevi and Mountainous Tusheti is forested.

3.4.5 Orobiomes

The high-mountain vegetation of the Caucasus and particularly of Georgia is distinguished by a great diversity which in the first place is conditioned by the location of the Caucasian range at the junction of Europe and Asia which have entirely different landscapes, also by the climatic contrasts, because of its interesting relief and other unique characteristics. Particular focus should be made on the vegetation of the upper border of the forest of sub-Alpine zone (2400–2750 m) pointed with its huge diversity of flora and *phitocenosis* and great abundance of endemic and relict species. The vegetation of the sub-Alpine zone is characterized by the following formations: (i) light forests, (ii) forests with deformed trees, (iii) creeping shrubs, (iv) tall grass, and (v) broad-leaved meadows (Nakhutsvrishvili, 2000).

3.4.5.1 The Orobiome of the Sub-Alpine Zone

The Alpine zone in Georgia is spread at the altitude of 2400–3000 m. Here, the following basic ecosystems occur: Alpine meadows seed bearing and multiform grass, Alpine shrubbery and microgroups (microsystems) of vocks and crumbled areas.

3.4.5.2   Sub-Nival and Nival Orobiome

The subnival zone (3000–3600 m) is distinguished by an entirely different setting. It is characterized by the abrupt fluctuations of temperature (especially in the topsoil), the high sun radiation with quite a wide specter of ultra-violet rays, low-partial pressure, rather a great mechanical pressure of a mobile substrate upon the plant. It is quite natural that only a particular group of plants is adapted to such extreme environment; it is the group, which for its long evolutionary process has worked out metabolic, rhythmic and morphological mechanisms of adaptation to the environment.

The Subnival belt in Georgia's highlands is not clearly revealed in the eastern and central parts of the Caucasian range. Under the influence of climatic changes the biome of the semidesert, that undergoes a strong anthropogenic stress, may turn into a desert biome (Nakhutsvrishvili, 2000).

3.5   Diversity of Forest

The forest is one of the leading providers of natural resources in Georgia. They can be characterized for their great diversity of geographical location as well as growing conditions. The forest creates unique, manifold and inter dependent biocenosis. The main factor that determines above mentioned diversity is the naturally formed surface or relief of the country.

The total area of Georgia's forested territories is 2,752,000 hectares. The supply of timber amounts to 434 million cubic meters. The annual average timber increase is 4.5 million cubic meters (Gigauri, 2000). Forest vegetation is most prominent in the mountains. Georgia's forests differ greatly depending on geographic location, environmental, biological or other factors. The forests are conditioned by the great diversity of natural conditions in which they grow and develop.

3.5.1   The Forest in a Balanced Ecological System

The change or disappearance of one of the components of this system brings about the partial or complete change of the ecosystem as a whole. Forests in the Caucasus and particularly in Georgia show the unique biological diversity regarding its origin, growth, development, and composition, as well as other characteristics. In comparison with European forests, forests in Georgia are distinguished by a wide variety of landscape. The vast majority of European forest cenosis is of artificial origin, and thus it possesses limited and uninvolved biodiversity in comparison to natural ecosystems which are rare in Europe.

One of the most important defining factors of the biological diversity of forests is the number of species they contain. Georgian forests are populated by a great many species of trees and bushes (approximately 400). Each one is an inseparable part of the ecosystem as a whole and encompasses its own microcenosis. On an individual level they cannot create an independent ecosystem on a wide scale. Most of them occur in the groves where they are mixed with the species that prevail there and make up separate biological groups. This is precisely the manifestation of the wide spectrum biological diversity of the forest composition.

According to Gigauri (2000), the plant forms of Georgia are divided into the following groups: trees, the number of species – 153; bushes – 202 species; semibushes – 29 species and lianas – 11 species. Georgia's 153 species of trees are divided according to height: tall trees (25+ meters) represent 51 species.

The most noteworthy of them are: Caucasian Abies, Oriental fir trees, Sosnovsky or bitovinda pine trees, Oriental beech trees, ordinary chestnut trees, oaks, Caucasian hornbeams, alder trees, ordinary ash trees, line trees, etc.

- mid-sized trees (from 7 to 25 m) – 56 species, among them the 5 varieties of birch, and wild apple trees, wild pear trees, sour plum trees, persimmon trees, box-trees, etc.
- small trees (less than 7 m) make up 46 trees species, among them red juniper trees, Pontian oak trees, buckthorn trees, birch bark trees, pomegranate trees, Cornelia cherry trees, etc.

One of the important indicators of the biodiversity of Georgia forests is the great number of relict and endemic trees and plants. It is impossible to enumerate all of them within the scope of the present chapter, but the following precious relict and endangered flora are noteworthy:

Hartvisi Oak tree, Supin tree, bitchvinta pine tree, Colchian box tree, yew-tree, Imeretian Oak tree, fig-tree, Pontic rhododendron, Georgian maple tree, strawberry, ordinary persimmon, juniper, etc. Most of these varieties are included into the Red Book of Georgia. (Japaridze, 2001).

The species that are found in Georgia's forests and adjacent specific territories vary greatly, even within small areas. Firstly, it depends on the relief of the territory, the variety of the soil and climate conditions. The interrelation of soil and climate conditions defines the character of the productivity and diversity of forest ecosystem. The forest vegetation and the representatives of fauna populating the forest are directly or indirectly depend on the depth of the soil, on its physical and chemical qualities, such as humidity, the quality of the food and the characteristics of the aboveground microclimate. It is true that these environmental factors define the general process of the formation of forest cenosis, but along with this, the forest introduces some direct factors to the character of the landscape, thus presenting vital force of environment formation. The existence of mixed and homogeneous groves in nature indicates the biodiversity and complex structure of the forests. Despite the presence of many species of trees and bushes in Georgia's forests, the forests are primarily represented by groves predominantly composed of timber species. For instance, groves of coniferous trees (spruce, fir or pine trees) occupy only 19% of the forest territory, while larch forests (beech, oak, chestnut, birch, etc., which intermittent bushes) occupy 81%.

The largest territories of Georgia's forests are occupied by Oriental beech groves. They are spread on 45% of the forested land; oak groves occupy 10.5%; Abies groves – 8.5%; hornbeam groves – 6.6%; fir-tree groves – 5.8%; pine-tree groves – 4.7%; birch-tree groves – 3.1%; chestnut tree groves – 2.5%, etc. (Japaridze et al., 2015; Gigauri, 2000).

The spread of forests in vertical zones and the degree of slant of the slopes define the biodiversity of the forests. Forests on vertical zones are not equally distributed, thus 26.8% of the forests are spread below the elevation of 1000 m a.s.l. and from 1000 m a.s.l. and higher 73.2% of forests occur, and that is, three fourths of the total territory of forests.

The difference in natural conditions in eastern and western Georgia influenced the spread of forests in vertical belts, which in its turn created the biodiversity of the forests. For instance, in western Georgia, the unforested, arid and semiarid vegetation belt does not occur, which is typical in eastern Georgia; at the elevation of 300–400 m to 500–600 m, a so-called arid or light forest belt is spread with a unique composition, origin and disposition of trees. Experts report that the zone is a transitional step from semidesert to forest and can be considered an analogous with subtropical forest-steppes. In Vashlovani, an aspen tree grows which dates back to the Tertiary period. Mixed subtropical forests growing in eastern Georgia are marked by a great abundance of biodiversity.

According to the inclination of the slopes, Georgia's forests are distributed as follows: on slopes with an inclination of 20–22% of the total amount of forests are situated. The greater part of the forests is situated on slopes with an inclination of 78°, 21–35°, 36°, that is on very steep slopes. The above-mentioned facts explain that the great varieties of biodiversity of Georgia's forests are conditioned by vertical zones and the degree of inclination of the mountain slopes. The most unique examples of Georgia's forest ecosystems and biodiversity are its virgin forests spread over approximately 500,000–600,000 hectares. Fortunately, such forests are preserved on large territories of Georgia. These forests have not only national but also broad regional significance because virgin forests no longer exist in European countries and even forests of natural origin are a great rarity on the continent. Virgin forests in Georgia primarily occur on protected areas and on areas situated along the upper points of rivers, on very steep slopes, and in the sub-Alpine zone. These territories are part of State forest fund. Upper and inaccessible points of the rivers are especially rich in virgin forests: those are the rivers Bzipi, Kodori, Enguri, Tskhenistskali, Rioni, Alazani, Liakhvi, and others (Gigauri, 2000).

Virgin forests are populated by groves of beeches, abies, firs, pines, and birches. The evolution and decay of organic substances is an ongoing process here. Virgin forests are represented by vertical united stands and by the complete spectrum of biodiversity. The constitution and phytocenosis of these stands is revealed in their underground or aboveground spread and in their formation.

The virgin forest varies in composition. It passes through all the biological stages during a lifetime of one generation. The change of generations over time has a cyclical character. The completion of one full biological cycle of a stand formation occurs over great many hundreds of years. At the final stages of forest development, the foundation for the emergence and development of a new generation is laid. The most essential changes take place when the oldest part of the stand begins to naturally disintegrate and die, thus returning, over a comparatively short period of time, a huge supply of biological substances to the soil that have been accumulated over hundreds of years. The supply is returned to the soil in the form of dead timber, thus enriching and intensifying the biological process in the soil, which, in its turn, demonstrates a wide range of biodiversity. The trees, reaching the height of 40–50–60 meters and the width of 2–2.5 meters are not rare in such forests. The average supply of timber is 1,000–1,200 cubic meters per 1 hectare, and in some stands it amounts to 1,800–2,000 cubic meters (in the stands where *Abies* dominate).

In the Akhmeta region, in particular in Mta-Tusheti, a unique natural phenomenon occurs: virgin pine forests and virgin birch forest. The pine forests are populated by trees of three generation. They occur at the elevation of 2,000–2,200 meters. Above them are birch forests at the elevation of 2,300–2,600 meters. Experts assume that the pine and birch forests on Mta-

Tusheti have no analogy, not only in Georgia and the Caucasus, but beyond the region as well.

Georgia's multiaged's, multistep (in height and in width as well) according to particular conditions, are presented by stands of varying types of composition.

To illustrate this fact, a scheme is offered which shows the distribution of multiaged *Abies* according to altitude. It also indicates the biodiversity of the stands.

The upper or first floor of the stands is represented by the tallest and thickest trees.

The second floor of the stand, where the trees experience the influence of taller and thicker trees, yet still tend to grow tall, and over a certain period of time reach the upper floor of the stand.

The middle parts or third floor of the stand: the number of the trees makes up 30% of the total.

The subordinate or lowest fourth floor of the stand, where the trees are situated below the trees of the upper floors and thus receive a deficit of light. The growth of these trees is significant.

The fir, spruce, beech, pine and other forests are composed of different biological and forestry components which indicate their biodiversity. Under the cover of these forests the continual process of creation and development of new forest generations takes place. In the vertical direction of the forest, a wall-like, open-work structure of assimilative parts of the plants is formed. They struggle against one another for the sunlight. Because of this, the trees of the lower part of the forest have coniferous leaves or needles primarily on their upper branches which are situated perpendicular to the rays of the sun, resulting in heliotropic deformation of the stem and branches of the upper parts of trees (Japaridze et al., 2015).

Another genetic trait among beeches, spruces and firs is that they can grow very slowly in the shade over tens and hundreds of years (250–300), and then after obtaining more favorable light conditions begin to grow intensely, pass through all the stages of biological development and achieve maximum height and thickness. Such trees occur in great quantity in Georgia's virgin forests on a genetic level have yet to be studied thoroughly (Gigauri, 2000).

Forests that are marked by a great range of biodiversity (composition due to species, structure, age structure, degree of denseness, etc.) are characterized by much higher and broader environmental functions than stands of much simpler composition and structure (Figure 3.2).

Pinus pithyusa   Pinus eldarica

Quercus imeretina   Betula megrelica

Figure 3.2 Georgian endemic trees (from www.wikipedia/georgia).

3.6 Diversity of Algoflora

In the continental waters of Georgia about 2605 taxa of water-plants are registered, out of which Chlorophyta – 1039, Bacillariophyta – 830, Cyanophyta

– 431, Euglenophyta – 161, Xanthopyta – 60, Pyrrophyta – 35, Chrysophyta – 26, Rhodophyta – 10, Charophyta – 13 (Gagnidze et al., 2000).

The majority of floristic ecological researches have shown that in the mountainous part of Georgia's more or less studied rivers and their tributaries, especially in the upper parts, are typical mountainous rivers. Their characteristic features are bentonic and epiphytic algae flora. Here, plankton forms are not developed and the algae flora rheopilic groups dominate, while the leading forms are *Hydrurus foetidus* from Chrysophyta section from Chlorophyta *Ulothrix zanata, Cladophora glomerata* and from Bacillariophyta section *Diatoma hiemale* var. *hiemale et* var. *mesodon, Ceratoneis arcus, Caloneis silicula, Navicula crytocephala, Nitzschia linearis, Cymbella hebridica, Gomphonema angustatum* var. *productum, Didymosphenia geminata*.

The exceptions are the slow currents of the rivers where algo flora is mostly developed. In this area, a great number of the species from Bacillariophyta section and fiber forms of green algae occurs. In these groups also take part blue-green algae flora. In the rivers of the valley is except benton and epypitic algae flora, especially in their slow currents develop plankton algae flora.

The flora of lakes and ponds differ greatly from that of rivers. Plankton algae flora is represented here in considerable quantity, though the flora of bentonic and epiphytic water plants is well developed, also in these ponds the main components of plankton are the types of Bacillariophyta and Chlorophyta and less from Cyanophyta section. In lake planktons, Euglenophyta occurs in small amounts except in dying lakes, where they represent one of the consisting components of algae groups. In various small ponds, pools, and holes an algo flora of different systematic groups is developed. Generally, in these ponds, any algo florist groups are not observed.

In mineral waters, mostly the bentons algae are present. Generally, the mineral waters algae flora is poor and mostly consists of the representatives of the sections of Cyanophyta and Bacillariophyta. According to their quantity of the species, Bacillariophyta is superior in number, but the leading forms are mostly blue-green algae flora. In the more or less investigated various types of swamps of Georgia, the algae is represented in different variety. In different parts of one and the same swamps in different ecological groups, algae flora is developed in different ways.

In the swamps planktons, mostly are met the representatives from Desmidiales and Chlorococcales orders. On the banks of the swamps, the representatives of blue-green and Euglenophyta are well-developed between waters-plants and mosses. There is a small amount of Bacillariophyta in swamps plankton, which better develops in bentons, or settle in stems of water-plants. On the banks of the swamps, among mosses, fiber forms of green algae are predominantly developed (Kukhaleishvili et al., 2000).

3.7 Diversity of Fungi

According to the mycological, phytopathological literature and other sources, up to 6500 species of fungi are recorded in Georgia. If considering unpublished but identified materials, the real figure exceeds 7000 (Gvritish-vili et al., 2000).

Peronosporales, Taphrinales, Erysiphales, Uredinales, Ustilaginales, Agaricales, Ganodermatales, Fistulinales, Hymenochaetales, Poriales, Russulales, Lycoperdales, Phallales, etc., as well as some genera of mitosporic fungi are groups of fungi that have been studied more thoroughly than other groups in terms of species composition.

Macrofungi are more important for mycobiotic complexes of forest ecosystems, since many of them are mycobionts, they are old forest indicator species and have conservation value. Among macrofungi, 'Agarics' (Agaricales, Boletales, Cortinariales, Poriales (Polyporales), Russulales, Schizophyllales) are the perfect example of this. Such ecological groups of macrofungi as mycorrizal (81 species), lignicolous ("xylophilic") (128), litter saprotrophs (132), humus saprotrophs (91), etc. are found in Georgia. The number of species of mycobionts connected with their phytobionts is as follows: *Fagus orientalis* – 81; *Picea orientalis* – 73; *Quercus iberica* and other species – 43; *Pinus kochiana* (mainly) – 33; *Abies nordmanniana* – 1; *Betula* spp. – 4; *Populus tremula* – 4. The lists of mycorrhizal fungi in beech forests, oak forests (with *Carpinus caucasia, C. orientalis*), dark coniferous forests (*Abies nordmanniana, Picea orientalis*), pine forests are given.

There are approximately 200 species of edible fungi reported in Georgia; more than 50 species belong to poisonous or are regarded as suspicious or conditionally suspicious. It is worth mentioning that over 80 macrofungi have original local names in Georgia. From the given number of edible fungi Georgians mainly consume approximately 30 species. Three species are known to be used in folk medicine: *Bovista nigrescens, Inonotus obliquus, Phalus impadicus* (Nakhutsvrishvili et al., 2000).

3.8   Diversity of Medicinal and Aromatic Plants

Recent years, interest in medicinal, aromatic, spicy, melliferous, poisonous and dye plants has been increased, and the potential of their use has been progressing, though, in modern medicine, cosmetology or cookery there are a lot of synthetic-chemical means (Table 3.1). It is natural because the use of the latter is often followed by side effects, like allergies, while medications produced from plants have no harmful effects. Primarily, the effectiveness of herbal means comes from their high biological activity and less toxicity. It is possible to use them for various chronic and acute diseases. The process mentioned above has a great importance because in metabolic processes taking place in ontogenesis period of plants they form very important and precious compounds, like essential oils, alkaloids, glycosides, tanning matters, vitamins or pharmaceutically active substances that have soft and long-term effects, as well as stable results on the human organism. They also have a positive physiologic effect on the organism (Kacharava and Korakhashvili, 2008; Kacharava, 2015) (Figure 3.2).

Georgian scientists studied and unified a single complex model of the following specialty and singularity, which are conditioned by the research and cataloging of genetic resources of the Medicinal, Aromatic, Spicy, Melliferous, Poisonous and Dye plants, including unique plants and those on the verge of extinction, conservation; diagnostics of the indigenous-endemic and collection material for the purposes of selection of the plants distinguished for their pharmacological and farming peculiarities; enrichment of seed bank and its inclusion into the international exchange programs; establishment of the database for the purposes of sustainable use and conservation of the aforementioned plants in certain regions of Georgia with different ecosystems (Kacharava, 2015).

3.9   Diversity of Agricultural Crops

3.9.1   Grapevine

Georgian species of the vine are one of the most ancient ones, and Georgia is acknowledged as the country where the first cultural grapevine forms emerged. In the result of "national selection," more than 500 aboriginal species of grapevine have been selected, domesticated and distributed; Also, a direct predecessor of modern cultural grapevine – *Vitis vinifera sativa,* a wild grapevine – *V. vinifera silvestris*, now included in the Red Book, was identified and described together with other 400 wild vines, including those

Table 3.1 List of Medicinal and Aromatic Plants

S. No.	Family	Species	Order	Genus	Notes
1	Asteraceae/Compositae: Number of Georgian endemics – 44; Number of Caucasian endemics – 88; (Kuchukhidze et al., 2016)	Achillea millefolium	Asterales	Achillea	There are 13 species of this genus identified in Georgia. One of them – Achillea sedelmeyeriana L – is endemic to Georgia. Medicinal and honey plants (Shetekauri et al., 2009).
		Artemisia absinthium	Asterales	Artemisia	There are 13 species of this genus identified in Georgia, including Artemisia fragrans Willd; Artemisia x scoparia Wald et Kit; Artemisia annua L; Artemisia vulgaris L; Artemisia caucasica Willd (Shetekauri et al., 2009).
		Artemisia annua	Asterales	Artemisia	(Kuchukhidze et al., 2016).
		Bidens tripartita	Asterales	Bidens	There are 3 species of this genus identified in Georgia – Bidens tripartita L, Bidens cernua L, Bidens bipinnata L. (Kuchukhidze and Jokhadze, 2012).
		Centaurea cyanus	Asterales	Centaurea	(Kacharava, 2009).
		Cichorium pumilum	Asterales	Cichorium	Only one wild perennial species occurs in Georgia – Cichorium intybus L. Melliferous plant (Kuchukhidze et al., 2012).
		Taraxacum officinale	Asterales	Taraxacum	There are 9 species of this genus identified in Georgia. 4 of them are Endemic to the Caucasus. Melliferous plant (Kuchukhidze et al., 2012).
		Senecio rhombifolius	Asterales	Senecio	Endemic to the Caucasus (Kacharava, 2009).

Table 3.1 (Continued)

S. No.	Family	Species	Order	Genus	Notes
2	Apiaceae/ Umbelliferae: Number of Georgian endemics – 21; Number of Caucasian endemics – 37; (Kuchukhidze et al., 2016).	Heracleum	Apiales	Heracleum	There are 25 species of this genus identified in Caucasus and 21 in Georgia. 6 Endemic to Georgia and 8 Endemic to the Caucasus 2 species – *Heracleum sosnowskyi* L. and *Heracleum mantegazzianum* L. – are widely spread in sub-Alpine zone of Georgia. Medical-aromatic plant (Kuchukhidze et al., 2016).
		Agasyllis latifolia	Apiales	Agasyllis	Endemic to the Caucasus. This species is widespread in Georgia (Kuchukhidze et al., 2012).
		Eryngium planum	Apiales	Eryngium	There are 5 species of this genus identified in Georgia (Shetekauri et al., 2009).
		Sanicula europaea	Apiales	Sanicula	Only this species is identified in Caucasus and Georgia (Shetekauri et al., 2009).
		Pastinaca sativa	Apiales	Pastinaca	There are identified 5 wild and 1 cultivated species of this genus in Caucasus. *Pastinaca sativa* L. is cultivated species (Shetekauri et al., 2009).
		Anthriscus cerefolium	Apiales	Anthriscus	There are 7 species of this genus identified In Georgia. One of them – *Anthriscus ruprechtii* L. – is endemic to the Caucasus 2 – *Anthriscus schmalhausenii* L. and *Anthriscus sosnowskyi* L. are endemic to Georgia. Medical-spicy plant (Kuchukhidze et al., 2012).
		Anethum graveolens	Apiales	Anethum	There are 3 species of this genus identified in Caucasus. Medical-spicy plant (Kuchukhidze et al., 2012).
		Conium maculatum	Apiales	Conium	Medical – poisonous plant (Kuchukhidze et al., 2012).

Table 3.1 (Continued)

S. No.	Family	Species	Order	Genus	Notes
3	Lamiaceae/ Labiateae: Number of Georgian endemics – 9; Number of Caucasian endemics – 17; (Kuchukhidze et al., 2016)	Thymus vulgaris	Lamiales	Thymus	There are 17 species of this genus identified In Georgia, from them 14 Endemic to the Caucasus and 2 – Thymus ladjanuricus L, Thymus tiflisiensis L – Endemic to Georgia. Medical –aromatic –spicy plant (Kacharava, Korokhashvili, 2008; Kacharava, 2015).
		Origanum vulgare	Lamiales	Origanum	Medical –aromatic –spicy – Dye (Kacharava, 2015).
		Mentha spicata	Lamiales	Mentha	There are 4 wild species of this genus identified in Georgia: Mentha pulegium L, Mentha arvenis L., Mentha aquatica L., Mentha longifolia L. Spice Plant (Kacharava, 2015).
		Satureja	Lamiales	Satureja	There are 3 species of this genus identified In Georgia: The cultivated species is Satureja laxiflora L. It is aromatic plant. Satureja spicigera and Satureja montana L., are spicy plants (Qvachakidze, 2009).
		Lamium album	Lamiales	Lamium.	Lamium album L. is widespread in Georgia. Honey plant (Shetekauri et al., 2009).
		Urtica dioica	Rosales	Urtica	Is widespread in Georgia Spicy and dye plant (Qvachakidze, 2009).
		Leonurus quinquelobatus	Lamiales	Leonurus	Two other species are also spread in Geogria and Caucasus – Leonurus glaucescens Bunge and Leonurus persicus Boiss (Gagnidze, 2005).
		Lavandula angustifolia	Lamiales	Lavandula	Two other species are also spread in Geogria and Caucasus Lavandula latifolia L., Lavandula vera L., Medical – aromatic plant (Gagnidze, 2005).

Table 3.1 (Continued)

S. No.	Family	Species	Order	Genus	Notes
		Nepeta cataria	Lamiales	*Nepeta*	There are 12 species of this genus identified in Georgia; 2 of them are endemic to Georgia; 4 of them are endemic to the Caucasus. Medical – aromatic – Melliferous plant (Shetekauri et. al., 2009).
		Rosmarinus officinalis	Lamiales	*Rosmarinus*	*Rosmarinus oficinalis* L. cultivated in Europe and Georgia (Eristavi et al., http://www.mobot.org/ MOBOT/research/georgia/checklist.pdf).
4	Boraginaceae	*Symphytum caucasicum*	Unplaced	*Symphytum*	There are 5 species of this genus identified In Georgia. Medical plant (Kacharava, 2009).
5	Solanaceae	*Atropa caucasica*	Solanoideae	*Atropa*	*Atropa caucasica* L. – is related to *Atropa belladonna* – Endemic to the Caucasus. Medical – poisonous plant (Kacharava, 2009).
6	Valerianaceae	*Valeriana officinalis*	Dipsacales	*Valeriana*	There are 24 species and 4 genera of this family identified in Georgia, including *Valeriana oficinalis* L. and *Valeriana alliarifolia*; One of them – *Valeriana colchica* Utk. is Endemic to the Caucasus. (Kacharava and Korakhashvili, 2006, 2008; Kacharava and Esvanjia, 2007)
7	Plantaginaceae	Greater Plantain	Lamiales	*Plantago*	There are 11 species of this genus identified in Georgia. 2 of them – *Plantago major* L., *Plantago media* L. and *Plantago lanceolata* L. are widespread species. Medical plant (Gagnidze, 2005).
8	Papaveraceae	*Chelidonium majus*	Ranunculales	*Chelidonium*	Medical – poisonous plant (Qvachakidze, 2009).

Table 3.1 *(Continued)*

S. No.	Family	Species	Order	Genus	Notes
		Papaver setigerum	Ranunculales	*Papaver*	There are 13 species of this genus identified in Georgia. One of them is endemic to the Caucasus Medical –food plant (Shetekauri et al., 2009).
9	Amaryllidaceae	*Galanthus caucasicum* (Baker) Grossh	Asparagales	*Amaryllis*	*Galanthus caucasicum* (Baker) Grossh – is endemic to the Caucasus and Vulnerable species. *Galanthus lagodechianus* Kem – Endemic to Georgia. *Galanthus nivalis* L. subsp. *angustifolius* (G.Koss) Artjushenko – endemic to the caucasus, Vulnerable species. *Galanthus platyphyllus* Traub. et Moldenke – Endemic to Georgia. *Galanthus woronowii* Losinsk. – Endemic to the Caucasus, rare species (Eristavi et al., http://www. mobot.org/MOBOT/research/georgia/checklist.pdf).
10	Berberidaceae	*Gymnospermium smirnowii*	Ranunculales	*Gymnospermium*	*Gymnospermium smirnowii* -Endemic to Georgia *Berberis iberica*-Endemic to the Caucasus; *Epimedium circinatocucullatum* – Endemic to Georgia. (Eristavi et al., http://www.mobot.org/MOBOT/research/georgia/ checklist.pdf).
11	Lythraceae	*Lythrum salicaria*	Myrtales	*Lythrum*	There are 7 species of this genus identified in Caucasus, 3 – in Georgia. 2 of them are endemic to the Caucasus (Kuchukhidze et al., 2016)
12	Primulaceae	*Cyclamen vernum*	Ericales	*Cyclamen*	Two other species are also identified in Caucasus. *Cyclamen colchicum* L. – Endemic to Georgia (Kuchukhidze et al., 2016).

Table 3.1 (Continued)

S. No.	Family	Species	Order	Genus	Notes
13	Ranunculaceae	*Helleborus orientalis*	Ranunculales	*Helleborus*	Two other species are also identified in Georgia *Helleborus caucasicus* L, – Endemic to the Caucasus and *Helleborus abchasicus* L – Endemic to Georgia (Eristavi et al., http://www.mobot.org/MOBOT/ research/georgia/checklist.pdf).
14	Apocynaceae	*Vinca herbacea*	Gentianales	*Vinca*	(Eristavi et al., http://www.mobot.org/MOBOT/ research/georgia/checklist.pdf).
15	Orchidaceae	*Orchis purpurea*	Asparagales	*Orchis*	There are 20 species of this genus identified in Georgia. Two of them are endemic to Georgia – *Orchis iberica* L. and *Orchis caucasica* L. (Kuchukhidze et al., 2012).
16	Equisetaceae	*Equisetum arvense*	Equisetales	*Equisetum*	There are 9 species of this genus identified in Georgia. Among them: *Equisetum salvaticum* L., *Equisetum pretense* L., *Equisetum fluviatile* L., *Equisetum paluster* L., etc., Some of them are poisonous (Kuchukhidze et al., 2016).

Original adapted by T. Kacharava.

Equisetum arvense L   *Artemisia absinthium L*

Valeriana alliarifolia L   *Leonurus cardiaca L*

Origanum vulgare L   *Calendula oficinalis L*

Figure 3.3 Georgian endemic medicinal plants (Photos by T. Kacharava, with permission).

which became wild in a course of time. The scientists have also depicted some transitional species, which indicate to the process of "transition" from wild grapevine species, to cultural ones.

Georgian aboriginal vine species, according to their morphological and biological characteristics, belong to the Black Sea basin (*Convar Pontie Negr.*), and Georgian region (*Sub. Convar Georgica Negr.*) subgroups. The majority of grapevine species of Western Georgian, and some of Eastern Georgia, such as: *Kachichi, Tsolikouri, Tsitska, Krakhuna, Otskhanuri Sapere, Chkhaveri, Rkatsiteli, Saperavi*, and others belong to this subgroup. The characteristics typical for those species are as follows: they have a middle-sized grape clusters which have an abundant number of leaves on the tip of the shoots. Some grapevine species of Easter Georgia belong to Eastern Asia ecological – geographical zone group (*Convar orientalis Negr.*) and Caspian Sea (*Sub.convar Caspica Negr.*) group, e.g., *Tavkveri and Shavkapito.* Special characteristics of the species are: the tips of the shoots are not fully covered with leaves, the grape clusters are bigger than average size and are quite juicy (Aleksidze and Chkhartishvili, 2015).

3.9.2   Wheat

Wheat is one of the most important crops grown in the country. In the result of monographic study of wheat spread in Georgia, the unique composition of species for their qualitative and agronomic characteristics has been identified.

Georgian wheat (*Triticum*) has been botanically differentiated and cultivated since ancient times, (neoliths period). It can be distinguished by its suitability for intensive cultivation, high productivity, and good digestibility. Wheat, as testified by archeological researches, is one of the most important food grain crops in Georgia. There have been identified 16 cultivated species of wheat in Georgia, namely: *T. eredvianum, T. monococcum, T. dicoccum, T. timopheevii, T. georgicum, T. carthlicum, T. durum, T. polonicum, T. turanicum, T. aestivum, T. compactum, T. spelta, T. macha*, out of which – 4 species are endemic to the area: *Triticum makha, T. timopheevii, T. zhukovski,* and *T. georgicum.*

The Georgian wheat has been widely used in the breeding of wheat as it represents a rich source of gene conferring resistance to diseases and drought; for example, *T. timopheevii,* is known for its resistance for scab and rusts, which have been incorporated into some improved varieties. Wheat variety *T. carthlicum* is cosmopolite and has spread beyond country borders.

Apart from the broad scope of varieties of wheat discovered and described in Georgia (152 varieties), there are indigenous species (more than 100) a great diversity. Such big number and rich diversity of species of wheat enable us to conclude that Georgia is one of the most important places of wheat origin. Thus, on the basis of comprehensive experimental material, it has been determined that Georgia is one of the breeding centers of the origin of cultivated wheat. The wheat variety (*Triticum* L.) is distinguished with its high-quality characteristics, with a high level of endemism and polymorphism (Khalikulov and Aleksidze, 2010; Naskidashvili et al., 2015).

3.9.3 Some Statistical Information About Agricultural Crops

- **Grapevine (species):** local – 444, introduced – 287, wild – 259 (Aleksidze et al., 2015a).
- **Fruits (seed and stone):** local – 326, introduced – 347 (Aleksidze et al., 2015d).
- **Subtropical crops:** local – 132, introduced – 78 (Aleksidze and Goliadze, 2015).
- **Cereals:** local – 156, introduced – 63 (Aleksidze et al., 2015b).
- **Vegetable crops:** local – 303 (Aleksidze and Kakabadze, 2015).
- **Legume crops:** local – 130 (Aleksidze et al., 2015c).
- **Technical crops**: local – 14 (Aleksidze et al., 2000; Aleksidze and Alpaidze, 2015).
- **Feed grasses**: local – 174 (Aleksidze and Sarjveladze, 2015).

3.10 Animal Diversity

The study of animal populations in Georgia has a long history. Regular studies of Georgian fauna began after the establishment of Caucasian Museum in the nineteenth century. Subsequently, the studies on species diversity of animal world were carried out by the scientific/research staff of Tbilisi State University and Institute of Zoology of the Georgian Academy of Sciences, and by the representatives of different higher education institutions.

At present, it is difficult to evaluate the knowledge on invertebrate animals, but it is possible to make a rough estimate of their great taxonomic diversity. The ratio of known and assumed species is 13553/26312. A relatively complete study has been conducted on nematodes, annelid worms, and several taxa of beetles, Hymenoptera, butterflies, and Diptera. Many groups of invertebrates have not been sufficiently studied. Until present, the vertebrates have been the most completely studied. The species' composition has been defined. It is presumed that approximately 700 species and 684 varieties have already been described.

Studies of dynamics of the spatial distribution of Georgian populations of some groups and endemic status of separate representatives have been conducted. As a result, endemic species have been revealed.

In the twentieth century, the number of species of animal world has been reduced dramatically. About 6 out of 18 species of predatory mammals are on the verge of extinction (CR) in Georgia. One out of eight species of ungulates inhabiting in Georgia became extinct, two species are endangered and one is on the verge of extinction (CR). The conditions of three species of *Cetaceans* have not been studied yet. As a result of human activities, habitats of wild animals were destroyed, and distribution range of many predatory animals and ungulates were fragmented. It is important to take effective, onsite protection measures in order to preserve the biological diversity of the animal world in Georgia. It is necessary to apply the methods of reintroduction and translocation to restore some endangered species. Taxonomic research to clarify diversity of species of Georgian fauna needs to be continued emphasizing the priority of protection and restoration of number of species (Badridze et al., 2000).

3.10.1 Invertebrates

A wide diversity of Georgian Nature determines large variety of vertebrate fauna, among which are many endemic species. Different groups of animals differ according to level of scientific studies carried out, though some separate taxa have been thoroughly investigated. Among these are: Lepidoptera, Geometridae, bugs (Coleoptera: Curculionidae, Carabidae), flatworm (Nemathelmintes), *Hymenoptera era, Hemiptera psylloidea.*

The following regions of Georgia stand out for the reason that there is a big number of endemic and other species of vertebrates facing the danger to become extinct, these regions are: The Big Caucasus Mountain range; Kolkheti, Borjomi Valley, Ivris Elevation, Southern front hills of Meskheti Mountain range (Rukhadze et al., 2005).

3.10.2 Vertebrates

3.10.2.1   Fish

There are more than 80 species of fish developed in Georgian freshwaters, out of which many are endemic. For example, out of 12 species of fish spread in Mtkvari River basin, 9 species are endemic of Mtkvari River and its estuaries. Some of the most widely spread species are: *Barbus lacer, Barbus mursa,* and *Barbus capito.*

Six species out of those existed in Black Sea are endemic; besides, five species of sturgeon family are presented, and one which belongs to sturgeon family – Atlantic Stergeon (*Acipenser sturio*) – is on the verge of extinction. In addition to local ones, there are nine introduced species and the most widespread is *Carasius carasius* (Rukhadze et al., 2005).

3.10.2.2 Amphibians

Twelve species of amphibians are common in Georgia among which Caucasus Salamander, *(Mertensialla caucasica)*, *Pelobates syriacus*, *Pelodytes caucasicus* are the most frequent. The forests of mountainous Kolkheti region represent a significant habitat of wide variety of amphibians. Gardabani Valley is known as the area where *Pelobates syriacus* is found (Arabuli, 2005).

3.10.2.3 Reptilians

More than 50 species of reptilians are known in Georgia, among which are: 3 species of tortoise, 27 species of lizards and 23 species of snakes; out of those species: 3 species which belong to family of Pelias, and 12 species of lizards which belong to the family of Archaeolacerta – are the representatives of Caucasus endemic species. Among the Caucasus endemics there are also the following species: *Elaphe hohenackeri, Pelias kaznakovi*. Those species worldwide are identified as endangered ones (Rukhadze et al., 2005).

3.10.2.4 Birds

There are more than 300 species of birds in Georgia. The territories around Kolkheti Valley including Black Sea coast and the lake of Paliastomi, also lake system of Javakheti highlands are important places where the migrant birds rest and hibernate. Among the bird species which are most prevalent in Georgia – three belong to Caucasus endemic species: *Tetrao mlokosiewiczi, Tetraogalus caspius* and *Phylloscopus lorenzi* (Arabuli, 2005).

3.10.2.5 Small-Size Vertebrates

About 79 species, which belong to four different order series, are found in Georgia. These are: insect-eaters – 10 species; hand-wings – 20 species; vermin – 39 species, and rabbit-like – 1 species.

Out of small-size vertebrates, the following Caucasus endemic species should be distinguished: *Sorex caucasica, Sorex volnuchini, Talpa caucasica, Neomis schelkovnikovi, Sicista caucasica, Sicista khlukhorica, Sicista kazbegica, Prometheomys schaposchnikovi, Chionomys gud*, and others. Also, a few which do not represent endemic species can be listed, such as: *Suncus etruscus, Sciurus anomalu, Allactaga elate, Rhinolopus euriale, Rhinolopus mehelyi, Myo emarginatus*, and others.

Besides the species named above, there are some which have been introduced into Georgia during different periods. These are the following: *Sciurus vulga, Myocastor coypus, Ondatra zibethicus* (Arabuli, 2005).

3.10.2.6   Large-Size Vertebrates

There are 60 species of vertebrates belonging to three order series of vertebrates in Georgia: wild, artiodactyls, and whale-type. Since the twentieth century, the areas where those animals inhabited have been reduced, also, the process of massive reduction in a number of large-size vertebrates started. Today, many of abovementioned species face the danger of becoming extinct. Only a few numbers of leopard and striped hyena have survived. A population of South Niamori (once inhabited on the territory of Trialeti Hill) and dorcas have disappeared totally.

Among large-size vertebrates, two species of goat should be mentioned: *Capra cylindricornis* and *C. caucasica,* which belong to Caucasus endemic species (Rukhadze et al., 2005).

3.10.2.7   Mammals

- Mediterranean horseshoe bat – *Rhinolophus euryale* Blasius – has become vulnerable in the whole world;
- Mehely's horseshoe bat – *Rhinolophus mehelyi* Matschie – Vulnerable in the whole world;
- Bechstein's bat – *Myotis bechsteinii* Kuhl – Vulnerable in the whole world;
- Barbastelle/Western barbastelle – *Barbastella barbastellus* Schreber – Vulnerable in the whole world;
- Caucasian squirrel, or Persian squirrel – *Sciurus anomalus Güldenstaedt* – Under stress as a result of being moved to other habitat;
- Eurasian beaver or European beaver – *Castor fiber* Linnaeus – Extinct in the nineteenth century;
- Caucasian birch mouse – *Sicista caucasica* Vinogradov – only very small area;
- Kluchor birch mouse – *Sicista kluchorica* Sokolov et al. – only very small area;

- Kazbeg birch mouse – *Sicista kazbegica* Sokolov et al. – only very small area;
- Nehring's blind mole-rat – *Nannospalax nehringi* Satunin – only very small fragmented area;
- Gray dwarf hamster, Gray hamster or Migratory hamster – *Cricetulus migratorius* Pallas – decrease of density of population;
- Mesocricetus – *Mesocricetus brandti* Nehring – only very small fragmented area;
- Long-clawed mole vole – *Prometheomys schaposchnikovi* Satunin – only very small fragmented area;
- Bank vole – *Clethrionomys glareolus ponticus* Schreber – Spotted area;
- *Meriones tristrami*, known as Tristram's jird – *Meriones tristrami* Thomas – only very small fragmented area;
- Harvest mouse – *Micromys minutus* Pallas – only very small fragmented area;
- Jungle cat/Reed cat or swamp cat – *Felis chaus* Schreber – Reduction of area;
- Eurasian lynx – *Lynx lynx* Linnaeus – only very small fragmented population;
- Leopard – *Panthera pardus* Linnaeus – Very small population, rare species;
- Tiger – *Panthera tigris* Linnaeus – Extinct in the twentieth century;
- Striped hyena – *Hyaena hyaena* Linnaeus – only very small population;
- Eurasian otter/European otter, Eurasian river otter, common otter, and Old World otter – *Lutra lutra* Linnaeus – Reduction of area;
- Marbled polecat – *Vormela peregusna* Güldenstadt – Reduction of area;
- Mediterranean monk seal – *Monachus monachus* Hermann – became extinct in the coastal waters of Georgia;
- Brown bear – *Ursus arctos* Linnaeus – only very small fragmented population;
- Harbor porpoise – *Phocoena phocoena* Linnaeus – Vulnerable in the whole world;
- Bottlenose dolphin or the Atlantic bottlenose – *Tursiops truncatus* Montagu – Sharp decrease in number in the Black Sea;
- Red deer – *Cervus elaphus* Linnaeus – Still exists in only two Geographical areas;

- Goitered/black-tailed gazelle – *Gazella subgutturosa* Güldenstaedt – Extinct in the twentieth century;
- West Caucasian tur – *Capra caucasica* Güldenstaedt et Pallas – Vulnerable in the whole world;
- East Caucasian tur or Daghestan tur – *Capra cylindricornis* Blyth – Vulnerable in the whole world;
- Domestic goat – *Capra aegagrus* Linnaeus – Entire population in one area;
- Chamois – *Rupicapra rupicapra* Linnaeus – Significant reduction in number in recent years.

3.10.2.8 Birds

- Murtali – *Podiceps grisegaena* Boddaert – Very small population;
- Great white pelican /Eastern white pelican, rosy pelican or white pelican – *Pelecanus onocrotalus* Linnaeus – Very small population;
- Dalmatian pelican – *Pelecanus crispus* Bruch – Vulnerable globally;
- White stork – *Ciconia ciconia* Linnaeus – very small population;
- Black stork – *Ciconia nigra* Linnaeus – very small population;
- Lesser white-fronted goose – *Anser erythropus* Linnaeus – very small population;
- Ruddy shelduck – *Tadorna ferruginea* Pallas – Vulnerable globally;
- Marbled duck, or marbled teal – *Marmaronetta angustirostris* Menetries – Vulnerable globally;
- Velvet scoter/velvet duck – *Melanitta fusca* Linnaeus – very small population;
- White-headed duck – *Oxyura leucocephala* Scopoli – Vulnerable globally;
- White-tailed eagle/eagle of the rain, sea gray eagle, gray eagle, and white-tailed sea-eagle – *Haliaeetus albicilla* Linnaeus – small population;
- Levant sparrowhawk – *Accipiter brevipes* Severtzov – small population;
- Long-legged buzzard – *Buteo rufinus rufinus* Cretzschmar – small population;
- Eastern imperial eagle – *Aquila heliaca* Savigny – Vulnerable globally;
- Greater spotted eagle – *Aquila clanga* Pallas – Vulnerable globally;
- Golden eagle – *Aquila chrysaetus* Linnaeus – small population;

- Egyptian vulture/white scavenger vulture or pharaoh's chicken – *Neophron percnopterus* Linnaeus – small population;
- Bearded vulture – *Gypaetus barbatus* Linnaeus – small population;
- Cinereous vulture – *Aegypius monachus* Linnaeus – very small population;
- Griffon vulture – *Gyps fulvus* Hablizl – small population;
- Saker Falcon – *Falco cherrug* Gray – vulnerable globally;
- Red-footed falcon – *Falco vespertinus* Linnaeus – very small population;
- Lanner falcon – *Falco biarmicus* Temminck – very small population;
- Lesser kestrel – *Falco naumanni* Fleischer – small population;
- Boreal owl – *Aegolius funereus* Linnaeus – globally and locally vulnerable;
- Barn owl – *Tyto alba* Scopoli – being reduced;
- Snowcocks – *Tetraogallus caspius* Gmelin – small population;
- Caucasian grouse – *Tetrao mlokosiewiczi* Taczanowski – very small population;
- Common crane /Eurasian crane – *Grus grus* Linnaeus – small population;
- Little bustard – *Tetrax tetrax* Linnaeus – small area and reducing number of habitat;
- Eurasian stone curlew/Eurasian thick-knee – *Burhinus oedicnemus* Linnaeus – very small population;
- Bearded reedling – *Panurus biarmicus* Linnaeus – Vulnerable in Europe;
- Güldenstädt's redstart/White-winged redstart – *Phoenicurus erythrogastrus* Güldenstadt – Vulnerable in Europe;
- Great rosefinch – *Carpodacus rubicilla* Güldenstadt – Decrease of population density;
- Radde's accentor – *Prunella ocularis* Radde – Small fragmented area.

3.10.2.9 Cartilaginous Fishes

- Beluga or European sturgeon – *Huso huso* Linnaeus – Vulnerable in the world;
- *Acipenser* – *Acipenser sturio* Linnaeus;
- Bastard sturgeon, fringebarbel sturgeon, ship sturgeon, spiny sturgeon, or thorn sturgeon – *Acipenser nudiventris* Lovetsky – Vulnerable in the world;

- Starry sturgeon /Stellate sturgeon or sevruga – *Acipenser stellatus* Pallas – Vulnerable in the world;
- Russian sturgeon /Diamond sturgeon or Danube sturgeon – *Acipenser gueldenstaedti* Brandt & Ratzeberg – Vulnerable in the world;
- Persian sturgeon – *Acipenser persicus* Orodin – Vulnerable in the world.

3.10.2.10   Bone Fishes

- Brown trout – *Salmo fario* Linnaeus – Significant reduction in recent years;
- *Rutilus frisii*, called the vyrezub, Black Sea roach, or kutum – *Rutilus frisii* Nordmann – Small fragmented area;
- *Capoeta sieboldii,* also called the Colchic khramulya or nipple-lip scraper – *Varicorhinus sieboldi* Steindachner – Small fragmented area;
- *Sabanejewia aurata* De Filippi – Small fragmented area;
- Monkey goby – *Neogobius fluviatilis* Pallas – Small fragmented area.

3.10.2.11   Crabs

- *Astacus colchicus* Kessler – Small and fragmented area;
- Pevtsov crab – *Pontastacus pylzowi* Skorikov – Small and fragmented area.

3.10.3   Diversity of Insects

Each of the Classes of arthropods, including the insects, can be split into a number of smaller groups, which reflect progressively more detailed structural similarities between the group members. Many types of arthropods are identified in Georgia; 128 – Elateridae, 90 – ladybugs Coccinelidae, 217 – Scolycidae, 350 – Chrisomelidae; Other groups of beetles are studied separately. Out of Lepidoptera Geometridae family – 434 species in total is fully studied (Figure 3.4).

Out of Hymenoptera group, Apoidea – 298 species are well studied, 100 – Aphelenidae, 210 – Encyrtidae, 160 – Braconoidea, 142 – Sirphydae, 185 – Formicidae.

From Diptera family around 500 species have been identified. From which 71 are Sarcophagidae and 29 are Califoridae. There are few notes about other classes of insects (Badridze et al., 2000).

Lyrurus mlokosiewiczi

Phylloscopus lorenzii

Tetraogallus himalayensis

Phasianus colchicus

Figure 3.4 Georgian endemic animals (Photo from www.bazieri.ge).

Keywords

- agricultural crops
- endemic animals
- grapevine
- insects
- invertebrates
- vertebrates
- wheat

Capra caucasica   Capra caucasica

Felis silvestris caucasicus   Gazella subgutturosa

Figure 3.4 (Continued).

References

Aleksidze, G., & Alpaidze, L., (2015). Genetic resources of technical crops in Georgia. In: *Agrobiodiversity of Georgia*, GAAS, Tbilisi, Georgia, pp. 169–172.

Aleksidze, G., & Chkhartishvili, N., (2015). Genetic resources of Grapevine in Georgia. In: *Agrobiodiversity of Georgia,* GAAS, Tbilisi, Georgia, pp. 10–31.

Aleksidze, G., & Goliadze, V., (2015). Genetic resources of subtropical crops in Georgia. In: *Agrobiodiversity of Georgia,* GAAS, Tbilisi, Georgia, pp. 87–106.

Aleksidze, G., & Kakabadze, N., (2015). Genetic resources of vegetables in Georgia. In: *Agrobiodiversity of Georgia*, GAAS, Tbilisi, Georgia, pp. 126–155.

Aleksidze, G., & Sarjveladze, I., (2015). Feed cereals and their genetic resources in Georgia. In: *Agrobiodiversity of Georgia*, GAAS, Tbilisi, Georgia, pp. 173–182.

Aleksidze, G., Akparov, Z., et al., (2000). *Biodiversity of Beta species in Caucasus Region (Armenia, Azerbaijan, Georgia, Iran)*. 3th Joint meeting of the ECP/GR *Beta* Working Group and *Beta* Network. pub. IPGRI, Tenerife, Spain.

Aleksidze, G., Korokhashvili, A., & Vacheishvili, P., (2015c). Genetic resources of legume crops in Georgia. In: *Agrobiodiversity of Georgia*, GAAS, Tbilisi, Georgia, pp. 156–168.

Aleksidze, G., Maghradze, D., & Chipashvili, R., (2015). Genetic resources of wild grapevine in Georgia. In: *Agrobiodiversity of Georgia*, GAAS, Tbilisi, Georgia, pp. 41–50.

Aleksidze, G., Naskidashvili, P., & Chkhutiashvili, G., (2015b). Genetic resources of wheat in Georgia. In: *Agrobiodiversity of Georgia*, GAAS, Tbilisi, Georgia, pp. 107–125.

Aleksidze, G., Ujmajuridze, L., Chkhartishvili, N., & Mamasakhlisashvili, L., (2015). Introduced cultivars of grapes in Georgia. In: *Agrobiodiversity of Georgia*, GAAS, Tbilisi, Georgia, pp. 32–40.

Aleksidze, G., Vasadze, I., & Maghradze, D., (2015). Genetic resources of fruits in Georgia. In: *Agrobiodiversity of Georgia*, GAAS, Tbilisi, Georgia, pp. 51–86.

Arabuli, G., (2005). Species and Habitats (Flora). In: *Georgia's Biodiversity Strategy and Action Plan*. Tbilisi, Georgia, pp. 27–33.

Badridze, I., Eliava, I., Kajaia, G., & Cholokava, A., (2000). Present condition of species diversity of fauna in Georgia. In: *Biological and Landscape Diversity*, Tbilisi, Georgia.

Didebulidze, A., (1997). *Agriculture and Rural Development in Georgia: Problems and Prospects*, United Nations Development Program, Tbilisi, Georgia.

Eristavi, M., Shulkina, T., Sikharulidze, S., & Asieshvili, L. *Rare, Endangered and Vulnerable plants of the Republic of Georgia*. http://www.mobot.org/MOBOT/research/georgia/checklist.pdf, pp. 91.

Gagnidze, R., (2000). Diversity of Georgia's Flora. In: *Biological and Landscape Diversity of Georgia*. Tbilisi, Georgia, pp. 21–32.

Gagnidze, R., (2005). *Vascular Plants of Georgia – A Pomenclatural Checklist*. Tbilisi.

Gigauri, G., (2000). *Biological and Landscape Diversity*. Tbilisi, Georgia, pp. 2–300.

Gvritishvili, M., Nakhutsvrishvili, I., Svanidze, T., Murvanishvili, I., & Dekanoidze, N., (2000). Fungal biodiversity of Georgia. In: *Biological and Landscape Diversity of Georgia*. Tbilisi, Georgia, pp. 97–114.

Japaridze, G., (2001). *Forest is a National Treasury*. Magazine 'Kvali," Tbilisi, Georgia, *vol. 1–2*.

Japaridze, G., Chagelishvili, R., & Gagoshidze, G., (2015). *Genetic Resources of Forest in Georgia*. Materials of the conference "Genetic Resources of Perennial Crops," GAAS, Tbilisi, Georgia.

Kacharava, T., & Esvanjia, V., (2007). The ways of increasing effectiveness and qualitative characteristics of *Valeriana officinalis* (*Valeriana officinalis* L). *Annals of Agrarian Science*, Tbilisi, 5(31), 75–79.

Kacharava, T., & Korakhashvili, A., (2006). *Catalog of Medicine and Aromatic Plants of Georgia*. Tbilisi, Georgia.

Kacharava, T., & Korakhashvili, A., (2008). *Catalog of Medicine, Aromatic, Spicy, Poisonous Plants of Georgia*. Tbilisi, Georgia.

Kacharava, T., (2009). *Medicine, Aromatic, Spicery and Poisonous Plants*, Tbilisi.

Kacharava, T., (2015). Sustainable use: Genetic resources if medicinal, Aromatic, spicy, poisonous plants, *International Conference Applied Ecology: Problems, Innovations*, Tbilisi, Georgia, pp. 241–246, http://icae-2015.tsu.ge.

Khalikulov., & Aleksidze, G., (2010). *Plant Genetic Resources in Central Asia and Caucasus*. 8th International Conference (abstracts), St. Petersburg, Russia.

Kuchukhidze, J., & Jokhadze, M., (2012). *Botany- Medicinal Plants*, Tbilisi.

Kuchukhidze, J., Gagnidze, R., Gviniashvili, T., & Jokhadze, M., (2016). *Endemic Flowering Plants of Georgian Flora*. Tbilisi.

Kukhaleishvili, L., & Kanchaveli, K., (2000). The biodiversity of Georgia's Algoflora. *In the Book: Biological and Landscape Diversity*. Tbilisi, Georgia.

Nakhutsvrishvili, T., Svanidze, T., & Murvanishvili, M., (2000). Fungal biodiversity of Georgia. In: *Biological and Landscape Diversity*. Tbilisi, Georgia, pp. 43–68.

National Biodiversity Strategy and Action Plan, 2005. Tbilisi.

Qvachakidze, R., (2009). *Vegetation of Georgia*, Tbilisi.

Rukhadze, A., Arabuli, G., Abdaladze, O., & Jorjadze, M., (2013). *Hyena, Levantine Viper*, and others, Biodiversity Conservation and Research Center. (NECRES), Tbilisi.

Sabashvili, M., (1995). *Soil of Georgia.*

Shetekauri, S., & Jakoby, M., (2009). *Mountain Flowers & Trees of Caucasia*, Istanbul, Bunebaprint.

CHAPTER 4

Biodiversity in Greece

ANASTASIOS LEGAKIS,[1] THEOPHANIS CONSTANTINIDIS,[2] and PANOS V. PETRAKIS[3]

[1]University of Athens, Faculty of Biology, Zoological Museum, 157 84 Athens, Greece, E-mail: alegakis@biol.uoa.gr

[2]University of Athens, Faculty of Biology, Department of Ecology and Systematics, 157 84 Athens, Greece, E-mail: constgr@biol.uoa.gr

[3]Hellenic Agricultural Organization Demeter, Institute for Mediterranean Forest Research, Laboratory of Entomology, Terma Alkmanos, 115 28 Athens, Greece, E-mail: pvpetrakis@fria.gr

4.1 Introduction

The text that follows is meant under the definition given by UNEP (Benn, 2010) and has been adopted by the majority of authors. According to this definition, "Biodiversity is the variety of life on Earth, [and] it includes all organisms, species, and populations; the genetic variation among these; and their complex assemblages of communities and ecosystems. It also refers to the interrelatedness of genes, species, and ecosystems and in turn, their interactions with the environment." This is a working definition and is used in all conversations, articles, and brochures referred to the variation of the globe life. Inherent in the definition of biodiversity is the classification into three main hierarchical components or levels. Namely, ecosystem 'sometimes called landscape diversity,' taxonomic 'sometimes called species or organismal,' and genetic diversity.

The three levels of biodiversity are commonly recognized – genetic, species and ecosystem diversity are actually umbrella terms typified below:

1. Genetic diversity is all the different genes contained in all the living species, including individual plants, animals, fungi, and microorganisms. Sometimes within the term gene are included all gene modifiers that regulate gene expression (Miko, 2008).
2. Species diversity is all the different species, as well as the differences within and between different species. All types of species can be used but usually the biological, morphological and taxonomic types of species are meant (Wilkins, 2003) in these discussions.
3. Ecosystem diversity is all the different habitats, biological communities, and ecological processes together with the services the ecosystem provides as well as variation within individual ecosystems.

The high level of biodiversity and endemism of Greece are noticeable and Greece is the most diverse, in proportion to its area, Mediterranean country in all levels of biodiversity (Table 2 in Georghiou and Delipetrou, 2010). All scientists working in Greece are familiar with the discovery of new taxa. This causes enthusiasm to scientists and laymen in other northern countries who are investigating Greece sometimes in a very consistent way. This has raised substantially our knowledge on the biodiversity of this country. Unfortunately, due to space and time limitations, the citation of many works on the biodiversity of Greece is impossible. For this reason, as a rule, only important books and large reference bibliographic sources will be included.

Many factors have played a role in the high biodiversity of the Hellenic space (Greek mainland, Aegean and Ionian Islands and Crete). The first consideration is the dramatic geological history of Greece. According to the compiled work of Higgins and Higgins (1996), Greece can be better described by means of *isopic* zones and *massif* geological entities. *Isopic* zones are considered as a widespread set of rocks that share a common history of sedimentation, faulting, and deposition; they were usually continental fragments, ocean floors, ridges, while they can be several kilometers long and thick. *Massifs* are plutonic and metamorphic rocks that are better considered as exposed lower parts of the continental crust. Greece is a set of 8 zones (Vardar, Pelagonian, sub-Pelagonian, Ionian, Gavrovo, Pindhos, Parnassos, pre-Apulian) and 2 *massifs* (Rhodope, Serbo-Macedonian), while the islands are lined by the Attica-Cycladic belt. Crete, on the other hand, suffered isolation into four islands in Miocene that later were connected and formed the present-day island of Crete. The extremely variegated contemporary picture is shaped by the grabens and horsts together with the eolian erosion of the formed alpine structures. The map of Creutzburg (1963) is also indicative of the Aegean Sea at Pleistostocene when a central Cycladian mass permitted extensive gene exchange between species. The Pleistocene separation of the grabens created the high number of islands (approx. 2700) that created the nowadays picture of high number of species [for lizards (Hurston et al., 2009), for land snails (Mylonas, 1981, 1984; Vardinoyannis and Mylonas, 1988; Giokas et al., 2000; Triantis et al., 2005), for spiders and land snails (Chatzaki et al., 1998), for isopods and land snails (Cameron et al., 2000), for terrestrial flowering plants (Turland et al., 1993; Strid, 2016), for the Aegean tortoise (Mantziou et al., 2004)]. For plants, the reader is directed to the work of Georgiou and Delipetrou (1990–2009, Chloris database).

Second, it is the suitability of some areas for the development and multiplication of biological species. Such areas are abundant in Greece and with this feature, the ecology and the past history (geology and paleoecology) of the country are connected. These areas are commonly used for conservation and are areas of exceptionally high biodiversity. They are generally named as *hotspots* and they are variously defined on specific criteria and inhabiting organisms (Sfenthourakis and Legakis, 2001). Sfenthourakis and Legakis have found on the basis of four criteria and five animal taxa (Oniscidea Isopods, Gastropoda Pulmonata, Orthoptera, Carabidae beetles, and Tenebrionidae beetles) that biodiversity hotspots of endemism are mostly at higher mountains. It is the high percentage of plant endemism of Greece as a Balkan country (see the nomogram in Major, 1988) that raises the Greek biodiversity at high estimates. The endemism rate of plants denotes a high endemism rate of animal sessile species in the sense that they receive the influence of the same factors (Myers, 1990). Moreover, it has been suggested (Baselga, 2008) that endemics are concentrated in southern Europe and Greece has the majority of them at least proportionally to the total area (for plants see Georgiou and Delipetrou, 2010). The same author has found also a positive and significant Pearson correlation between log-transformed endemism and species richness ($R = 0.51$, $t = 3.48$, $p = 0.001$) although different spatial and environmental variables were found to be significantly related with endemism and species richness.

A third causative factor is the existence of glacial refugia (areas that remained unglaciated surrounded by glaciated ones in the last glacial maximum [LGM] and were the basis of the extension of geographic limits of certain taxa (e.g., dung beetles) (Hortal et al., 2011). The same authors postulate that "... current climate and Pleistocene climatic changes are both known to be associated with geographical patterns of diversity." Many recent phylogeographic studies of European taxa show that the three Mediterranean peninsulas are not only the cradle for genetic differentiation, but also a species repository for the northern latitudes of Europe after the Pleistocene glaciation (Gómez and Lunt, 2006).

On the other hand, it is known that the post-glacial expansion of the taxa is expected to have created several entailed successive bottlenecks, especially for northern expanded taxa causing thus their low biodiversity (Dapporto, 2009). These cases seem to be exceptional and restricted to forest trees and some insects. For instance, it is probable that *Polyommatus icarus* (Lepidoptera: Lycaenidae) was widely distributed in the Mediterranean region during the last ice age and expanded into central Europe in the postglacial period without substantial genetic reduction (Schmitt et al., 2003). But for some forest trees, the situation was different resulting in higher mid-European diversity than in the Mediterranean refugial areas. For example, working with chloroplast DNA variation in 22 species of forest trees and shrubs Petit et al. (2003) have found that the genetic diversity of the taxa in these refugia was not homogenized but remained higher in mid-latitudes than in the three European peninsulas (Greek, Italian and Spanish).

In general, it is accepted that the advent of DNA technology provides most suitable markers to examine crucial details and has drastically changed our ideas on the biodiversity of areas into which the post-glacial expansion occurred. It became clear that the present genetic structure of populations, species, and communities has been mainly formed by Quaternary ice ages (reviewed by Hewitt, 2000). It is now believed that Greece remained an area of high biodiversity not only because of its Quaternary past as refugium of many taxa but also because of the variety of habitats and excessive geographic terrain.

Biodiversity studies commonly use some sort of meta-analysis on large numbers of ecosystems across different geographic areas. These studies, with a few exceptions, are involving certain well-known taxa like birds and mammals as predictors of biodiversity since the inclusion of all resident taxa is an impossible task (e.g., Soininen et al., 2007). The inclusion of invertebrates in the set of biodiversity predictors strengthens the estimation of species numbers since several studies doubt the generality of the patterns revealed by large-scale studies or have found very loose correlation among the taxa used as predictors (e.g., Wolters et al., 2006; for a review see Willig et al., 2003). However, the dust for this debate is far from settled and European researchers presented a better picture by writing that arthropod assemblages – not species or species numbers – are best predicted by plant composition (Schaffers et al., 2008). This picture seems to suit the natural environment in Greece and is connected to the following fourth causative force of the Greek biodiversity.

Fourth, there is a temporal component in the alpha diversity which gives an augmentative estimate in various surveys (Petrakis, 2015) and in the long run, it causes an increase of the richness by means of allochronic speciation (Alexander and Bigelow, 1960). Also Petrakis (2017) decomposed the beta-diversity – species turnover between sites – of a set of seven community types in a flag biodiversity area in Greece and in world history (Schinias, Marathon, Attica) and showed that the temporal component was a substantially higher percentage of the entire biodiversity than the spatial beta-diversity and richness together (Figure 4.1). Of course high beta-diversities do

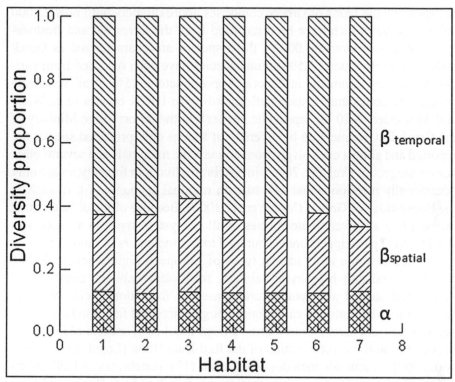

Figure 4.1 Stacked bar diagram showing the proportion of total (γ) diversity that is partitioned to α, $\beta_{spatial}$, $\beta_{temporal}$ diversity. The lower part of each column corresponds to the mean bimonthly number of species in each plot which is α diversity (after Petrakis, 2017).

not increase the number of species in a particular area but do show the difference of responsiveness of various taxa groups along environmental gradients (McCune et al., 2002) and finally show how many and different habitats exist in a landscape and how many and different landscapes exist in an area and in this way it is used in the ecological literature (Magurran, 2004).

4.2 Ecosystem Diversity in Greece

Since ecosystems are better described by the fundamental species (flag species in many cases) occurring and function in them, then we suppose that a coarse description of the ecosystem diversity necessarily passes through the general account of animal and plant species. In other words, the ecosystem biodiversity is a reflection of the number of occurring species. For the time being the flora comprises 5,800 species and the percentage endemism is

15.6% while the fauna of Greece comprises approximately until now 23,000 animal species which have been recorded from the terrestrial and freshwater ecosystems. Almost 4,000 of these species are characterized as Greek endemics. Furthermore, 3,500 animal species have been recorded from various marine environments in Greece. Approximately, 15% of all vertebrate species are considered endangered according to IUCN criteria (e.g., Nieto and Alexander, 2010 for saproxylic beetles). In the report of the Ministry of the Greek Environment on biodiversity of forests only protected species are reported and almost exclusively from animals the mammals and several other vertebrate groups (Vakalis, 2006). Invertebrates living in forest types are only occasionally mentioned and as a rule on regional accounts of taxa such as the Buprestidae of Greece (Mühle et al., 2000). Because of the difficulty met in recording all invertebrate species in all ecosystem types in all countries (CBD, 2011) participants are encouraged to use indicators of biodiversity.

Greece as home to 85 habitat types of European importance, including forest, coastal and halophytic habitats, freshwater habitats, coastal sand dunes, and natural and semi-natural grassland formations, sclerophyllous scrubs, rocky habitats and caves, raised bogs, mires and fens and lastly, temperate heath and scrub. Greece hosts a large number of species of European importance and the 2009 edition of the Red Data Book (Legakis and Maragou, 2009) for animals includes 468 species (171 vertebrates and 297 invertebrates) of animals as variously threatened. Two Red Data Books on Plants have also been published (Phitos et al., 1995, 2009). The books evaluate 263

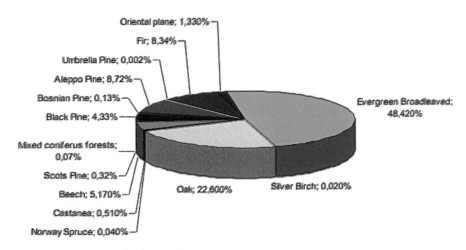

Figure 4.2 Percentage cover of the main forest types in Greece (redrawn from Vakalis, 2006).

and 300 plant taxa, respectively, and assign them under the IUCN categories following the IUCN universal criteria (older and recent). Evaluation status and threat categories of additional Greek plants, particularly those provided legal protection on European scale can be found on the relevant IUCN web pages. The conservation status of several species and habitat types were assessed in 2007 and 2015 for the period 2000–2015, where a rather disappointing picture is revealed with several knowledge gaps, which prevented the assessment in a number of habitat types and species. New assessments of the conservation status of these habitat types and species are expected to be regularly done in the next years.

The following sites must be consulted for those interested in all reported species inhabiting a particular ecosystem type:

1. **www.biol.uoa.gr.** For technical inquiries zoolmuse@biol.uoa.gr; or http://en.uoa.gr/about-us/museums/ for all museums at the University of Athens or www.fria.gr and www.fri.gr for insects and plants of Greek forests.

   These sites correspond to the 'Zoological Museum and the Botanical Museum of the University of Athens' and to the 'The Collection of Insects, Plant Species, and Wood types met in Greek Forests' in the Institute of Mediterranean Forest Ecosystems for southern Greece. These sites can direct interested persons in other collections or sites.

2. **http://natura2000.eea.europa.eu/Natura2000/.** This is the Standard Data Form for Natura 2000 sites. The contents of the website are: "Site identification; Site location; Ecological information; Site description; Site protection status; Site management; and Map of the site."

3. Site and premises of the Greek Taxon Information System (GTIS) which is an initiative of the LifeWatchGreece Research Infrastructure (Bailly et al., 2016). According to the authors, this system will include over 50,000 species, and in this system, a database will be constructed in the near future which will be the umbrella of several other specialized databases for various groups.

The long list of species and habitats listed there is impossible to be included in a short chapter like this.

4.2.1 Forest Ecosystems

Greece has many forest types (Table 4.1) which are subdivided in a number of subtypes according to the criteria given in Dimopoulos et al. (2012).

Table 4.1 Habitat Types in Greece (According to Anex I, Directive 92/43/EEC). The Text is a Modification of the Work of Dafis et al., 1996; and Legakis, 1990)

Type name	Subtype name	Natura 2000 description	How many Natura 2000 sites exist in Greece	Area (km²)
Coastal and halophytic habitats			28	54,957
	Open sea and tidal areas	Invertebrates like polychaetes, Ciicindellinae beetles Plants: *Cymodocea nodosa* and *Posidonia oceanica*, *Ruppia maritima*, *Sarcocornia* spp., *Najas* spp., algae species	8	46,574
	Sea cliffs and shingle or stony beaches	Insects (Diptera, Hymenoptera, Coleoptera) Plants: *Cakile maritima*, *Salsola* spp., *Atriplex* spp., *Polygonum* spp., *Euphorbia peplis*, *E. paralias*, *Elymus repens*, *Glaucium flavum*, *Matthiola sinuata*, *M. tricuspidata*, *Eryngium maritimum*, endemic *Limonium* spp.	5	1,381
	Atlantic and continental salt marshes and salt meadows	Invertebrate communities of marshes tolerating NaCl and sand-gravel dwelling organisms Plants: *Salicornia* spp., *Suaeda maritima*, *Frankenia pulverulenta*, *Salsola soda*, *Cressa cretica*, *Parapholis incurva*, *Hordeum marinum*, *Sphenopus divaricatus*, *Sagina maritima*, *S. nodosa*, *Crypsis* spp., *Spergularia* spp., etc.	4	2,054

Table 4.1 (Continued)

Type name	Subtype name	Natura 2000 description	How many Natura 2000 sites exist in Greece	Area (km²)
	Mediterranean and thermo-Atlantic salt marshes and salt meadows	*Halonitrophilus plant and animal taxa*	3	1.759
		Plants: *Juncus maritimus, J. acutus, Trifolium pannonicum, Samolus valerandi, Hordeum maritimum, Trifolium squamosum, Carex divisa, Ranunculus ophioglossifolius, Plantago crassifolia, Aeluropus littoralis, Puccinellia* spp., *Eleocharis palustris, Arthrocnemum macrostachyum, Cressa cretica, Glinus lotoides, Limonium echioides, Schoenoplectus litoralis,* etc.		
	Salt and gypsum inland steppes	Molluscs; Spectacular and locally important insects: *Euplagia quadripunctaria, Lycaena dispar;* Mammals: *Microtus oeconomus mehelyi;* Birds: *Aythyanyroca, Ardea purpurea, Panurus biarmicus*	3	2,716
		Plants: *Limonium* spp., *Lygeum spartum, Sphenopus divaricatus.*		
	Boreal Baltic archipelago, coastal and land upheaval areas	Mammals: *Careta careta;* Birds: *Larus* sp., *Stercorarius parasiticus, Sterna caspia, Uria aalge;* Crustaceans: *Balanus improvisus, Idothea* sp; Molluscs: *Mytilus edulis*	5	471
Coastal sand dunes and inland dunes	Sea dunes of the Atlantic, North Sea and Baltic coasts	Invertebrate communities of embryonic, white and grey dunes as well as dune slacks	21	5,230
		Plants: *Elytrigia juncea, Sporobolus pungens, Euphorbia peplis, Achillea maritima, Medicago marina, Eryngium maritimum, Pancratium maritimum,* etc.	10	3,264

Table 4.1 (Continued)

Type name	Subtype name	Natura 2000 description	How many Natura 2000 sites exist in Greece	Area (km²)
	Sea dunes of the Mediterranean coast	Plants: *Juniperus macrocarpa, Crucianella maritima, Pancratium maritimum, Silene subconica*, etc.	7	1,484
	Inland dunes, old and decalcified	Mammals: *Lutra lutra*; Serpentia: *Mauremys caspica, Emys orbicularis, Elaphe quatuorlineata*; Pisces: *Aphanius fasciatus*; Insects: *Euplagia quadripunctaria*	4	479
Freshwater habitats			19	22,674
	Standing water	Insects: The beetles *Agonum lugens, A. livens, Badister meridionalis*, and *Pelophila borealis* (dry phase)	10	17,552
		Plants: *Isoetes* spp., *Pilularia minuta, Potamogeton polygonifolius, Eleocharis acicularis, Sparganium minimum, Juncus tenageia, Cyperus fuscus, C. michelianus, Limosella aquatica, Juncus bufonius, Centaurium pulchellum, Cicendia filiformis, Lemna* spp., *Spirodela polyrhiza, Wolffia arrhiza, Hydrocharis morsus-ranae, Utricularia* spp., species of *Chara* and *Nitella*.		
		In temporary ponds many rare species in addition to the above: *Myosurus heldreichii, Juncus minutulus, J. pygmaeus, J. sphaerocarpus, Illecebrum verticillatum, Radiola linoides, Callitriche pulchra*, etc.		
	Running water	Plants: species of *Ranunculus* sect *Batrachium, Myriophyllum* spp., *Callitriche* spp., *Potamogeton* spp., *Zannichellia palustris, Paspalum distichum, Polypogon viridis, Salix* spp., *Populus alba*.	9	5,121

Table 4.1 (Continued)

Type name	Subtype name	Natura 2000 description	How many Natura 2000 sites exist in Greece	Area (km²)
Temperate heath and scrub		Plants: *Astragalus angustifolius, Acantholimon graecum, Rindera graeca, Aster alpinus, Globularia stygia, Minuartia stellata, Erysimum pusillum, Thymus teucrioides, Anthyllis aurea, Achillea ageratifolia, Sideritis scardica, Linum flavum, Thymus boissieri, Sesleria caerulans, Arctostaphylos uva-ursi* and in Crete: *Atraphaxis billardieri, Berberis cretica, Chamaecytisus creticus, Daphne oleoides, Prunus prostrata, Euphorbia acanthothamnos, Verbascum spinosum, Sideritis syriaca, Satureja spinosa, Asperula idaea, Pimpinella tragium, Acinos alpinus,* etc.	12	41,954
Sclerophyllous scrub (matorral)			13	14,693
	Sub-Mediterranean and temperate scrub	Plants: *Buxus sempervirens, Prunus spinosa, P. mahaleb, Cornus mas, Crataegus* spp., *Berberis vulgaris, Ligustrum vulgare, Viburnum lantana, Amelanchier ovalis, Geranium sanguineum, Dictamnus albus*	4	3,331
	Mediterranean arborescent matorral	Plants: *Juniperus oxycedrus, J. phoenicea, J. excelsa, J. foetidissima, J. communis, Arbutus unedo, Ceratonia siliqua, Fraxinus ornus, Laurus nobilis, Olea europaea* var. *sylvestris, Phillyrea latifolia, Quercus ilex, Smilax aspera, Viburnum tinus*	3	3,772
	Thermo-Mediterranean and pre-steppe brush	Plants: *Helichrysum italicum, Pistacia lentiscus, Thymelaea hirsuta, T. tartonraira, Genista fasselata, Euphorbia dendroides*	3	4,857

Table 4.1 (Continued)

Type name	Subtype name	Natura 2000 description	How many Natura 2000 sites exist in Greece	Area (km²)
	Phrygana	Plants: Many cushion-forming, often thorny small shrubs, as *Sarcopoterium spinosum*, *Centaurea spinosa*, *Satureja thymbra*, *Thymbra capitata*, *Genista acanthoclada*, *Anthyllis hermanniae*, *Euphorbia acanthothamnos*, *Ballota acetabulosa*, *Erica manipuliflora*, *Rhamnus oleoides*, *Fumana arabica*, *F. thymifolia*, *Cistus creticus*, *C. parviflorus*, *C. salvifolius*, *Pistacia lentiscus*, *Teucrium brevifolium*, *T. divaricatum*, *T. polium*, *Calicotome villosa*, *Micromeria graeca*, *M. juliana*, *M. nervosa*, *Salvia fruticosa*, *Phagnalion graecum*, *Phlomis cretica*, *Hypericum empetrifolium*, *H. aegyptiacum*, etc.	3	2,732
Natural and semi-natural grassland formations				42,422
	Natural grasslands	Plants: *Alyssum alyssoides*, *A. montanum*, *Cerastium* spp., *Hornungia petraea*, *Saxifraga tridactylites*, *Sedum* spp., *Sempervivum* spp., *Allium schoenoprasum*, *Dianthus deltoides*, *Euphorbia seguieriana*, *Dryas octopetala*, *Anthyllis vulneraria*, *Helianthemum nummularium*, *Geum montanum*, etc.	31 9	10,379
	Semi-natural dry grasslands and scrubland facies on calcareous substrates (Festuco-Brometalia) (* important orchid sites)	Plants: *Anthyllis vulneraria*, *Arabis hirsuta*, *Brachypodium pinnatum*, *B. distachyon*, *B. retusum*, *Bromus inermis*, *Campanula glomerata*, *Carex caryophyllea*, *Carlina vulgaris*, *Eryngium campestre*, *Leontodon hispidus*, *Medicago sativa* ssp. *falcata*, *Ophrys* spp., *Orchis* spp., *Primula veris*, *Sanguisorba minor*, *Scabiosa columbaria*, *Bromus erectus*, *Fumana procumbens*, *Hippocrepis comosa*, *Festuca valesiaca*, *Silene otites*, etc.	12	16,101

Table 4.1 (Continued)

Type name	Subtype name	Natura 2000 description	How many Natura 2000 sites exist in Greece	Area (km²)
	Sclerophillous grazed forests (dehesas)	Plants: *Quercus ilex, Q. coccifera.*	1	6,514
	Semi-natural tall-herb humid meadows	Plants: *Molinia caerulea, Sanguisorba officinalis, Viola palustris, Luzula multiflora, Potentilla erecta, Carex pallescens, Glechoma hederacea, Epilobium hirsutum, Filipendula ulmaria, Petasites hybridus, Chaerophyllum hirsutum, Aegopodium podagraria, Alliaria petiolata, Lyhrum salicaria, Aconitum lycoctonum, Geranium sylvaticum, Trollius europaeus, Adenostyles alliariae, Digitalis grandiflora, Calamagrostis arundinacea,* etc.	6	5,026
	Mesophile grasslands	Plants: *Arrhenatherum elatius, Trisetum flavescens, Centaurea jacea, Crepis biennis, Knautia arvensis, Tragopogon pratensis, Daucus carota, Leucanthemum vulgare, Alopecurus pratensis, Anthoxanthum odoratum, Campanula patula, C. glomerata, Leontodon hispidus, Narcissus poeticus, Linum bienne, Malva moschata, Serapias cordigera,* etc.	3	4,399
Raised bogs and mires and fens			12	28,630
	Sphagnum acid bogs	Plants: *Drosera rotundifolia, Eriophorum vaginatum, E. gracile, Carex lasiocarpa, C. rostrata, C. limosa, Rhynchospora alba, Epilobium palustre, Sphagnum* spp.	6	13,403
	Calcareous fens	Plants: *Cladium mariscus, Schoenus nigricans, Carex* spp., *Eriophorum latifolium.*	4	2,533

Table 4.1 (Continued)

Type name	Subtype name	Natura 2000 description	How many Natura 2000 sites exist in Greece	Area (km²)
	Boreal mires	Insects: *Lepidoptera: Apamea maillardi.* Plants: *Eriophorum vaginatum, Carex rostrata, C. lasiocarpa, Molinia caerulea, Dactylorhiza incarnata,* various mosses.	2	15,226
Rocky habitats and caves		This type of habitats bears many endemic plants and animals	14	15,016
	Scree	Plants: *Drypis spinosa, Ranunculus brevifolius, Senecio thapsoides, Aethionema saxatile, Campanula hawkinsoniana, Arenaria conferta, Cardamine glauca, Alyssum montanum, Silene haussknechtii,* etc.	6	4,643
	Rocky slopes with chasmophytic vegetation	Plants: *Silene lerchenfeldiana, S. congesta, Sempervivum heuffelii, Potentilla haynaldiana, Saxifraga pedemontana, Haberlea rhodopensis, Symphyandra wanneri, Carex kitaibeliana, Campanula* spp., *Sedum* spp., fern species	4	8,486
	Other rocky habitats	Plants: *Myosotis speluncicola*	4	1,886
Forests			81	107,558
	Forests of Boreal Europe	Plants: *Picea abies, Pinus sylvestris, Vaccinium myrtillus, Actaea spicata, Carex remota, Epipogium aphyllum, Geranium sylvaticum, Impatiens noli-tangere, Paris quadrifolia,* etc.	8	32,432
	Forests of temperate Europe	Many different woody species, mostly representing the genera *Pinus, Abies, Fagus, Quercus, Carpinus,* etc.	37	36,917

Table 4.1 (Continued)

Type name	Subtype name	Natura 2000 description	How many Natura 2000 sites exist in Greece	Area (km²)
	Mediterranean deciduous forests	Saproxylic insects are found in this subtype forest: *Cerambyx cerdo, Lucanus cervus, Rosalia alpina, Morimus asper* subsp. *funereus, Lacon* spp., *Osmoderma eremita* [sensu lato] was found by PVP on Mt. Parnon. Plants: *Quercus trojana, Q. frainetto* and *Q. pubescens, Castanea sativa, Salix* spp., *Alnus glutinosa, Cercis siliquastrum, Celtis australis, Populus alba, P. nigra, Juglans regia, Fraxinus ornus, Crataegus monogyna, Cornus sanguinea, Ruscus aculeatus, Vitex agnus-castus, Nerium oleander, Rubus* spp., *Rosa sempervirens, Hedera helix, Clematis vitalba, Platanus orientalis, Liquidambar orientalis, Tamarix* spp.	13	13,082
	Mediterranean sclerophyllous forests	Plants: *Quercus pubescens, Q. ilex, Q. ithaburensis* subsp. *macrolepis, Olea europaea* var. *sylvestris, Ceratonia siliqua, Pistacia lentiscus, Myrtus communis*, etc.	10	12,574
	Temperate mountainous coniferous forests	Plants: *Picea abies, Vaccinium* spp.	3	3,923
	Mediterranean and Macaronesian mountainous coniferous forests	Plants: *Pinus nigra, P. pinea, P. peuce, P. halepensis, P. brutia, Juniperus foetidissima, J. drupacea*	10	8,628

The same criteria are also used in the modified classification system EUNIS (Level III). More specifically the list of criteria is the following:

1. Hydrological status, for example, flooding, riverine, alluvial or dry and/or seasonally wet ecotopes.
2. Alluvial or riverine zone, which comprises the linear thickets along river and torrent banks.
3. Dominant plant species, which harbors specific animal assemblages
    a. Dominant plant species is thermophilous, for example, *Populus* sp., *Platanus* sp., *Quercus* sp.
    b. Dominant plant species is oligotrophous, for example, onoligotrophous soils.
    c. Dominant plant species is a Mediterranean tree/shrub, for example, *Quercus ilex, Q. coccifera, Ilex* sp.*, Laurus* sp., *Olea europaea, Ceratonia silqua, Erica arborea.*
    d. Dominant plant species is *Abies* sp., *Picea* sp, *Pinus nigra, Cupressus sempervirens, Juniperus* sp.
4. Biogeographical region and dominant species, for example, *Pinus* sp., *Juniperus* sp.
5. Altitudinal zone, for example, *Pinus heldreichii, P. peuce, P. nigra, P. halepensis, P. brutia, P. sylvestris, P. pinea.*

Among the 31 forest types listed in the Directive 92/43EEC Annex I, can be augmented by 6 Hellenic (Greek). All forest types in Greece are grouped in 5 *Hellenic habitat types*, namely,

1. Temperate forest types of Europe;
2. Mediterranean deciduous forests;
3. Mediterranean evergreen forests;
4. Alpine and subalpine conifer forests;
5. Mediterranean mountain conifer forests.

Only a limited number of flag animal species will be given for each forest type. Especially species that are considered endangered and need conservation according to European Legislation because only for these species and their habitats exists comprehensive information (CBD, 2011). Also, only forest types in areas authorized by management bodies are presented; if such a body does not exist the faunistic information coming from various sources (student works, enthusiasts, local experts, not networked researchers) is doubtful.

1. **Temperate forest types of Europe.** We do not know exactly the invertebrate fauna of these forests, which is a common problem

for all Greek ecosystems, but we know that they are inhabited by many butterfly species (Lep., Rhopalocera) like *Euphydryas aurinia, Lucanus cervus, Morimus funereus* and Odonata. Vertebrate species *Ursus arctos, Rupicapra rupicapra, Rhinolophus ferrum-equinum, Myosotis emarginatus, Citellus citellus, Bombina variegata, Mauremys caspica, Elaphe quatuorlineata, E. situla, Triturus cristatus, T. alpestris, Testudo graeca, Salmo macrostigma* are also found.

2. **Mediterranean deciduous forests.** Dominant plant species are *Populus tremula, Quercus trojana, Castanea sativa*. The most impressive invertebrates are *Lucanus cervus, Morimus funereus, Rosalia alpina, Cerambyx cerdo* and among vertebrates *Ursus arctos, Rupicapra rupicapra, Bombina variegata, Lutra lutra, Testudo hermanni, T.graeca, Elaphe quatuorlineata, E. situla, Rhinolophus ferrum-equinum, Myosotis bechsteinii, Miniopterus schreibersii, Emys orbicularis.*

3. **Mediterranean evergreen forests.** Many plant species are included in this forest type. A prominent example is the small endemic tree of Crete *Zelkova abelicea* on Mt Cedros, Lasithi; Omalos plateau, Gerakari Rethymnon, Katharoplateau and other sites. Dominant animal species are the vulture *Gyps fulvus*, the predatory bird *Hieraetus fasciatus*, and the very rare *Gypaetus barbatus*. Other animals are *Rhinolophus ferrum-equinum, R. hipposideros, Elaphe situla, Mauremys caspica*. Also in the Natura 2000 sites where this forest type predominates have been also recorded *Monachus monachus* and *Careta caretta.*

4. **Alpine and subalpine conifer forests.** This forest type is quite restricted in Greece with only two Natura 2000 sites, namely Dasos Fractou and Elatia village on Mt Rhodopi. The other forest type met on this mountain is the birch pine mixed forest (*Betula pendula* and *Pinus sylvestris*).

   Invertebrates are mainly *Lucanus cervus, Rosalia alpina*, and may Buprestidae. Vertebrate animal species occurring in this forest type include *Ursus arctos, Rupicapra rupicapra, Bombina variegata, Lutra lutra, Testudo hermanni, Elaphe quatuorlineata, E. situla, Rhinolophus ferrum-equinum, Myosotis bechsteinii.*

5. **Mediterranean mountain conifer forests.** Invertebrates in this forest type are *Eriogaster catax, Lucanus cervus, Euplagia quadripunctaria*. Vertebrates met are *Ursus arctos, Rupicapra rupicapra balcanica, Bombina variegata, Lutra lutra, Testudo hermanni, T. marginata, T. graeca, Elaphe quatuorlineata, E. situla, Rhinolo-*

phus ferrum-equinum, R. hipposideros, Myosotis bechsteinii, Myotis emargnatus.

4.2.2 Other Non-Forest Terrestrial Ecosystems

Many Greek ecosystems characteristic of the East Mediterranean climate cover Greece, especially the central and southern sites and islands. Table 4.1 shows in a descriptive scale the types of these ecosystems which are fluctuating between sand dunes and sclerophyllous scrub to rocky habitats, crevices, and caves. More detailed view of these ecosystems can be obtained in Dafis et al. (1996), and Dimopoulos et al. (2012).

4.2.3 Marine, Littoral, and Freshwater Ecosystems

The Greek marine ecosystems, in general, can be conveniently separated to the northern eutrophic seas and the southern oligotrophic waters. This diversity of the environment should be added the local diversity which makes Greek seas very heterogeneous habitats. For instance, the eutrophication of northern water is enhanced by the increased eutrophication of the major gulfs such as Saronic, South and North Euboeic, Argolic, Corinthian, Patraic, Laconic, Messenian, Pagasetikos, and the gulfs and recesses of the Asia Minor coasts and the East Aegean islands. To this must be added the eutrophication of the waters near the coasts; which are also inhabited by humans since ancient times because of the ease of fishing of a great variety of fishes. In these waters, 447 autochthonous (=indigenous) fish species are found (Papaconstantinou, 1988; Papathanasiou and Zenetos, 2005). Also, 34 fish species are allochthonous (H alien) migrations, mainly Lessepsian from the Red Sea (Corsini-Foka and Economidis, 2007). The continuum of discontinuities, as it is usually called, of the Greek sea waters is enhanced by the existence of lagoon in the estuaries of rivers (e.g., Nestos River in northern Greece).

Also, the freshwaters in Greece are surprisingly numerous. In general in Greece except for the large wetlands of the northern and the western areas, there are 200 small wetlands distributed in 40 islands which at first glance give the impression of an arid island (Katsadorakis and Paragamian, 2006). Unfortunately, since the beginning of the 20th century, more than 60% of these wetlands have been dried up, for agricultural purposes and the reduction mosquito populations, by special irrigation works made by humans (e.g., Copais lake at Boeotia, Agoulinitsa lake at Ilia, Lapsista lake at Epirus, Giannitsa lake and Elç Philippon [= Phlipppon swamps] at Macedonia). The reduction of wetlands triggered a series of biodiversity reduction events affecting waterfowl, aquatic invertebrates, mammals, through the disappearance of many habitats, the hydrodynamic balance of the areas, illegal hunting, and the invasion of alien taxa.

4.2.4 Ecosystem Services and Products

A human-oriented way to assess the biodiversity of a country is the account of the services and products that local ecosystems provide. The pollination services and the quality of the produced honey are the first mentioned in the literature services and products that ecosystems provide. An important aspect of high biodiversity is the augmentation of fruit setting in agriculture because of abundant pollinators in a highly diverse situation. Also high biodiversity in one component of the ecosystem, for example, insects promotes high biodiversity in the plant component, which increases the feeding, resting, hiding, and nesting sites of more insects perpetuating this positive feedback circle. For instance, Cho et al. (2017) have found that despite the fact that forest floor herbaceous plants account of less than 1% of the total biomass of a forest ecosystem, they contain most of the floral resources of a forest. The diversity of understory honey plants determines visitation rate of pollinators such as honey bee (*Apis mellifera*), which increases the plant visitation of wild pollinators offering increased services in nearby agricultural feeds. It also maintains the populations of pollinators that decrease because of habitat destruction and various diseases that usually accompany.

Importantly, Greek ecosystems produce high-quality honey from various herbs. Honey derived from thyme (Aliferis et al., 2010) is appreciated not only in Greece but all around the globe. The main plant of this honey dominated low bush formations (phrygana and pre-steppe brush). Another product derived from coniferous forest, is the special honey derived from insects, mainly *Mindarus* and *Dreyfusia* (Hemiptera, Sternorrhyncha) living on Greek fir trees in central Peloponnese. However, the pine honey derived from the pine scale *Marchalina hellenica* all over Greece, is of very low quality and the pine scale damages the pine tree in a slow and irreversible way (Petrakis et al., 2010; Petrakis et al., 2011c). It is speculated that beekeepers mix this honey with essential oils in order to add flavor to it.

Another valuable product of Greek ecosystems is the blends of secondary metabolites, derived from plants and animals, which are used by several pharmacopeias either for the construction of drugs or the chemical synthesis of new ones by using these compounds as chemical structures. The entire

Greek and Roman history is full of myths about the introduction of many plants, and animals in a lesser degree, that affected human health or acted as healing agents (Bauman, 1982; Ruck, 1995). Many authors stated that the basis of our medicines is the set of natural products, and the paucity of new drugs started at 1900 when the combinatorial chemistry replaced natural products (Ji et al., 2009). Indeed Greece was the westward station of the numerous natural products derived from plants, animals, fungi, and microbes arrived from India, Malaysian peninsula, Indochina, and Far East at the times of Minoan and Mycenaean civilization (Ruck, 1995). This was made possible because the ecosystems in Greece were already rich in naturally derived plants and the inhabitants were predisposed for their curing and flavoring benefits, which at those times was a property already sprung from gods. This last property proved very useful in tracking the route of ecosystem products from Asia to the western world. A paradigm of such a transfer and relationship between the uses is the chemical affinity and similarity of uses of the Alkannin and Shikonin. These compounds were derived for many centuries as extracts from the roots of *Alkanna tinctoria* in Europe and *Lithospermum erythrorhizon* in the Orient (e.g., China). They have been used independently as natural dyes and as crude drugs in wound healing. It has been proven that the second chemical, that is, *shikonin* is the *R* enantiomer (Papageorgiou et al., 1999). All the aforementioned arguments constitute the reasons for which the high biodiversity of Greece made this country the ideal place for the introduction of the services and products of Asian ecosystems.

The medicinal uses of plants in Greece were studied by many scientists and are routinely used in drug formulation (amongst others Roussis et al., 2000a, b; Daferera et al., 2000; Stamatis et al., 2003; Liolios et al., 2010). Attempts have been recently made for exploiting the insect repellency of the rich amphibian fauna of temperate forests in an ecologically sound way instead of the routine habit of using frogs and toads as delicacies in restaurants at northern Greece (Ioannina, Epirus). An example is the construction of farms for the derivation of mosquito repellents from frog skins in a nondestructive way. This attempt failed for political reasons without regard to the ecologically sensitivity of farmers, investors and local scientists.

Finally, the flagship of ecosystems particular to Greece is the olive grove agro-ecosystem. Since antiquity, the live grove based olive culture was practiced in the East Mediterranean geographical area (Petrakis, 2000; Petrakis et al., 2008; Dais et al., 2012). Because of this, it is possible that native cultures biosynthesized thousands of chemical compounds (Petrakis et al.,

2008; Garcia-Gonzalez and Aparicio, 2010). Several of these compounds configure the special taste of the extra virgin olive oil (EVOO) of Greek olive cultures. These chemical compounds are believed to be the result of the selection of olive trees through a special breeding system. Olive seeds were planted within arborescent matorral (Table 4.1) or pine forests; the seedlings that survived the adverse abiotic and biotic conditions were replanted in the olive groves which were, as a rule, in proximity to pine stands (Petrakis, 2000; Agiomyrgianaki et al., 2012). The Greek EVOOs are so rich in secondary compounds as to enable researchers to detect and authenticate the geographical origin of various EVOOs from the cultivar Koroneiki by using NMR data (e.g., Petrakis et al., 2008). This detection is very detailed and permitted the separation of the temporal effect in the composition of the oil from the geographical effect even at very local scales (e.g., Chania, Heraklion, Sitia in Crete, and Laconia, Messenia in Peloponnese, and Zakynthos in the Ionian Sea).

4.3 Species Diversity in Greece

4.3.1 Historical Notes on Animals

Our knowledge of the animals of Greece begins in the past, essentially by Aristotle, who, over 2,300 years ago, wrote his *Historia Animalium*, where he described in detail about 600 species. The knowledge of the Greek fauna has increased with the contribution of various naturalists and scholars during the Hellenistic, Roman and Byzantine periods, and it has gained a more complete form with the help of various foreign travelers in the 17th, 18th and 19th centuries. The first large organized mission was carried out by French naturalists in the Peloponnese between 1832 and 1836 (Bory de Saint-Vincent 1832–1836). From then until the 1970s the vast majority of scientific work on the fauna was carried out by foreign, mainly German-speaking researchers. Since the '70s, the number of Greeks involved in the fauna of Greece has increased (Legakis, 1983). Each year over 150 works are published for all animal groups and for subjects ranging from physiology to ecology and conservation.

4.3.2 General Constitution of the Greek Fauna

The fauna of Greece includes representatives of most of the animal groups. According to recent censuses (de Jong et al., 2004), 23,130 species of land

and freshwater species have been recorded (Legakis, 2004). To this, we can add another 3,500 species of marine species. If we add a number of species that have been recorded but are not included in the present catalogs, we reach a total of about 30,000 species. However, it is well known that Greece's fauna is not well studied. The best-studied groups are the vertebrates, while larger gaps exist in invertebrate species. If we compare the number of species of Greece with the number of species in Europe, we see that there are groups where the Greek species make up 40% of the European fauna, while others have 10% (Table 4.2). Therefore it is believed that if the Greek fauna is fully studied, it should include more than 50,000 species.

As expected, the majority of animal species belong to the arthropods (92%), followed by the mollusks and the chordates. Within the arthropods, the insects are very abundant, especially Coleoptera, Lepidoptera, Hymenoptera, and Diptera, while Homoptera, Heteroptera, and Orthoptera have relatively high numbers of species. Among the other groups of arthropods are the spiders, the Chilopoda and the Diplopoda.

The fauna of Greece has more affinities with the fauna of the eastern Mediterranean, a region affected by Europe, Central Asia, Anatolia, the Middle East and Africa. In many regions of Greece, the prevailing faunal elements are East-Mediterranean. The southern European elements are also important and are strengthened as we go to the north of the country. In the northern regions, we see more common European as well as Palearctic elements.

Table 4.2 Number of Animal Species Per Log (Number of km²)

Coleoptera		Terrestrial Gastropoda		Orthoptera		Trichoptera	
Italy	1832.8	Greece	149.0	Greece	67.6	Italy	66.5
France	1552.0	Italy	77.6	Spain	55.0	Switzerland	65.2
Ισπανία	1440.3	Spain	67.4	Italy	48.5	France	60.9
Greece	1309.5	France	67.2	Bulgaria	39.2	Germany	56.4
Slovakia	1176.6	Croatia	65.2	Yugoslavia	38.1	Greece	56.2
Terrestrial Isopoda		**Reptiles**		**Dictyoptera**		**Raphidioptera**	
Italy	58.3	Greece	11.9	Spain	12.3	Greece	6.6
Greece	48.2	Spain	8.6	Sicily	7.5	Italy	3.3
Rumania	41.3	Albania	8.1	Italy	7.1	France	2.6
France	33.3	Italy	7.8	Greece	6.8	Switzerland	2.6
Spain	32.1	Portugal	7.3	Croatia	6.1	Bulgaria	2.6

(Legakis, 2004)

Using some animal groups, Greece has been divided into 11 zoogeographical regions, each of which has its own characteristic fauna. These areas are most affected by neighboring regions outside Greece. For example, the region of the eastern Aegean is affected by the neighboring Asia Minor, with which most islands were once connected.

The shaping of the fauna of Greece is due both to historical and ecological factors. The main historical factors are the presence of glaciers and, in general, the cold climate of the Pleistocene, the long connections of islands with the continental regions and between them, the existence of barriers that prevented dispersal and changes in vegetation, especially during the last 20,000 years.

Ecological factors include climatic factors such as temperature and humidity, soil, vegetation, soil relief, altitude, mosaicity of Mediterranean ecosystems, human activities, and others. There are, for example, species that cannot withstand the extremely cold winter temperatures or the very high temperatures of the summer. Other species have a preference for limestone soils, but they cannot survive at high altitudes, while species with special preferences successfully exploit the specific microenvironments of the Mediterranean ecosystems.

There are some groups of animals that show a significant number of species compared to the corresponding number of other European countries. Groups such as land mollusks, isopods and reptiles have the greatest abundance in relation to the area in Greece (Sfenthourakis et al., 1999). This species density can be due to many reasons. On the one hand, there is the mosaic of ecosystems. Within a short distance you can find coastal, phryganic, bushy, forest and subalpine ecosystems, each of which has different forms and subdivisions. On the other hand, as mentioned, Greece lies on biogeographic crossroads between different dispersal corridors. The fragmentation of Greece's surface in many marines and "mountainous" islands has led to the creation of many endemic species. The phenomenon of high abundance of biodiversity is not limited to animals or the Greek area. The flora of Greece is equally more abundant than the flora of most European countries. More generally, all of Southern Europe, and in particular the three major peninsulas, Iberian, Italian and Balkan, are richer in plant and animal species from central and northern Europe.

The other peculiarity of the Greek fauna is the high rate of endemism. Recent data show that up to now 3,956 endemic species of land and freshwater have been recorded, a percentage of 17.1. There are some groups such as terrestrial isopods and Orthoptera with endemism higher than 30% (64% and

32% respectively). The main reasons for the existence of these high rates are the long-term isolation of islands and the existence of Pleistocene shelters in the mountainous areas.

Some areas of Greece have a particularly high number of species and especially endemic species. This high number can come from the long-term isolation and the vigorous follow-up that followed. Such areas in southern Greece are the tops of the mountains such as Mt. Taygetos and Mt. Psiloritis and islands like many Cycladic ones (Legakis and Kypriotakis, 1994; Sfenthourakis and Legakis, 2001).

Unfortunately, we have relatively few data on the status of the populations of species living in Greece. A general estimate is that populations are relatively small if compared to those of Central Europe. There are groups that reach very high levels, such as certain types of lizards, Orthoptera and Coleoptera that are adapted to warm climates. In general, vertebrates do not have populations with high densities.

4.3.3 *Plant Species*

4.3.3.1 Fungi

Fungi constitute a large and diverse group of organisms inadequately and fragmentary investigated in Greece. Research up to current days has covered mainly fungi of phytopathological importance and macrofungi. Existing figures indicate a total of about 2,500 recorded species out of more than 25,000 fungi that are expected to occur in country's territory, taking into account its size and diversity of habitats (Zervakis, 2001). Pantidou (1973) produced a comprehensive fungus-host index that needs updating and Zervakis et al. (1998, 1999) compiled a checklist of Greek macrofungi and ascomycota. Altogether, the group comprises at least 935 species, of which 815 belong to Basidiomycota and 120 to large Ascomycota. Furthermore, at least 138 species of Myxomycota exist in the country (Zervakis et al., 1998). The above numbers are indicative only; recent research in various parts and different habitats of Greece constantly enriches its mycodiversity at a stable rate.

4.3.3.2 Lichens and Lichenicolous Fungi

Lichens and lichenicolous fungi is another group not fully investigated in Greece. Abbott (2009) summarizes 1,296 accepted taxa in the country and estimates that the final total may well exceed 1,500 taxa.

4.3.3.3 Cyanobacteria and Algae

Cyanobacteria and algae have been investigated since the early 19th century in Greece but comprehensive checklists or reviews that summarize existing diversity have appeared recently and referred to particular target groups only. The checklist of Greek Cyanobacteria comprises 543 species, classified in 130 genera, 41 families and 8 orders (Gkelis et al. 2016). Synechococcales and Oscillatoriales have the highest number of species (158 and 153, respectively), whereas these two orders along with Nostocales and Chroococcales cover 93% of the known Greek cyanobacteria. A checklist of the red seaweeds (Rhodophyta) of the order Ceramiales along the Greek coasts is provided by Tsiamis and Panayotidis (2016).

The total number of genera, species and infraspecific taxa currently accepted is 60, 118 and 2, respectively. Still, surveys in unexplored areas and particularly the deeper habitats will further increase the number of taxa. Tsiamis et al. (2013) revised the Phaeophyceae of Greek coasts and the total number of species and infraspecific taxa accepted is 107, with 17 taxa pending confirmation. Knowledge of the green seaweeds (Ulvophyceae) of the Greek coasts is summarized by Tsiamis et al. (2014), with the total number of species and infraspecific taxa currently accepted being 96. For the last two groups, the authors agree that the number of species will increase with additional surveys. Finally, a useful list of toxic and harmful algae of Greek coasts is provided by Ignatiades and Gotsis-Skretas (2010) accounting for 61 species.

4.3.3.4 Bryophytes

Important contributions to Greek bryophytes have appeared during the last decades. Preston (1981, 1984b) compiled a checklist of liverworts (Anthocerotophyta and Marchantiophyta) from Greece, together with a checklist of Greek mosses (Bryophyta; Preston, 1984a). Düll (1995, 2014) summarized knowledge on bryophytes in Greece and the Aegean Islands, respectively, and Blockeel (2013) presented a brief summary of Greek bryophytes with species estimations. According to the latter, 536 mosses and 154 liverworts grow in Greece, totaling 690 species of bryophytes.

4.3.3.5 Vascular Plants

Knowledge of the Greek vascular flora has much increased during the last years, resulting in the publication of the Vascular Plants of Greece Checklist

(Dimopoulos et al., 2013, 2016) and the Atlas of the Aegean Flora (Strid, 2016). Summarized data up to October 2013 indicate that the vascular flora of Greece comprises 5,758 species and 1,970 subspecies (native and natural-ized), representing 6,620 taxa, belonging to 1,073 genera and 185 families.

Among them, there are 75 species of ferns (pteridophytes), 24 species of gymnosperms and the remaining 5,659 species represent angiosperms.

The most species-rich family of the Greek vascular flora is Asteraceae, comprising 749 species in 113 genera, and accounting for 13% of all Greek species. The second most species-rich and the most genus-rich family is Poaceae, which contains 439 species in 125 genera (7.6% of all Greek spe-cies). Other species-rich families, in descending order, are Fabaceae, Caryo-phyllaceae, Brassicaceae, Lamiaceae, Apiaceae, Rosaceae, Boraginaceae and Rubiaceae (Dimopoulos et al., 2013). *Centaurea* (Asteraceae) is the most species- and taxon-rich genus in Greece (110 species and 141 taxa). *Hieracium* (also Asteraceae) ranks second in taxon richness (137 taxa) but tenth in species richness (66 species). Other genera that differ considerably in numbers of species and taxa are *Dianthus* and *Silene* (both Caryophyl-laceae). A few critical and taxonomically difficult genera of the Greek flora (e.g., *Allium, Hieracium, Limonium, Ophrys*) may exhibit contrasting treat-ments by different authors and a full consensus is still pending and desirable. For example, the related *Allium cupani* Raf. and *A. hirtovaginatum* Kunth, considered as members of the Greek flora are both excluded by Brullo et al. (2015) and replaced by three species described as new: *A. balcanicum* Brullo, Pavone & Salmeri (2015), *A. cephalonicum* and *A. tzanoudakisanum* Brullo, Pavone & Salmeri (2015). Crete comprises 25 *Limonium* species according to Brullo and Erben (2016) but 19 species are mapped by Strid (2016). A very recently published Atlas of the Greek Orchids (Tsiftis and Antonopoulos 2017; Antonopoulos and Tsiftsis 2017) enumerates 91 *Oph-rys* taxa (species and subspecies), while 43 taxa are accepted by Dimopoulos et al. (2013) with small differences in Dimopoulos et al. (2016).

A new initiative, the publication of a complete Flora of Greece, started in 2017. This is a joint project between the Hellenic Botanical Society and the Botanic Garden and Botanical Museum Berlin-Dahlem of Germany. Publica-tion will take place in both printed and electronic formats. The Flora of Greece Web already serves as a focal point for searches on the Greek vascular flora.

4.3.3.6   Endemism of Vascular Plants

According to present knowledge (Dimopoulos et al., 2016), the endemic vascular flora of Greece comprises 1,459 taxa (species and subspecies, 22%

of the total number of taxa in Greece), corresponding to 1,274 endemic species (22.1% of the total number of Greek species). There are no endemic families, but there are eight endemic genera: *Hymenonema* (Asteraceae), with two species, and with one species each: *Horstrissea* (Apiaceae), *Jankaea* (Gesneriaceae), *Leptoplax* (Brassicaceae, recently suggested as a member of *Bornmuellera*, see Rešetnik et al., 2014), *Lutzia* (Brassicaceae), *Petromarula* (Campanulaceae), *Phitosia* (Asteraceae) and *Thamnosciadium* (Apiaceae).

The endemic richness in absolute numbers and the rate of endemism are not uniformly distributed across Greece. As a general pattern, south Greece and east Greece are richer in absolute numbers of endemics. The highest number of Greek endemic species and subspecies is observed in Peloponnisos (464 taxa), while the second highest numbers are in the Cretan region (part of southern, insular Greece that includes Crete, Karpathos, Kasos, Saria and several smaller islands). While the Cretan region is second highest among the floristic regions in its absolute number of Greek endemic taxa, its endemism rate is the highest (21.1% for subspecies, 17.1% for species and 17.7% for taxa). The high endemism rate, the long isolation of Crete from the mainland and its diverse topography (see Trigas et al., 2013), are the main reasons for often considering the Cretan area as an independent floristic region indicated by the abbreviation Cr as opposite to Gr for the remaining part of Greece.

In the eastern parts of Greece, the East Aegean Islands (abbreviated as EAe) present important floristic affinities with neighboring western Anatolia and serve as the westernmost part of the Asiatic flora. Indeed, the imaginary line between the East Aegean Islands and the Islands of Cyclades to the west serves as the phytogeographical border between Europe and Asia; it has been proposed to be given the name '*Rechinger line*' by Strid (1996).

Dimopoulos et al. (2013) introduced the concept of '*range-restricted species,*' for those species that occupy a limited area of distribution not exceeding a linear distance of 500 km between the farthest points of occurrence, regardless of whether the distribution area crosses a national border or not. The range-restricted vascular flora of Greece consists of 1,972 taxa (29.8% of the total taxa number in Greece), corresponding to 1,703 species accounting for 29.6% of the total number of species, and 611 subspecies (31% of the total number of subspecies). Again, Peloponnesus is the richest floristic region in Greece with 505 range-restricted taxa (Dimopoulos et al., 2016).

Although most endemic or range restricted Greek taxa are herbaceous perennials, a few are trees or tall shrubs. Among them, *Abies cephalonica*

Loudon (Pinaceae) grows in the southern and central Greek mainland and the islands of Kefallinia and Evvia, *Quercus trojana* Webb subsp. *euboica* (Papaioannou) K.I. Chr. (Fagaceae) is endemic to Evvia and *Zelkova abelicea* (Lam.) Boiss. (Ulmaceae) is restricted to Crete.

4.4 Major Threatened Plants and Animals

4.4.1 Threats

In recent years, especially since the 1960s, the expansion and intensification of human activities have greatly increased the pressures and threats to the fauna, leading to a severe decline in populations of many species. The main threats facing the fauna and its habitats today in Greece are either indirect or direct.

4.4.1.1 Alterations and Destructions of Habitats

It is the world's most serious problem for the fauna and is almost exclusively due to economic development. It appears today in many forms such as wetlands drainage, extensive logging, fires, extension of settlements and tourist facilities to beaches, interventions in mountains, etc.

Since the late 19th century, Greece has lost many large wetlands due to extensive drying out for agricultural land or the fight against malaria such as the lakes of Kopaida, Giannitsa, Achinos, Lesini, Xyniada, Sharigiol, Philippi, and others. Many of these wetlands were of particular importance to birdlife resulting in a dramatic decline in populations of many aquatic birds and other animals such as amphibians and waterfowl, freshwater fish, otters, etc.

The forests, which currently cover 19% of Greece, appear to have been much more extensive in the recent past, reaching 45% in the early 19th century (Spanos et al., 2015). This shrinking is almost exclusively due to anthropogenic factors such as extensive logging for the expansion of agriculture, progressive intensification of forest exploitation, overgrazing for residential or tourist use, fires, etc.

The result of the interventions was the degradation of almost all the large herbivorous or carnivorous mammals such as deer, wolf, bear, etc.

The opening of forest roads and fires are two more interventions that have increased in recent years. In the 1950s–1960s the total length of the forest roads was 1,672 km while in 1985 it reached 15,310 km. By 1973 the fires burned an average of 115,000 acres. In 1985, 1,050,000 acres were burned, while the rate of reforestation barely reached 50–55,000 acres a year (Dimitrakopoulos and Mitsopoulos, 2006; Mitsopoulos and Mallinis, 2017).

4.4.2 *Invasive Alien Species*

Invasive alien species are species intentionally or accidentally introduced into areas other than their natural habitats where they have the ability to settle, prevail over indigenous species and dominate their new environment. For many countries, this is the number two ecological problem after the loss of habitats.

Unfortunately, no detailed studies have been carried out on the species that have invaded Greece, except for some plant species, the magnitude of the damage they have caused and the measures are taken. Occasionally, we know a number of cases, but we cannot be sure of their impact. An example of mammals is the coypu, a mammal introduced for the exploitation of its fur and which has created several natural populations in wetlands in northern Greece. Among the birds, we can mention the introduction of the island partridge from the islands of the Aegean where it lives, to mainland Greece for the enrichment of the game. Many areas have already been installed (e.g., Corfu) causing serious problems to the rock partridge population due to competition, while the transfer of animals from one island to another can cause genetic erosion (Handrinos and Akriotis, 1997). Among the reptiles, we can refer to the case of the American pond slider *Trachemys scripta* which is used as a pet and is released very often in small or large wetlands. The introduction of this species has been banned across the European Union because it causes problems for most of the indigenous aquatic organisms. This is also the case with the American bullfrog *Rana catesbeiana*, which was introduced for food production and has created free local populations (such as in Crete). Many species of foreign fish such as the American rainbow trout *Oncorhynchus mykiss* and the mosquito fish *Gambusia affinis* have been introduced into lakes and rivers for the purpose of exploiting or fighting insects and causing problems in native species due to competition (Economidis, 1992). Finally, we can mention the transfer of bees from mainland Greece to Crete where there is an endemic subspecies, *Apis mellifera adami*, which seems to have almost disappeared due to competition.

Other insects that invaded Greece are *Leptoglossus occidentalis* (Heteroptera, Coreidae) (Petrakis, 2011) and *Zelus renardii* (Heteroptera, Reduviidae, Harpactocorinae) (Petrakis and Moulet, 2011). The majority of potential invaders are intercepted at the points of entry in the country (ports, stations, border stations). However, some insect species, especially those that are preadapted for the invading process successfully avoided the control and treatments (e.g.,

Sinoxylon anale and *S. unidentatum* (Coleoptera, Bostrichidae) (Lykidis et al., 2016)]. These insect species can be easily concealed under the 'enemy release' umbrella since their natural enemies can hardly follow them and new enemies in the invaded country are very slowly recruited from the local species pool. Other insects that invaded Greece are *Megastigmus* sp. (PVP personal observation), *Drosophila suzuki* (Papachristos et al., 2013), *Tuta absoluta* (Lepidoptera, Gelechiidae) (Roditakis et al., 2010), *Harmonia axyridis* (Coleoptera, Coccinellidae) (Brown et al., 2008), and some other insects, mainly hosted by exotic species like *Phoracantha semipunctata* and *P. recurva* (Coleoptera, Cerambycidae) have been repeatedly collected by us and are listed in EPPO (= European Plant Protection Organization) (entered in 2003 but deleted in 2006). By employing inspections in all prefectures of Greece, the 'Plant Protection Service' of 'The Hellenic Ministry of Agriculture' intercepts several organisms that are considered as quarantine pest organisms by EPPO.

The alien flora of Greece has been investigated (Arianoutsou et al. 2010) and at least those taxa permanently established somewhere in the country are incorporated in the national checklist (Dimopoulos et al. 2013). Still, new cases of alien plants have been recorded and their invasion capacity evaluated. Dispersal of most aliens is mediated by mammals and birds (43%) and by the wind (28%), that is by agents which provide the chance for long-distance dispersal. A few highly invasive alien plants have been recorded in Greece: the seaweed *Caulerpa taxifolia* and the vascular plants *Ailanthus altissima, Ambrosia artemisiifolia, Carpobrotus edulis, Opuntia ficus-indica, Oxalis pes-caprae. Paspalum distichum.* Artificial habitats, especially cultivations and road networks, host the highest numbers of alien plants. The natural habitats that host the highest numbers of aliens are the coastal zones and the inland surface waters. Artificial habitats host the highest numbers of naturalized taxa, as well.

4.4.3 Pollution

The gradual restructuring and intensification of Greek agriculture have created additional problems at the expense of the fauna. Extensive monocultures and the need to produce high yields have led to a steep increase in the unreasonable use of chemicals, mainly pesticides, and fertilizers. These substances are channeled to water and soil, increasing eutrophication, causing physiological changes in organisms and reducing their resistance to diseases and pests. In addition, direct airborne pollution results in massive deaths. Until now, no comprehensive study has been done on the effect of pesticides on the fauna of Greece. Nevertheless, the assessment of certain incidents or phenomena in conjunction with more general observations provides strong indications that even in our country the problem is real, albeit not as intense as in other European countries.

4.4.4 Overexploitation

One of the most important nuisances for the fauna comes from hunting. Hunting in Greece has grown rapidly in recent years (Table 4.3). In 1962,

Table 4.3 Animal Species That Are Subject to Exploitation in Greece

Mammals
- Hare
- Wild rabbit
- Wild boar

Birds
- Hunted species

Amphibians
- Marsh frog

Fish
- Fished species

Invertebrates
- Sponges
- Corals
- Leeches
- Polychaetes (fishing baits)
- Sipuncula (fishing baits)
- Marine bivalves
- Terrestrial gastropods
- Acari (predators of harmful insects)
- Crustaceans (lobsters, prawns, crabs, shrimps, baits)
- Silkworms
- Bees
- Bumblebees (pollinators)
- Parasitic Hymenoptera (predators of harmful insects)
- Sea urchins
- Ascidians (food, baits)

there were 165,339 hunters in Greece while 322,882 hunting permits were issued in 1987 (Paraschi and Handrinos, 1992). At the same time, Greece has a wide variety of fauna, but because of its small size and habitat fragmentation, available hunted populations are very small and vulnerable. The hunting problems start from the ignorance of most game hunters, with the result that thousands of rare, protected or forbidden birds and mammals are killed or seriously injured each year. Another parallel problem is the great nuisance caused by hunting and which in many cases causes worse effects on animals than the hunting itself. Poaching is also particularly strong using prohibited hunting methods at the expense of large mammals.

In addition to hunting, a common practice in some areas of Greece, such as Chios, is the massive capture of fauna species either for food or for cage keeping. Today, the overwhelming majority of illegally captured birds are birds that are kept in captivity for their singing. The most common species are the various finches, especially the goldfinch, the siskin, the greenfinch, the serin, the chaffinch, the calandra lark, the black-headed bunting, and others. Other species included in this category are various snails and many seafood invertebrates. Embalming and trafficking of embalmed species is also a serious incentive for the murder of many impressive birds and mammals.

The presence of many rare and endemic animal species in Greece has raised the interest of several Greek and mainly foreign collectors. These species are used as pets by many Europeans and market prices are extremely high. Typical examples are various species of reptiles such as the Milos viper and the land turtles.

4.5   Conservation Strategy and Protection Measures

The measures for the protection and preservation of fauna are divided into in situ and ex situ measures. In situ measures include the existence of protected areas, some of which have been created specifically to protect particularly endangered species.

The first protected area was established in 1938 with the opening of the Olympus National Park. Today there are several categories of protected areas:

1. National parks and aesthetic forests
2. Marine parks
3. Monuments of nature
4. Wildlife refuges

5. Game reserves
6. Controlled hunting areas
7. Protected fishing grounds
8. Areas of world cultural heritage

Several protected areas have been included in the European Natura 2000 network (Dafis et al., 1996). These areas fall into two categories: Sites of Community Interest (SCI) established in accordance with the Habitats Directive (92/43) and the Special Protected Areas (SPAs) defined by the Birds Directive (79/409). Up to now, 419 sites have been included in the proposed Sites of Community Interest and in the Special Areas of Protection. The total area covered by these areas exceeds 21% of the terrestrial area and 6% of the marine area of Greece.

Some of the protected areas were established for the protection of specific species and include the Samaria National Forest, with main objective the protection of Cretan wildlife, the National Marine Park of Zakynthos to protect the marine turtles, and the National Marine Park of Alonissos-Northern Sporades to protect the Mediterranean monk seal and some mountain areas for the protection of bears. The protected areas include 11 wetlands of the Ramsar Convention mainly concerned with the protection of birds. Many other Natura 2000 sites were proposed for the conservation of sustainable populations of the species included in the European Birds and Habitats Directives.

The ex situ protection of the Greek fauna is virtually non-existent. The few zoos and aquariums that exist serve only for the recreation and provide very little information to the visitors. There are also a few wildlife care centers that are insufficient for the needs of the country and are not well-staffed, since they belong to non-governmental organizations that are financially dependent on sponsorships and offers. There are no organized programs for raising endangered species in captivity with a view to reintroducing or enriching populations. The enrichment of game species by hunting organizations is not intended to protect natural populations but rather to hunt them.

4.5.1 Animal Agro-Diversity

Besides wildlife, the diversity of farm and domestic animals is also important, as several indigenous breeds have been recorded. The current list includes 3 cattle races, 1 buffalo race, 22 sheep races, 5 horse races, 1 pig race and 1 donkey race.

4.5.2   Animals as Indicators of Environmental Quality

Although several species are used as indicators of the state of the environment, this sector has not yet begun to be systematically implemented in Greece. The only legislative instrument that requires the use of animal indices is the European Water Framework Directive requiring the existence of specific species whose presence shows a satisfactory state of inland waters. The implementation of this directive has just begun in Greece and programs for the identification of Greek species-indices, adapted to local conditions, are currently in progress.

4.5.3   The Cultural Value of the Fauna

Man treats animals not only as a source of food, for the provision of other goods and services, or as a danger or aversion, but also gives them other characteristics by likening them to situations unknown to him. Animals also often have a cultural value representing superior ideas.

Ancient Greek mythology as well as the present traditions that have been passed on to us from generation to generation are littered with animal examples, real or mythical. Harpies, griffins, sphinxes, hydras, chimeras, mermaids and many other mythical animals appear in Greek mythology. The gods, when they want to descend into the world of mortals, they are transformed into animals like swans and bulls. They also transform those who want to avenge or protect from the wrath of other gods such as Circe who transformed Odysseus' comrades into pigs or Jupiter who transformed Europe into a cow.

Folklore is also scattered in references to animals possessing human qualities: evil and threatening (wolves, snakes), evil (foxes, crows), goods (sheep, cows), clever (ferrets, dolphins), etc.

Some species have, over time, acquired a special cultural value for the inhabitants of the areas where they live. The Cretan wild goat is the emblem of Crete as well as the fallow deer for Rhodes and the eagle for Epirus. Sea turtles are often met in ancient coins of coastal areas and the owl was and still is the symbol of Athens.

4.5.4   The Value and Uses of Greek plants

4.5.4.1   Uses of Greek Plants

A number of Greek native plants have been cultivated (and selected) since ancient times and certain wild herbs still participate as dietary components in traditional recipes, particularly in rural areas. The Mediterranean diet, rich in herbs and vegetables, has been shown to offer protection against the occurrence of major chronic degenerative diseases (Trichopoulou et al, 2009; Sofi et al, 2010). Some wild trees, shrubs and perennial or annual plants are genetically related to widely cultivated crops or may offer future challenges for agricultural exploitation and new product resources. Medicinal and aromatic plants are still harvested from wild populations in various parts of Greece. Wood is provided by forests mainly on the Pindos mountain chain and northern Greece.

Some common Greek native species of economic importance or useful in various ways are provided below. For additional information see Hanlidou et al (2004), Skoula et al. (2009), Psaroudaki et al. (2012).

4.5.4.2 Timber Species

Abies cephalonica, A. borisii-regis (fir trees), *Pinus nigra, P. sylvestris, P. heldreichii, P. halepensis* and *P. brutia* (pine trees; some species also provide resin), *Picea abies* (Norway spruce), *Fagus sylvatica* (common beech), various species of *Quercus* (oak trees), species of *Populus* (poplar), *Juglans regia* (walnut) and *Castanea sativa* (sweet chestnut).

4.5.4.3 Trees or Shrubs With Edible Fruit/Seed

Pinus pinea (maritime pine), *Juglans regia* (walnut) and *Castanea sativa* (sweet chestnut), *Corylus* spp. (hazel, filbert), *Ficus carica* (fig tree), *Rubus* spp. (blackberry, raspberry), various *Prunus* species, *Ceratonia siliqua* (carob tree), *Vitis vinifera* (grapevine), *Olea europaea* (olive), *Morus* spp. (mulberries).

4.5.4.4 Edible Wild Herbs

Cichorium (particularly *Cichorium spinosum* in Crete) and *Taraxacum* species, *Sonchus oleraceus, Centaurea raphanina,* various species of *Crepis, Picris* and *Hedypnois, Urospermum picroides, Cynara cardunculus, Silybum marianum, Scolymus hispanicus* (all Asteraceae), *Capsella bursa-pastoris, Raphanus raphanistrum, Brassica* spp., *Eruca sativa, Sinapis alba* (all Brassicaceae), *Tordylium apulum, Scandix pectin-veneris, Daucus carota* (all Apiaceae), *Asparagus* spp., *Capparis spinosa, Portulaca oleracea, Amaranrthus* spp., *Muscari comosum, Urtica* spp.

4.5.4.5 Flavoring and Tea Plants

Origanum vulgare, O. onites (oregano herbs) and *O. dictamnus, Thymbra capitata, Satureja thymbra,* various perennial *Sideritis* species (mountain tea), *Salvia fruticosa* and *Mentha* spp., *Rosmarinus officinalis* (all Lamiaceae), *Foeniculum vulgare, Matricaria chamomilla, Tilia* spp., *Laurus nobilis*, and many others.

4.5.4.6 Dye Plants

Alkanna tinctoria (alkanet or dyer's bugloss), *Rubia tinctorum* (dyer's madder), *Genista tinctoria* (dyer's greenweed), *Isatis tinctoria* (woad), *Carthamus tinctorius* (safflower), *Juglans regia* (walnut), and others.

4.5.4.7 Wild Relatives of Cultivated Species

Beta vulgaris (beet), *Brassica* species, *Malus dasyphylla* and *M. sylvestris* (wild apples), *Pyrus* spp. (pear relatives), several species of *Prunus, Lupinus albus* (white lupin), *Lens* spp. (lentil), *Pisum sativum* (field peas), *Daucus carota* (variable relatives of cultivated carrot), various *Allium* species (wild leek), *Crocus cartwrightianus* (the closer relative of cultivated *Crocus sativus*, the saffron crocus), *Aegilops, Hordeum* and *Avena* species (relatives of cultivated cereals), and *Phoenix theophrasti* (related to cultivated date palm).

Keywords

- conservation strategy
- ecosystem diversity
- protection measures
- species diversity

References

Abbott, B. F. M., (2009). Checklist of the lichens and lichenicolous fungi of Greece. *Bibliotheca Lichenologica, 103.*

Agiomyrgianaki, A., Petrakis, P. V., & Dais, P., (2012). Influence of harvest year, cultivar and geographical origin on Greek extra virgin olive oils composition: A study by NMR spectroscopy and biometric analysis. *Food Chemistry, 135*, 2561–2568.

Alexander, R. D., & Bigelow, R. S., (1960). Allochronic speciation in field crickets, and a new species, *Acheta veletis. Evolution, 14*, 334–346.

Aliferis, K. A., Tarantilis, P. A., Harizanis, P. C., & Alissandrakis, E., (2010). Botanical discrimination and classification of honey samples applying gas chromatography/mass spectrometry fingerprinting of headspace volatile compounds. *Food Chemistry, 121*, 856–862.

Antonopoulos, Z., & Tsiftsis, S., (2017). *Atlas of the Greek Orchids,* Mediterraneo editions, Rethymno, vol. 2.

Arianoutsou, M., Bazos, I., Delipetrou, P., & Kokkoris, Y., (2010). The alien flora of Greece: Taxonomy, life traits and habitat preferences, *Biological Invasions, 12,* 3525–3549.

Bailly, N., Gerovasileiou, V., Arvanitidis, C., & Legakis, A., (2016). Introduction to the Greek Taxon Information System (GTIS) in LifeWatch Greece: The construction of the Preliminary Checklists of Species of Greece. *Biodiversity Data Journal, 4,* 7959.

Baselga, A., (2008). Determinants of species richness, endemism and turnover in European longhorn beetles, *Ecography, 31,* 263–271.

Baumann, H., (1982). *Die grichische Pflanzenvelt I Mythos, Kunst und Literatur,* Hirmer Verlag.

Benn, J., (2010). *What is Biodiversity?,* UNEP.

Blockeel, T. L., (2013). Mountains and Islands: In search of bryophytes in Greece. *Field Bryology, 109,* 16–25.

Bory de Saint-Vincent, J., (1832–1836). *Expédition Scientifique de Morée Enterprise et Publié Par Ordre du Gouvernement Français.* Travaux de la section des sciences physiques sous la direction de M. le colonel Bory de Saint-Vincent. Tome III. 1ere Partie. Zoologie.

Brown, P. M. J., Adriaens, T., Bathon, H., Cuppen, J., Goldarazena, A., et al., (2008). *Harmonia axyridis* in Europe: Spread and distribution of a nonnative coccinellid. *BioControl, 53,* 5–21.

Brullo, S., & Erben, M., (2016). The genus *Limonium* (Plumbaginaceae) in Greece. *Phytotaxa, 240.*

Brullo, S., Pavone, P., & Salmeri, C., (2015). Biosystematic researches on *Allium cupani* group (Amaryllidaceae) in the Mediterranean area. *Flora Mediterranea, 25,* 209–244.

Brussell, D. E., (2004). Medicinal plants of Mt. Pelion, Greece. *Economic Botany, 58*(1), 174–202.

Cameron, R. A. D., Mylonas, M., & Vardinoyannis, K., (2000). Local and regional diversity in some Aegean land snail faunas. *J. Moll. Stud. 66,* 131–142.

CBD, (2011). Convention on Biological Diversity, 5th National Report of Greece, Global Environment Facility (GEF) under the management of the United Nations Environment Program (UNEP), Washington DC.

Chatzaki, M., Trichas, A., Markakis, G., & Mylonas, M., (1998). Seasonal activity of the ground spider fauna in a Mediterranean ecosystem (Mt Youchtas, Crete, Greece). In: *Proceedings of the 17th European Colloquium in Arachnology,* Selden, P. A., (edt.), pp. 235–243.

Cho, Y., Lee, D., & Bae, S. Y., (2017). Effects of vegetation structure and human impact on understory honey plant richness: Implications for pollinator visitation. *J. Ecology and Environment, 41,* 2.

Corsini-Foka, M., & Economidis, P., (2007). Allochthonous and vagrant ichthyofauna in Hellenic marine and estuarine waters. *Mediterranean Marine Science, 8,* 67–89.

Creutzburg, N., (1963). Paleogeographic evolution of Crete from Miocene till our days, *Cretan Annals. 15*(16), 336–342.

Daferera, D. J., Ziogas, B. N., & Polissiou, M. G., (2000). GC-MS analysis of essential oils from some greek aromatic plants and their fungitoxicity on *Penicillium digitatum. J. Agricult. Food Chem., 48,* 2576–2581.

Dafis, S., Papastergiadou, E., Georghiou, K., Babalonas, D., Georgiadis, T., Papageorgiou, M., Lazaridou, T., & Tsiaoussi, V., (1996). *Directive 92/43/EEC. The Greek 'Habitat' Project NATURA 2000: An Overview.* The Goulandris Natural History Museum – Greek Biotope/Wetland Centre, pp. 1–917.

Dais, P., Agiomyrgianaki, A., & Petrakis, P. V., (2012). Influence of harvest year, cultivar and geographical origin on Greek extra virgin olive oils composition: A study by NMR spectroscopy and biometric analysis. *Food Chemistry, 135,* 2561–2568.

Dapporto, L., (2009). Speciation in Mediterranean refugia and postglacial expansion of *Zerynthia polyxena* (Lepidoptera, Papilionidae), *J. Zool. Syst. Evol. Res., 48,* 229–237.

De Jong, Y., et al., (2014). Fauna Europaea – All European animal species on the web. *Biodiversity Data Journal, 2,* e4034, accessed in 10 Dec 2016.

Dimitrakopoulos, A. P., & Mitsopoulos, I. D., (2006). Global forest resources assessments 2005 – Report on fires in the Mediterranean region. In: *Fire Management Working Paper 8.* doi www.fao.org/forestry/site/fire-alerts/en, accessed in 25 Apr 2010.

Dimopoulos, P., Bergmeier, E., Eleftheriadou, E., Theodoropoulos, K., Gerasimidis, A., & Tsiafouli, M., (2012). *Monitoring Guide for Habitat Types and Species in the Natura, 2000.* Sites with management institutions, University of West Greece.

Dimopoulos, P., Raus, T., Bergmeier, E., Constantinidis, T., Iatrou, G., Kokkini, S., Strid, A., & Tzanoudakis, D., (2013). Vascular plants of Greece: An annotated checklist, Botanic garden and botanical museum Berlin-Dahlem, Berlin and Hellenic Botanical Society, Athens. *Englera, 31.*

Dimopoulos, P., Raus, T., Bergmeier, E., Constantinidis, T., Iatrou, G., Kokkini, S., Strid, A., & Tzanoudakis, D., (2016). Vascular plants of Greece: An annotated checklist. Supplement. *Willdenowia, 46,* 301–347.

Düll, R., (1995). Moose Griechenlands (Bryophytes of Greece). *Bryologische Beiträge, 10,* 1–229.

Düll, R., (2014). *A Survey of the Bryophytes Known From the Aegean Islands.* Weissdorn-Verlag, Jena.

Economidis, P., (1992). Fish. In: *The Red Book of the Threatened Vertebrates of Greece,* Karandeinos, M., & Legakis, A., (eds.). Hellenic Zoological Society-Hellenic Ornithological Society, pp. 41–81.

García-González, D. L., & Aparicio, R., (2010). Research in olive oil: Challenges for near future. *J. Agric. Food Chem., 58,* 12569–12577.

Georghiou, K., & Delipetrou, P., (1990–2009). *Database 'Chloris': Endemic, Rare, Threatened and Protected Plants of Greece.* Synonyms, distribution, conservation and protection status, biology, ecology, bibliography. Electronic Database in MS Access and in ORACLE for WINDOWS NT. University of Athens.

Georghiou, K., & Delipetrou, P., (2010). Patterns and traits of the endemic plants of Greece. *Bot. J. Linn. Soc., 162,* 130–422.

Giokas, S., Mylonas, M., & Sotiropoulos, K., (2000). Gene flow and differential mortality in a contact zone between two *Albinaria* species (Gastropoda, Clausiliidae). *Biol. J. Linn. Soc., 71,* 755–770.

Gkelis, S., Ourailidis, I., Panou, M., & Pappas, N., (2016). Cyanobacteria of Greece: An annotated checklist. *Biodiversity Data Journal., 4,* 10084.

Gómez, A., & Lunt, D. H., (2006). Refugia within refugia: Patterns of phylogeographic concordance in the Iberian Peninsula. In: *Phylogeography of Southern European Refugia* Weiss, S., & Ferrand, N., (eds). Springer. *The Netherlands,* pp. 155–188.

Handrinos, G., & Akriotis, T., (1997). *The Birds of Greece*. Christopher Helm.

Hanlidou, E., Karousou, R., Kleftoyanni, V., & Kokkini, S., (2004). The herbal market of Thessaloniki (N Greece) and its relation to the ethnobotanical tradition. *J. Ethnopharmacol.*, *91*(2–3), 281–299.

Hewitt, G., (2000). The genetic legacy of the Quaternary ice ages. *Nature*, *405*, 907–913.

Higgins, M. D., & Higgins, R., (1996). *A Geological Companion to Greece and the Aegean*. Cornell University Press, Duckworth Publishers, Ithaca.

Hortal, J., Diniz-Filho, J. A. F., Bini, L. M., Rodrıguez, M. A., Baselga, A., Nogues-Bravo, D., Rangel, T. F., Hawkins, B. A., & Lobo, J. M., (2011). Ice age climate, evolutionary constraints and diversity patterns of European dung beetle, *Ecology Letters*, *14*, 741–748.

Hurston, H., Foufopoulos, J., Voith, L., Bonanno, J., Pafilis, P., Valakos, E., & Anthony, N., (2009). Effects of fragmentation on genetic diversity in island populations of the Aegean wall lizard *Podarcis erhardii* (Lacertidae, Reptilia). *Mol. Phyl. Evol.*, 395–405.

Ignatiades, L., & Gotsis-Skretas, O., (2010). A review on toxic and harmful algae in Greek coastal waters (E. Mediterranean Sea). *Toxins*, *2*, 1019–1037.

Ji, H. F., Li, X. J., & Zhang, H. Y., (2009). Natural products and drug discovery. *EMBO*, *10*, 194–200.

Katsadorakis, G., & Paragamian, K., (2006). *Aegean Wetlands*. Athens, Greece.

Legakis, A., & Kypriotakis, Z., (1994). A biogeographic analysis of the island of Crete (Greece). *J. Biogeogr.*, *21*, 441–445.

Legakis, A., & Maragou, P., (2009). *Greek Red Data Book of Threatened Fauna*. Hellenic Zoological Society.

Legakis, A., (1983). Recent trends in the study of the Greek fauna. *Biol. Gallo-Hellen.*, *10*, 17–20.

Legakis, A., (1990). The status of the Bern invertebrates in Greece. In: *Colloquy on the Bern Convention, Invertebrates and Their Conservation*, No 10. Council of Europe, Environmental Encounters Series, pp. 17–19.

Legakis, A., (2004). *How Many Animal Species Exist in Greece*. Hellenic Congress of the Union of Greek Ecologists & Hellenic Zoological Society, Mytilini.

Liolios, C. C., Graikou, K., Skaltsa, E., & Chinou, I., (*2010*). Dittany of Crete: A botanical and ethnopharmacological review. *J. Ethnopharmacol.*, 131, 229–241.

Lykidis, C. T., Nardi, G., & Petrakis, P. V., (2016). First record of *Sinoxylon unidentatum* and *S. anale* (Coleoptera: Bostrichidae) from Greece, with an updated account on their global distribution and host plants. *Fragmenta Entomologica*, 48, 101–121.

Magurran, A. E., (2004). *Measuring Biological Diversity*. Blackwell Publishing. Malden, MA.

Major, J., (1988). Endemism: A botanical perspective. In: *Analytical Biogeography: An Integrated Approach to the Study of Animal and Plant Distributions*. Myers, A. A., & Giller, P., (eds.), Chapman & Hall, London, UK, pp. 118–146.

Mantziou, G., Poulakakis, N., Lymberakis, P., Valakos, E., & Mylonas, M., (2004). The inter and intraspecific status of Aegean *Mauremys rivulata* (Chelonia, Bataguridae) as inferred mitochondrial DNA sequences. *Herpet. J.*, *14*, 35–45.

Mayden, R. L., (1997). A hierarchy of species concepts: The denouement in the saga of the species problem, In: *Species: The Units of Diversity*, Claridge, M. F., Dawah, H. A., & Wilson, M. R., (eds.), London: Chapman and Hall, pp. 381–423.

McCune, B., Grace, J. B., & Urban, D. L., (2002). *Analysis of Ecological Communities*. MjM, Gleneden Beach, Oregon.

Miko, I., (2008). Genetic dominance: Genotype-phenotype relationships. *Nature Education*, *1*, 140.

Mitsopoulos, I., & Mallinis, G., (2017). "A data-driven approach to assess large fire size generation in Greece," *Natural Hazards*, *88*(3), 1591–1607. doi: 10.1007/s11069–017–2934-z.

Mühle, H., Brandl, P., & Niehuis, M., (2000). *Catalogus Faunae Graeciae, Coleoptera: Buprestidae*. Selbstverlag, Augsburg, Germany.

Myers, N., Mittermeier, R. A., Mittermeier, C. G., Da Fonseca, G. A. B., & Kent, J., (1990). Biodiversity hotspots for conservation priorities. *Nature, 403*, 853–858.

Mylonas, M., (1981). *The Zoogeography and Ecology of the Terrestrial Molluscs of Cyclades*. PhD thesis, University of Athens, Athens.

Mylonas, M., (1984). The influence of man: A special problem in the study of the zoogeography of terrestrial molluscs on the Aegean islands. In: *World-Wide Snails*, Solem, A., & Van Bruggen, A. C., (eds.). Biogeographical studies on nonmarine Mollusca. E. J. Brill. Leiden, pp. 249–259.

Nieto, A., & Alexander, K. N. A., (2010). *European Red List of Saproxylic Beetles*. Luxembourg, Publications Office of the European Union.

Pantidou, M., (1973). *Fungus-Host Index for Greece*. Benaki Phytopathological Institute, Kiphissia.

Papachristos, D., Matakoulis, C., Papadopoulos, N. T., Lagouranis, A., Zarpas, K., & Milonas, P., (2013). First Report of the Presence of *Drosophila Suzukii* (Diptera: Drosophilidae) in Greece. Abstracts of the 15th entomology meeting of the Hellenic Entomological Society.

Papaconstantinou, C., (1988). *Check-List of Marine Fishes of Greece. Fauna Graeciae*. National Centre for Marine Research, Athens, vol. 4.

Papageorgiou, V. P., Assimopoulou, N. A., Couladouros, E. A., Hepworth, D., & Nicolaou, K. L., (1999). The chemistry and biology of alkannin, shikonin, and related naphthazarin Natural Products. *Angewandte Chemie International Edition*, *38*, 270–300.

Papathanassiou, E., & Zenetos, A., (2005). *State of the Hellenic Marine Environment*. Hellenic Centre for Marine Research, Institute of Oceanography, Athens, Greece.

Paraschi, L., & Handrinos, G., (1992). Introduction. In: *The Red Book of the Threatened Vertebrates of Greece*, Karandeinos, M., & Legakis, A., (eds.). Hellenic Zoological Society – Hellenic Ornithological Society, pp. 1–20.

Petit, R. J., Aguinagalde, I., De Beaulieu, J. L., et al., (2003). Glacial refugia: Hotspots but not melting pots of genetic diversity. *Science, 300*, 1563–1565.

Petrakis, P. V., & Moulet, P., (2011b). First record of the Nearctic *Zelus renardii* (Heteroptera, Reduviidae, Harpactocorinae) in Europe. *Entomologia Hellenica.*, *20*, 75–82.

Petrakis, P. V., (2000). Larval performance in relation to oviposition site preference in olive kernel moth (*Prays oleae* BERN., Yponomeutidae, Praydina), *Agricultural and Forest Entomology.*, *2*, 271–282.

Petrakis, P. V., (2011a). First record of *Leptoglossus occidentalis* (Heteroptera, Coreidae) in Greece. *Entomologia Hellenica, 20*, 83–92.

Petrakis, P. V., (2015). Insects in agroforestry and urban waste management. In: *Advances in Bio-Informatics and Environmental Engineering – ICABEE*, Rome, Italy 18–19.

Petrakis, P. V., (2017). Ecosystem management: The roles of insects in ecosystems and their control. In: *ARENA- Human Health and the Environment*. Postgraduate course of the Medicine School in The National and Capodistrian University of Athens.

Petrakis, P. V., Agiomyrgianaki, A., Christoforidou, S., Spyros, A., & Dais, P., (2008). Geographical characterization of Greek virgin olive oils (Koroneiki cv) using ^{1}H and ^{31}P NMR fingerprinting with canonical discriminant analysis and classification binary trees. *J. Agric. Food Chem.*, *56*, 3200–3207.

Petrakis, P. V., Roussis, V., Vagias, C., & Tsoukatou, M., (2010). The interaction of pine scale with pines in Attica, Greece. *European, J. Forest Res.*, *129*, 1047–1056.

Petrakis, P. V., Spanos, K., Kalapanida, M., Lahlou, E., & Feest, A., (2011c). Insect biodiversity reduction of pinewoods in southern Greece caused by the pine scale (*Marchalina hellenica*). *Forest Systems, 20*, 27–41.

Phitos, D., Constantinidis, T., & Kamari, G., (2009). *The Red Data Book of Rare and Threatened Plants of Greece, (A-D) and 2 (E-Z)*. Hellenic Botanical Society, Patras, Greece, vols. 1.

Phitos, D., Strid, A., Snogerup, S., & Greuter, W., (1995). *The Red Data Book of Rare and Threatened Plants of Greece*. World Wide Fund for Nature (WWF), Athens.

Prendegast, J. R., Quinn, R. M., Lawton, J. H., Eversham, B. C., & Gibbons, D. W., (1993). Rare species, the coincidence of diversity hotspots and conservation strategies. *Nature, 365*, 335–337.

Preston, C. D., (1981). A check-list of Greek liverworts. *J. Bryology, 11*, 537–553.

Preston, C. D., (1984a). A check-list of Greek mosses. *J. Bryology, 13*, 43–95.

Preston, C. D., (1984b). A check-list of Greek liverworts: Addendum. *J. Bryology, 13*, 97–100.

Psaroudaki, A., Dimitropoulakis, P., Constantinidis, T., Katsiotis, A., & Skaracis, G. N., (2012). Ten indigenous edible plants: Contemporary use in Eastern Crete, Greece. *Culture, Agriculture, Food and Environment, 34*(2), 172–177.

Rešetnik, I., Schneeweiss, G. M., & Liber, Z., (2014). Two new combinations in the genus *Bornmuellera* (Brassicaceae). *Phytotaxa, 159*(4), 298–300.

Roditakis, E., Papachristos, D., & Roditakis, N. E., (2010). Current status of the tomato leafminer *Tuta absoluta* in Greece. *Bulletin OEPP/EPPO Bulletin, 40*, 163–166.

Roussis, V., Couladis, M., Tzakou, O., Loukis, A., Petrakis, P. V., Dukic, N. M., & Jancic, R., (2000b). A comparative study on the needle volatile constituents of three *Abies* species grown in South Balkans. *J. Essent. Oil Res., 12*, 41–46.

Roussis, V., Tsoukatou, M., Petrakis, P. V., Chinou, I., Skoula, M., & Harborne, J. B., (2000a). Volatile constituents of four *Helichrysum* species growing in Greece. *Biochem. Syst. Ecol., 28*, 163–175.

Ruck, C. A. P., (1995). Gods and plants in the classical world. In: *Ethnobotany-Evolution of a Discipline*, Dioscorides Press.

Schaffers, A. P., Raemakers, I. P., Sýkora, K. V., & Ter Braak, C. J. F., (2008). Arthropod assemblages are best predicted by plant species composition. *Ecology, 89*, 782–794.

Schmitt, T., Giessl, A., & Seitz, A., (2003). Did *Polyommatus icarus* (Lepidoptera: Lycaenidae) have distinct glacial refugia in southern Europe? Evidence from population genetics. *Biol. J. Linn. Soc., 80*, 529–538.

Sfenthourakis, S., & Legakis, A., (2001). Hotspots of endemic terrestrial invertebrates in southern Greece. *Biodiversity and Conservation, 10*, 1387–1417.

Sfenthourakis, S., Giokas, S., & Mylonas, M., (1999). Testing for nestedness in the terrestrial isopods and snails of Kyklades islands (Aegean archipelago, Greece). *Ecography, 2,* 384–395.

Skoula, M., Dal Cin D'Agata, C., & Sarpaki, A., (2009). Contribution to the ethnobotany of Crete, Greece. *Bocconea., 23,* 479–487.

Sofi, F., Abbate, R., FrancoGensini, G., & Casini, A., (2010). Accruing evidence on benefits of adherence to the Mediterranean diet on health: An updated systematic review and meta-analysis. *Amer. J. Clinical Nutrition, 92,* 1189–1196.

Soininen, J., McDonald, R., & Hillebrand, H., (2007). The distance decay of similarity in ecological communities. *Ecography, 30, 3–12.*

Spanos, I, Meliadis, I., Mantzanas, K., Platis, P., Samara, T., & Meliadis, M., (2015). *Forest Land Ownership in Greece COST Action FP1201 FACEMAP*, European Forest Institute, Vienna, pp. 31.

Stamatis, G., Kyriazopoulos, P., Golegou, S., Basayiannis, A., Skaltsas, S., & Skaltsa, H., (2003). In vitro anti *Helicobacter pylori* activity of Greek herbal medicines. *J. Ethno-pharmacol., 88,* 175–179.

Strid, A., (1996). Phytogeographia Aegaea and the Flora Hellenica database. *Annalen des Naturhistorischen Museums in Wien. Serie B, 98,* 279–289.

Strid, A., (2016). *Atlas of the Aegean Flora. Part 1: Text & Plates. Part 2: Maps.* Botanic Garden and Botanical Museum Berlin, Freie Universität Berlin, Berlin. *Englera, 33* (1&2).

Triantis, K. A., Mylonas, M., Weiser, M. D., Lika, K., & Vardinoyannis, K., (2005). Species richness, environmental heterogeneity and area: A case study based on land snails in Skyros archipelago (Aegean Sea, Greece*). J. Biogeogr., 32,* 1727–1735.

Trichopoulou, A., Bamia, C., & Trichopoulos, D., (2009). Anatomy of health effects of Mediterranean diet: Greek EPIC prospective cohort study. *British Medical Journal, 338,* 2337.

Trigas, P., Panitsa, M., & Tsiftsis, S., (2013). Elevational gradient of vascular plant species richness and endemism in Crete – The effect of post-isolation mountain uplift on a continental island system. *Pls One., 8*(3), 59425.

Tsiamis, K., & Panayotidis, P., (2016). Seaweeds of the Greek coasts: Rhodophyta: Ceramiales. *Acta Adriatica., 57*(2), 227–250.

Tsiamis, K., Panayotidis, P., Economou-Amilli, A., & Katsaros, C., (2013). Seaweeds of the Greek coasts. I. Phaeophyceae. *Mediterranean Marine Science, 14*(1), 141–157.

Tsiamis, K., Panayotidis, P., Economou-Amilli, A., & Katsaros, C., (2014). Seaweeds of the Greek coasts. II. Ulvophyceae. *Mediterranean Marine Science, 15*(2), 449–461.

Tsiftis, S., & Antonopoulos, Z., (2017). *Atlas of the Greek Orchids,* Mediterraneo Editions, Rethymno, vol. 1.

Turland, N. J., Chilton, L., & Press, J. R., (1993). *Flora of the Cretan Area: Annotated Checklist and Atlas.* The Natural History Museum, London, UK.

Vakalis, D., (2006). *Review of Forestry Service Activities in the Year 2005,* Ministry of Rural Development and Food, General Directorate for the Development and Protection of Forests, Directorate of Forest Resources Development. Athens, pp. 78.

Vardinoyannis, K., & Mylonas, M., (1988). The differentiation and distribution of the genus *Mastus* (Gastropoda, Enidae) in the Aegean Archipelago. *Rapp. Comm. Int. Mer Medit., 31*(2), 133.

Wilkins, J. S., (2003). How to be a chaste species pluralist-realist: The origins of species modes and the Synapomorphic Species Concept, *Biology and Philosophy*, *18*, 621–638.

Willig, M. R., Kaufman, D. M., & Stevens, R. D., (2003). Latitudinal gradients of biodiversity: Pattern, process, scale, and synthesis, *Annu. Rev. Ecol. Evol. Syst.*, *34*, 273–309.

Wolters, V., Bengtsson, J., & Zaitsev, A., (2006). Relationship among the species richness of different taxa, *Ecology*, *87, 1886–1895*.

Zervakis, G. I., (2001). Biodiversity in Greece. *Bocconia.*, *13*, 119–124.

Zervakis, G. I., Dimou, D., & Balis, C., (1998). A check-list of the Greek macrofungi, including hosts and biogeographic distribution: I. Basidiomycotina. *Mycotaxon.*, *66*, 273–336.

Zervakis, G., Lizon, P., Dimou, D., & Polemis, E., (1999). Annotated check-list of the Greek macrofungi. II. Ascomycotina. *Mycotaxon.*, *72*, 487–506.

Plate 4.1 A new record of *Osmoderma lassallei* subspecies from Mt. Parnon (Peloponnese) in July 2017 (photo by P. V. Petrakis).

Plate 4.2 Megatree (*Quercus ilex*) in Kouri Forest, Spilia, Mt Ossa, Larissa. *Osmoderma lassallei* (Col., Scarabaeidae, Cetoniinae) was found within the hollow of this tree. (photo by P. V. Petrakis).

Plate 4.3   A mixed forest with fir (*Abies borissii-regis*) and deciduous trees (*Quercus* spp., *Fagus* spp., *Acer* spp., *Prunus* sppp., *Sorbus* spp., *Pyrus* spp. near Spelea, Mt Ossa, Larissa. The high biodiversity of this vegetation type provides the necessary niches for a great variety of insects and vertebrates. (photo by P. V. Petrakis)

Plate 4.4 Beehives at Anopolis, Sfakia, Crete. The pine trees of the open pine forest are full of the cottony masses of *Marchalina hellenica* (Hemiptera, Coccoidea) and are introduced by beekeepers. This insect finally causes the death of the pine tree by sucking the plant juice in the onset of the dry season of the year (May-June) and providing the necessary food source to the bees. It is considered beneficial in pine-honey production and for this it is spread by beekeepers causing severe biodiversity losses in the otherwise taxonomically poor pine forests. (Petrakis et al., 2010, 2011). The mountain tops above are summits of Lefka Ori. They are considered as a mountain desert but they harbor many endemic plants (e.g. *Nepeta sphaciotica* on Svourichti summit in the NATURA2000 site with code GR4340008, inlet photo) (photo by P. V. Petrakis).

(a)

(b)

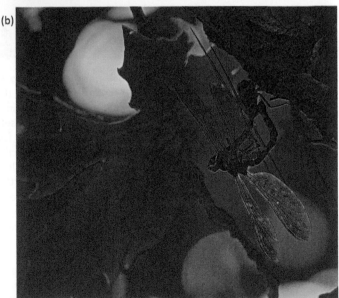

Plate 4.5 A Cretan endemic *Boyeria cretensis* (Odonata, Aeshnidae) known as 'Cretan spectre'. It is an endangered species (IUCN categorize it as C2a). It has an unknown population density and has been found only in 11 streams. (a) ovipositing female (b) mating on *Platanus orientalis* tree. (photo by F. Samaritakis).

Plate 4.6 A Cretan endemic subspecies *Calopteryx splendens* subsp. *cretensis* (Odonata, Calopterygidae) common name 'Banded demoiselle'. It is bigger than the nominal species and the markings on its wings go up almost to the end of the wing. It also shares the same habitat with the above species. (photo by F. Samaritakis).

Plate 4.7 *Monticola solitaria* (Blue rock thrush) at the cliffs of Oenoe gorge, Marathon lake, Attica. The bird is categoriezed as 'Least Concern' by IUCN. (a) Male bird and (b) its biotope. (photo by G. Vasilakis).

Plate 4.8   (a) Collecting saproxylic insects in the hollow of a huge 600 years old plane tree (*Platanus orientalis*) at Paraskevorema river, Prokopi, Euboea). The tree is a 'monument of nature'. (photo by P. V. Petrakis). (b) Not only humans are looking for insects (a *Bufo bufo*, 'common toad') in the same hollow and as a rule they are more successful than humans.

Plate 4.9 (a) *Hypericum fragile* (a rare Greek endemic of West Aegean calcareous cliffs) on a cliff at Chiliadou coast, Euboea. IUCN categorizes this species as 'vulnerable-VU'. (b) Cliff habitat from where the plant at pane (a) was photographed. (photo by P. V. Petrakis).

Plate 4.10 Typical habitat of *Cerambyx cerdo* (Coleoptera, Cerambycidae; common name 'great capricorn beetle') at Milea, Mt Taygetos, Peloponnese. The IUCN category is NT (=near threatened) but the estimated calculation of veteran tree loss in the next years is almost probable that will cause a decline of the existing populations of the insect. This is a saproxylic beetle leaving predominantly on *Quercus* spp. trees. This is a Q. *frainetto* veteran tree cut by local people with a machine-saw at a depth of 1-2 cm. This depth is enough to create many suitable sites for saproxylic insects on the tree though in the long run it causes dieback of the foliage and finally the death of the tree. This action is directed towards the removal of the area from the Natura 2000 network. However, saproxylic insects reverse this process and the area is retained in the Natura 2000 network of ecotopes because of the biodiversity of insects. (photo by P. V. Petrakis).

Plate 4.11 Sparse trees of *Pinus heldreichii* (common name 'Bosnian pine' and Greek vernacular name 'robolo'). IUCN categorized the species as of 'Least Concern (LC). (photo by P. V. Petrakis).

CHAPTER 5

Biodiversity in Hungary

ZOLTÁN VARGA

Department of Evolutionary Zoology, University of Debrecen, Hungary,
E-mail: varga.zoltan@science.unideb.hu

5.1 Introduction

Hungary is situated in the Carpathian basin, in a topographically discrete unit of southeastern Central Europe. This area belongs to the regions of Europe having the highest biodiversity (Williams et al., 1999). Its geographically transitional position resulted in an outstanding mixture of floristic and faunal elements of diverse origins and geographical histories. Especially the hilly areas surrounding the Pannonian lowland with varied orography, hydrography and also characterized by transitional climatic conditions, are populated by many biogeographically important floristic and faunal elements. The southern slopes and foothills of the Hungarian Middle Range served as refuges for thermo-xerophilous elements during several cold and cool-humid climatic phases of the Quaternary period and postglacially as centers of their dispersal (Varga, 2011; Schmitt and Varga, 2012). Thus, many thermophilous elements probably populated postglacially the Carpathian basin not only by long-distance colonization from southern "classical" glacial refuges, but also from numerous meso- or microclimatically favorable sites, lying at the fluctuating borderlines of the Mediterranean refugial areas and periglacial belts. The biostratigraphical structure of the Hungarian young Pleistocene, often characterized by a coexistence of forest and nonforest faunal elements provides evidence which demonstrated the transitional character of this region during the whole time-span of the Quaternary period (Willis et al., 1995; Magyari et al., 2010, 2014).

Hungary occupies the central, Pannonian part of the Carpathian basin. Despite the country's relatively small area (93,030 km²) and low altitudes (highest point is 1,015 m, Mts. Mátra) it is characterized by moderately humid continental climate, substantially influenced by Atlantic and sub-Mediterranean climatic regimes, but also with some alpine influences (Fekete et al., 2014; Lóczy, 2015). The diversity of flora and fauna is especially high due to the multiple biogeographic effects and the species dispersal processes during and after the glacial period (Varga, 1995; Fekete et al., 2014). The uniqueness of the flora and fauna contributed to the forma-

tion of a particular biogeographical unit within Europe named as Pannonian Region (or Pannonicum), covering the whole of Hungary, and smaller
areas from the neighboring countries and the Czech Republic as well
(Fekete et al., 2014, 2016). There are several unique endemic habitat types
of the Pannonian steppes and steppic forests as the plant communities of
the lowland alkaline, sandy and loess steppes, the lowland steppic forests
and meadow steppes on alkali, sand and loess; on calcareous sand dunes
the perennial *Festuca-Stipa* grasslands and juniper-white poplar forests; in
hilly areas the different types of xerothermic lanuginose oak forests, the
rupicolous and ravine forests and different types of rupicolous grasslands
of calcareous and volcanic hills. The deep floodplain of the basin also
shows peculiar types of wet habitats as softwood gallery forests, oxbow
lakes with rich pondweed vegetation, saline marshes, etc. (Fekete et al.,
2014). These habitat types are also characterized by numerous endemic or
Ponto-Pannonian plant species, as *Seseli leucospermum, Dianthus plumarius* subsp. *regis-stephani, Vincetoxicum pannonicum, Ferula sadleriana* on
dolomitic grasslands, *Festuca vaginata, Colchicum arenarium, Dianthus
diutinus, Iris humilis* on calcareous sand, *Suaeda prostrata, S. pannonica,*
and *S. salsa,* on alkali soil, etc.

Concerning the fauna, there are some widely distributed, pan-European or Eurasian species of priority importance for the EU that are present in remarkable population sizes in Hungary, i.e., the Imperial eagle
(*Aquila heliaca*), Saker falcon (*Falco cherrug*), Eurasian otter (*Lutra
lutra*). However, the most important speciality of the Carpathian Basin is
that this area is the most Western outpost of the Palaearctic steppe zone
(Wesche et al., 2016) and supports a mixture of biogeographic elements
from numerous geographical regions. Several faunal elements originated
from the Siberian, the Mediterranean, the Balkan, the Alpine or Atlantic
regions, are now the endemics of the Pannonian Biogeographical Region
and as such are unique natural assets of Hungarian nature conservation. The proportion of endemisms is high, e.g., in molluscs, diplopods,
orthopterans and trichopterans (Varga, 2003). Of the vertebrate species,
the Carpathian and Biharian barbel (*Barbus carpathicus, B. biharicus*),
the Hungarian meadow viper (*Vipera ursinii rakosiensis*), the Pannonian
birch mouse (*Sicista trizona*) and four Blind mole-rat species (of the genera *Nannospalax* and *Spalax*) can only be found in the Carpathian Basin.
Conservation regulations under the EU legislation are putting, therefore,
most of the responsibility to conserve the biodiversity of the Pannonian
region on Hungary.

5.2 Protection of Biodiversity: General Recent Situation in Hungary

The first nature conservation area in Hungary was declared in 1939 (the so-called "Great Forest" near Debrecen). The first designation was followed by some others, and until the end of the World War II about 200 nature conservation areas and natural monuments were declared protected. The declarations were suspended for some years after the World War II. The creation of the first landscape protection area in Tihany peninsula near the lake Balaton in 1952 was a real milestone in the national nature conservation. From 1962, the National Nature Conservation Authority (OTvH) became the responsibility for designations. Between 1950 and 1965 more than 100 areas were declared protected, on approximately 10,000 ha. The first national park of Hungary, the Hortobágy National Park, was founded by the decree No. 1850/1972 of the National Nature Conservation Authority. The process of declaring protected status got a new impetus in the early 70s (1972). The OTvH created two other national parks until 1977: the Kiskunság National Park in 1974 and the Bükk National Park in 1977. Fifteen landscape protection areas were also founded: *i.e.,* the Pilis, the Őrség, the Börzsöny, the Buda and the Zemplén landscape protection areas. The fourth national park of Hungary, the Aggtelek National Park was created in 1984. After the change of the political system the Ministry of Environment and Regional Development (KTM) was operating for eight years during which period it established five new national parks beside the already existing four. The Fertő-Hanság National Park was created in 1991, the Duna-Dráva National Park in 1996, the Körös-Maros National Park, the Balaton-felvidéki National Park and the Duna-Ipoly National Park were inaugurated in 1997 (Figure 5.1).

In September 2008, 9% of Hungary's territory was protected by national law, 16% of which was accorded 'strictly protected' status. All caves (4,077), springs (2,479), bogs (837), alkali lakes (317), tumuli (1,732) and earthen fortifications (378) are protected by the Act on Nature Conservation. In addition, 63 forest reserves have been designated. Caves have been protected by law since 1961. More than 4,000 caves are known to exist in the country today, of which 147 have received 'strictly protected' status. The total length of all known passages is 234 km. Highlights include the large hydrothermal caves of the Budapest thermal karst with their extraordinarily rich formations, as well as the caves of the Bükk Mountains and the Aggtelek Karst (Figure 5.2).

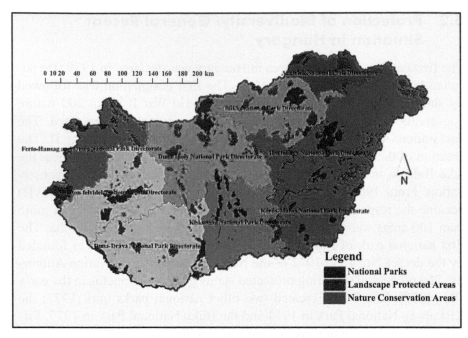

Figure 5.1 National parks, landscape protected areas and nature conservation areas in Hungary (*Source:* Ministry of Agriculture).

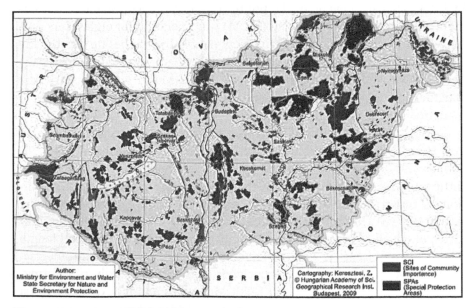

Figure 5.2 The NATURA 2000 network of Hungary (*Source:* Ministry for Environment and Water State, with permission).

Accession to the European Union in 2004 resulted in many changes in the legal-institutional setting as well as in financing nature conservation. Following the requirements of the Birds and Habitats Directives, 525 Natura 2,000 sites were designated in 2004 as an obligatory step: 56 Special Protection Areas (SPAs) and 479 Special Areas of Conservation (SACs) (21.39%, 14.77% and 15.51% of the country area, respectively) (CBD Report, 2014). Thus, recently 22% of the country's area is under different level of protection.

The incoming EU funds (e.g., LIFE, structural and rural development funds) started to play a key role in financing nature conservation activities for the rehabilitation of degraded habitats, species protection projects, nature friendly land management, construction of visitor centers, and establishing of nature schools ("Schools in Forests," Kovács et al., 2014). Hungary introduced the Common Agricultural Policy, and gained access to EU payment schemes for supporting agricultural production (Potori et al., 2013). The increase in available resources for both conservation and agricultural led to mixed effects in terms of habitat and species conservation. Several species and habitat conservation programs significantly contributed to conservation. A comprehensive guidebook has been recently published (Haraszthy, 2014) providing species and habitat-specific descriptions and management options for nearly all Natura 2000 species and habitats.

In Central Eastern European (CEE) countries political, legal and regulatory systems changed dramatically in a short period making a substantial impact on biodiversity conservation (Berkes, 2016). Despite of forced collectivization and intensification trends, extensive farming practices and extended seminatural and natural habitats survived in Hungary during socialism (Báldi and Batáry, 2011). These habitats contributed substantially to the increase of biodiversity-rich areas of the European Union when these countries joined (Henle et al., 2008; Young et al., 2007; Stoate et al., 2009; Sutcliffe et al., 2015). As a socioeconomic driver, a decline in economic performance of agriculture accompanied the transition in the early 1990s in almost all CEE countries, with a GDP fell by 15% between 1990 and 1993 in Hungary (e.g., Burger, 2001). Agricultural production has declined even more, its contribution to GDP fell from 18% in the 1980s to 5% by 1997 (Báldi and Faragó, 2007). Animal husbandry declined substantially as a result of the collapse of the former socialist market and land fragmentation during privatization after transition (Potori et al., 2013), as in almost every CEE country intensive privatization increased the share of land owned by private farms (Liira et al., 2008).

5.2.1 Diversity and Conservation Status of Species

As mentioned above, the largest part of the Pannonian biogeographical region belongs to Hungary. Compared to other member states of the European Union, biodiversity in Hungary is relatively well preserved. For example, the number of nesting bird species and their abundance is remarkably high. More than 53,000 described species occur in Hungary, 82% of them are animals. About 3% of the species are protected by national law (Table 5.1).

The number of protected species has grown by 6% (from 1,660 to 1,760), since 2003. About 36 plant, 91 bird and 105 other animal species listed on the European Union Habitats Directive occur in Hungary, some of them are protected by national law as well. Next to the legal protection, the species action plans are also important regarding the long-term conservation of the threatened species. Since 2004, action plans for 22 animal and 21 plant species were published. These plans can be found on the website of the State Secretary for Nature Protection.

In the 2008 IUCN Red List of Threatened Species, 520 species occurring in Hungary have been evaluated regarding their global endangerment status. 17% of them were found being somehow endangered (categories: Critically Endangered, Endangered, Vulnerable, Near Threatened, Lower Risk: Conservation Dependent) in the global level. Country-level red lists were compiled in 1989 for animal and plant species and in 2007 for vascular plant species (Table 5.2). The number of native vascular plant species of Hungary was estimated by Soó (1980) as 2148, by Simon (2000) as 2183. The proportion of vascular plants endangered at some level is 27.5%; numbers have grown by 30% between 1989 and 2007. 69% of these species are protected by national law.

Table 5.1   Number of Species Occurring in Hungary and the Share of Species Protected by National Law

Taxonomic group	Number of species detected in Hungary	Number of species protected by national law	% protected
Animals	43,560	997	2%
Plants	6,860	720	10%
Fungi	2,000–2,500	35	1–2%
Lichens	800	8	1%
Total	53,200–53,700	1 760	3%

Source: Standovár and Primack (2001), Hungarian Ministry of Environment and Water (2009).

Table 5.2 Number of Threatened Vascular Plant Species in Hungary According to Red Lists Compiled in 1989 and 2007

Categories Red list	1989 Red list	2007 Red List	Proportion of vascular flora	Changes 1989–2007 in %
Extinct or disappeared	36	47	1.7%	31%
Critically endangered	41	115	4.1%	180%
Endangered	127	162	5.8%	28%
Potentially endangered	386	441	15.8%	14%
Not classified (data deficient)		178	6.4%	
Total		943	33.9%	
Total (except data deficient)	590	765	27.5%	30%

Source: Németh (1989) and Király (2007).

About 25% of the 211 species of European importance are in a favorable status, according to their monitoring between 2001 and 2006. Unfortunately, the status of another 59% is inadequate or bad. These changes are the consequences of the habitat degradation and fragmentation.

5.3 The Biogeographical Composition of Flora and Fauna

5.3.1 Endemic Plant Species of the Pannonian Region

The distribution of endemics is usually restricted to some part of a region and is defined in terms of special topographic features, such as a mountain range or an island, a cave, or other natural elements, such as a specific habitat of relict character.

Vascular plant species known only from Hungary include *Dianthus diutinus* (calcareous sandy grasslands), *Vincetoxicum pannonicum* (xeric calcareous and dolomitic rupicolous grasslands), *Linum dolomiticum* (dolomitic rupicolous grasslands), *Achillea horanszkyi* and *A. tuzsonii* (volcanic rupicolous grasslands), *Ferula sadleriana* (calcareous rupicolous grasslands), *Seseli leucospermum* (dolomitic rupicolous grasslands), *Puccinellia pannonica* (Solonchak saltmarshes), and several Whitebeam trees of relict forest and shrub types incl. *Sorbus bakonyensis, S. ulmifolia, S. pseudolatifolia, S. gerecseensis, S. degenii, S. pelsoensis, S. redliana, S. udvardyana,* and *S. barabitsii*. Most of these species are connected to some special, most often xerothermic and/or rupicolous habitat types (syntaxa, see: Table 5.3).

This review shows that most Pannonian (or Pannonian-Transylvanian) endemic species are restricted to lower hilly ("peri-Pannonian") habitats

Table 5.3 The Connection of Pannonian Habitats With the Pannonian Endemic and Pontic-Pannonian Species

Syntaxa (sensu lato)	Pannonian endemisms	Pontic-Pannonian species
Dry-semidry rupicolous and steppe grasslands	*Dianthus giganteiformis* subsp. *pontederae*	*Cirsium pannonicum, Pulsatilla grandis, Thymus odoratissimus*
Dry subcontinental grasslands	*Carduus collinus, Centaurea sadleriana, Koeleria majoriflora, Seseli osseum*	*Campanula macrostachya, Chamaecytisus austriacus, Crepis pannonica, Inula ensifolia, Iris pumila, Linum flavum, Minuartia glomerata, Minuartia setacea, Thymus pannonicus, Viola ambigua*
Northern Hungarian (and Transylvanian) rocky grasslands on limestone	*Astragalus vesicarius* subsp. *albidus, Campanula sibirica* subsp. *divergenti-formis, Erysimum odoratum* subsp. *buekkense, Erysimum pallidiflorum, Ferula sadleriana, Sempervivum marmoreum, Sesleria heufleriana* subsp. *hungarica, Onosma tornense*	
Sub-Mediterranean limestone-dolomite rocky grasslands	*Bromus pannonicus, Dianthus plumarius* subsp. *regis-stephani, Galium austriacum, Hieracium kossuthianum, Linum dolomiticum, Seseli leucospermum, Thalictrum minus* subsp. *pseudominus, Vincetoxicum pannonicum*	*Paronychia cephalotes*
Carpathian silicate rocky grass-lands	*Achillea horánszkyi, Achillea tuzsonii, Poa pannonica* subsp. *scabra, Minuartia hirsuta* subsp. *frutescens*	
West Balkanic rocky grasslands	*Bromus pannonicus, Sedum neglectum* subsp. *sopianae, Vincetoxicum pannonicum*	

Table 5.3 *(Continued)*

Syntaxa (sensu lato)	Pannonian endemisms	Pontic-Pannonian species
Pannonian steppe grasslands	*Festuca stricta, Thlaspi jankae*	*Ajuga laxmannii, Astragalus dasyanthus, Bupleurum affine, Carduus hamulosus, Crambe tataria, Echium maculatum, Inula germanica, Inula oculus-christi, Lappula heteracantha, Lathyrus pallescens, Salvia austriaca, Salvia nutans, Serratula radiata, Taraxacum serotinum, Vinca herbacea*
Pannonian steppe grasslands and sand steppes		*Astragalus asper, Linum hirsutum*
Pontic-pannonian perennial sand grasslands	*Iris humilis* subsp. *arenaria, Colchicum arenarium, Dianthus diutinus, Dianthus serotinus, Festuca vaginata, Festuca wagneri, Pulsatilla pratensis* subsp. *hungarica, Sedum hillebrandtii, Tragopogon floccosus*	*Achillea ochroleuca, Centaurea arenaria, Corispermum canescens, Corispermum nitidum, Echinops ruthenicus, Erysimum canum, Peucedanum arenarium, Polygonum arenarium, Stipa borysthenica*
Puccinellia and *Camphorosma* stands on alkalic soil	*Lepidium crassifolium, Puccinellia limosa, Suaeda maritima* subsp. *pannonica*	*Camphorosma annua, Pholiurus pannonicus, Salicornia perennans, Aster tripolium* subsp. *pannonicus*
Alkali steppes and meadows	*Cirsium brachycephalum, Limonium gmelini* subsp. *hungaricum, Plantago schwarzenbergiana, Achillea asplenifolia*	*Gagea szovitsii*
Sub-Mediterranean-subcontinental xerothermic forests	*Epipactis bugacensis, Galium abaujense, Pyrus magyarica, Sorbus x danubialis, Thlaspi kovatsii* subsp. *sudichii*	
Subcontinental oakwoods, mixed xerothermic forests		*Chamaecytisus albus, Melica picta*
Pannonian white oak forests		*Serratula lycopifolia, Euphorbia epithymoides*
Fringe thickets		*Iris aphylla* subsp. *hungarica, Iris variegata*

Table 5.3   *(Continued)*

Syntaxa (sensu lato)	Pannonian endemisms	Pontic-Pannonian species
Mesophilous deciduous forests	*Scilla vindobonensis, Scilla drunensis*	*Polygonatum latifolium*
Southeast European white oak forests	*Paeonia banatica*	
Mountain beech forests	*Hesperis vrabelyiana*	
Tall herb communities, Molinia meadows	*Koeleria javorkae*	*Carex buekii*
Reeds, fens	*Armoracia macrocarpa*	*Urtica kioviensis, Leucanthemum serotinum*
Segetal weed communities	*Melampyrum barbatum, Ornithogalum* x *degenianum*	*Polycnemum verrucosum*

Source: Fekete et al., (2014) (shortened).

while the bulk of the more extended Ponto-Pannonian species typify some lowland steppic habitat type, of course with some overlaps and transitions (Fekete et al., 2014).

A significant number of interesting species is further represented by the endemic and subendemic Eastern Carpathian ("Dacian") floristic elements (*Aconitum moldavicum, Symphytum cordatum, Onosma pseudoarenaria, Helleborus purpurascens, Centaurea sadleriana*), which often also occur in the Middle Range of Northern Hungary. Another important group of plant species is formed by relict species of cold phases of the younger Quaternary period, often simply called as "glacial relicts." They are associated with two main routes of westwards migration. The relict species associated with the Sarmatian migration route, North of the Carpathians are represented, e.g., by *Astragalus arenarius, Allium strictum, Helictotrichon desertorum, Jurinea cyanoides, Stipa tirsa* and species of *Stipa dasyphylla* agg. However, some of them occur in the Pannonian region and in the Transylvanian basin. Other species of continental steppe spread along the Pannonian migration route south of the Carpathians and crossing the western gate of the Carpathian basin ("porta Hungarica") reached Lower Austria and South Moravia but not Bohemia. The species as *Agropyron pectinatum, Bassia prostrata, Crambe tataria, Crepis pannonica, Jurinea mollis, Onosma arenaria, Phlomis tuberosa, Prunus tenella, Scorzonera austriaca* and *Taraxacum serotinum* are well-known examples.

5.3.2 Endemic Zootaxa and Autochthonous Evolution in the Carpathian Basin

The level of endemism generally correlates with the geological age of the refugia where relict-like taxa have been evolved and/or could survive. The Carpathian Basin belongs to the geologically young areas of Europe. Its relief developed under the influence of the Alpine orogenesis and by retreat of the Paratethys and the Pannonian inland sea. Moreover, the phylogeography of some freshwater invertebrates (e.g., Neritidae snails; see, Bunje, 2007; Fehér et al., 2007) is clearly connected with the evolution of the Ponto-Pannonian water basin and of the Danube catchment area. In addition, there are several taxonomical groups with considerable proportion of endemic species, e.g., the land gastropods (Soós, 1943), the earthworms (Lumbricidae: Csuzdi and Pop, 2007) or some soil arthropods (e.g., Opiliones, Diplopoda: Korsós, 1994; Collembola: Dányi and Traser, 2007). Their core areas clearly coincide with the younger tertiary landmasses within and near to the Carpathian Basin.

Most endemic species are narrow specialists, inhabiting extreme habitats, such as thermal springs, karstic caves and karstic springs. Several endemic troglobionta have been described in gastropods, pseudoscorpions, harvestmen, spiders and springtails, often occurring within a single or a few caves of karstic mountains. Several species of earthworms, millipedes, centipedes and assels can be considered as holo-endemic species of the western Transylvanian (Apuseni) mountains, in the Carpathian basin, East of Hungary (Csuzdi and Pop, 2007; Varga and Rakosy, 2008). A bulk of endemic taxa is confined to the Eastern and Southern Carpathians, to the Mţi. Apuseni and to the mountains of Banat, which could preserve some endemic and several relict species of Isopoda and Diplopoda in refugia without permafrost phenomena during the last glaciations.

A family of nematodes, Lucionematidae, has been described as parasiting on pike-perch *Lucioperca lucioperca* in Lake Balaton. Other endemic invertebrates include the cavernicole earthworms *Allolobophora gestroides* (Bükk Mts.) and *A.mozsaryorum* (Baradla cave in Aggtelek), the snails *Bythiospeum hungaricum* (caves of Mecsek Mts.), *Hauffenia kissdalmae* (Börzsöny Mts. springs), *Kovacsia kovacsi* (gallery forests of Körös rivers), a ladybird spider *Eresus hermani* occurring in rupicolous habitats, a pre-Alpine ground spider *Parasyrisca arrabonica*, a cave-dwelling springtail *Protaphorura kadici*, a millipede *Hungarosoma bokori* (caves of Mecsek Mts.), the amphipods *Niphargus gebhardti* and *Niphargus molnari* (both in caves of Mecsek Mts. in South of Hungary). Insects unique to Hungary include the tiny ground beetle *Poecilus kekesiensis* (alkaline grasslands of Hortobágy), cave-dwelling ground beetles *Duvalius gebhardti* (Mecsek Mts.) and *D. hungaricus* (Hungarian and Slovakian Karst), the weevils *Brachysomus tenuicollis* and *B. hegyessyi* (NE Hungary), etc.

Endemic terrestrial insects of the Carpathians are, as a rule, short-winged, flightless such as the bush-crickets *Isophya, Poecilimon* spp.; some stenotopic relict grasshopper species (*Capraiuscola ebneri, Podismopsis transsylvanica, Uvarovitettix transsylvanica, Zubovskia banatica*; Kis 1965, 1980); numerous species of the ground beetles (*Duvalius, Trechus, Patrobus, Morphocarabus* spp.) and weevils (e.g., *Otiorrhynchus* spp.). A bulk of these endemic taxa is confined in the Carpathian basin – but outsite of Hungary – to the eastern and southern Carpathians, the Apuseni Mts. and the mountains of Banat, which could preserve relict species (e.g., the tertiary relict gastropods *Chilostoma banaticum, Pomatias rivulare,* however, these occur also in Hungary), or some narrow endemic species of Isopoda and Diplopoda (Table 5.4) in refugia without permafrost phenomena during the last glaciations (Bennett et al., 1991; Krolopp and Sümegi, 1995; Willis et al., 1995).

In the mobile insect groups, the proportion of endemism lies rather low (e.g., in dragonflies no endemic taxa occur in the Carpathian Basin, in butterflies and larger moths only at subspecific level). Most endemic Lepidoptera of the Carpathian Basin belongs to Microlepidoptera, which have flightless females and are strictly specialized to some food plants living on halophyta in the saline grasslands of the Fertő-Neusiedlersee area, and those of the Great Hungarian plain (Kiskunság and Hortobágy). Endemic subspecies of Geometridae and Noctuidae evolved as peripheric isolates of turano-eremic species from the late-glacial, *kryoxerotic* periods, e.g., *Narraga tessularia kasyi, Saragossa porosa kenderesensis* (on food plants: *Artemisia santonicum, A. pontica*) and *Hadula dianthi hungarica* (on *Gypsophila muralis*). Some endemic taxa in the sandy areas of the Pannonian lowland are specialized predators or parasitoids, e.g., the spider *Dictyna szaboi* and the pompilid wasp *Cryptocheilus szabopatayi*.

Majority the endemics of the lower hilly parts of the Carpathian basin, however, represent thermophilous post(inter?-) glacial relicts with connections to the Balkan Peninsula, Asia Minor or Southern Russia as, e.g., the noctuid moths *Apamea syriaca tallosi* in warm-humid alluvial areas, *Dioszeghyana schmidtii schmidtii* and *Asteroscopus syriacus decipulae* in Pannonian xerothermic oak forests, *Polymixis rufocincta isolata* in the Villányi Mts.; *Chersotis fimbriola fimbriola, Euxoa vitta vitta,* and *Cucullia mixta lorica* in

Table 5.4 Examples of Stenotopic Species of Extreme Habitats Endemic to the Carpathians and the Carpathian Basin (Some of Them Outside of Hungary, But in the Pannonian Region)

Taxonomic group	Species	Habitat type	Occurrence
Gastropoda	*Melanopsis parreysi*	Thermal springs	Baile Felix (Oradea)
Gastropoda	*Theodoxus prevostianus*	Thermal springs	Hungarian Middle Range
Gastropoda	*Paladilhia hungarica*	Karstic water in caves	Mecsek Mts. (Abaliget)
Gastropoda	*Paladilhiopsis transsylvanica*	Karstic water in caves	Mţi. Apuseni
Amphipoda	*Niphargus tatrensis*	Karstic water in caves	Calcareous mts. in the Northern Carpathians
Gastropoda	*Bythinella pannonica*	Karstic springs	Calcareous mts. in the Northern Carpathians
Palpigradi:	*Eukoenenia vagvoelgyii*	Karstic caves	Aggtelek karst, Slovakian karst
Diplopoda	*Haasea hungarica*	Karstic caves	Mecsek Mts. (Abaliget)
Diplopoda	*Hungarosoma bokori*	Karstic caves	Mecsek Mts. (Abaliget)
Isopoda	*Mesoniscus graniger*	Karstic caves	Aggtelek karst, Slovakian karst
Collembola	*Pumilinura dudichi*	Karstic caves	Aggtelek karst
Collembola	*Protaphorura kadici*	Karstic caves	Aggtelek karst
Collembola	*Arrhopalites dudichi*	Karstic caves	Aggtelek karst
Collembola	*Arrhopalites hungaricus*	Karstic caves	Aggtelek karst
Collembola	*Arrhopalites buekkensis*	Karstic caves	Bükk Mts. (N. Hungary)
Coleoptera, Carabidae	*Duvalius bokori*	Karstic caves	Slovakian karst
Coleoptera, Carabidae	*Duvalius gebhardti*	Karstic caves	Bükk Mts.
Coleoptera, Carabidae	*Duvalius hungaricus*	Karstic caves	Aggtelek karst, Slovakian karst

the dolomitic areas of the Transdanubian Middle Range, *Chersotis fimbriola baloghi* in the Aggtelek Karst. Balkanic connections have also been observed in butterfly species, which are restricted to special, Pontic-Pannonian steppic food plants, e.g., *Kretania sephirus* (feeding on *Astragalus exscapus, A. dasyanthus*), *Melitaea ornata kovacsi* (on *Cirsium pannonicum*).

A few endemic taxa are widespread in the Carpathians and in the neighboring mountainous areas, e.g., the Brown Argus blue *Aricia artaxerxes issekutzi*, while many others are confined to the southern and eastern Carpathians of Romania, often with Balkanic connections (*Erebia cassioides neleus*, subspecies of *Erebia melas, Coenonympha rhodopensis schmidtii*, etc.) (Varga, 2014).

5.3.3 *The Carpathian-Balkanic Connections*

The close geological and faunal connections of the Carpathians suggest the existence of highly dynamic contacts with the mountains of the Balkan Peninsula during the Upper Pleistocene. These connections show a contrasting picture compared to the refugia of the Iberian and the Appenine peninsula which have been much more sheltered by the glaciated mountains of the Pyrenées and the Alps, respectively. At least two major arboreal refugia can be traced here: the Illyrian refugium related to the Dinarids and its foothills and the Carpatho-Dacian refugium related to the Carpathians and its foothills. Some areas attached to these refuges served as periodic habitats over climatically favorable periods. These are regarded as fluctuation zones (Varga, 1995; Sümegi et al., 1998; Deli et al., 1997). Since the Carpathian Basin occupied a transitional position between the Balkanic refugia and the cold-continental tundro-steppe zone during the glacial periods, the postglacial repopulation of the Carpathian Basin proceeded (1) by long-distance dispersal from the more remote (atlanto- and ponto-) Mediterranean and southern Continental refugia, and (2) also from some adjacent local survival areas, e.g., from northwestern Balkanic ("Illyrian") versus South Transylvanian ("Dacian") arboreal refugia. In such cases, the arrows of the northwards dispersal of the southwestern and southeastern populations surround the arid central part of the basin.

These components of the flora and fauna extend northwards through the foothills of the eastern Alps and southwest-Pannonian hilly regions on the one hand, and through the hilly regions of the Banat area and the western foothills of the Transylvanian "Island" mountains (Apuseni Mts.), on the other. In some cases, the populations of the southwestern and southeastern "strains" do not display any significant taxonomical differentiation, e.g., the silver lime (*Tilia tomentosa*), the black bryony (*Tamus communis*), or some butterflies and moths (*Pyronia tithonus, Aplasta ononaria, Idaea nitidata,* etc.). Much more evidence is provided by the repopulation of the Carpathian Basin from different directions in the cases of vicarious pairs of closely related species or in subspecies of polycentric species. Such cases can mostly be mentioned in land gastropods, e.g., *Pomatias elegans – P. rivulare, Chilostoma illyricum – Ch. banaticum*, or in flightless insects, e.g., short-winged Orthoptera: *Odontopodisma schmidti – O. rubripes, Isophya modestior – I. stysi* (Orci et al., 2005). Other Balkanic-Pannonian connections are shown by the isolated occurrence of relict-like endemic floristic elements which are sister species of some Balkanic montane species as *Linum dolomiticum* in Central Hungary (sister species of *Linum elegans* in Greece, Olympos), *Onosma tornense* in the northern Karst mountains (sister species of *O. heterophylla in* northern Greece).

The western Balkanic ("*Illyrean*") influences are most significant in the southern and southwestern parts of Transdanubia. These areas are characterized by a humid sub-Mediterranean climate and do not have a significant rainfall deficit in the summer period. They belong to the belt of mesophilous zonal forests of *Fagion illyricum, Querco–Carpinion illyricum* and the Illyrean-Pannonian hardwood gallery forests (*Fraxino pannonicae–Ulmetum*) characterized by a richness in tertiary/interglacial relict, often geophytic plant species (Horvat et al., 1974).

The Transylvanian ("Dacian") influences are connected with the Eastern Carpathians and are often transmitted by the western Transylvanian mountains (Mahunka, 1993, 2007; Varga, 1995, 2003). The occurrence of some Dacian elements is typical for the eastern part of the Hungarian Middle Range. Eastern Balkanic influences also appear in the Hungarian Middle Range by relict-like occurrences of some Balkanic and Balkanic-Anatolian elements (e.g., noctuid moths: *Asteroscopus syriacus* and *Dioszeghyana schmidtii*), especially in the warm foothill zone with significant sub-Mediterranean influences Relict occurrences of Dacian elements (bush-crickets: *Isophya stysi, Leptophyes discoidalis, Pholidoptera transsylvanica*; the ground-beetle *Carabus hampei ormayi*) have been recently discovered on the small, island-like volcanic hills of the Bereg lowland.

The influences of the northern Carpathians are significant in the NE part of the Hungarian Middle Range. There is a characteristic difference between the Eperjes-Tokaj volcanic chain on the one hand, and the limestone plateau of the Bükk Mts. and the N Hungarian karst on the other. The biotic contact of the Eperjes-Tokaj range with the Carpathians is young, obviously postglacial, and can be characterized mostly by the presence of species, which are either typical of the montane forest belts of the Carpathians (e.g., land snails: *Bielzia coerulans, Vestia gulo* and ground-beetles: *Abax schueppeli, Carabus obsoletus, C. zawadszkyi*) or are widely dispersed in the northern part of Central Europe. The Bükk Mts. with its old Dinaric connections in the Tertiary, has an insular character, however. Its Carpathian and de-Alpine elements are isolated relics (e.g., land snails: *Spelaeodiscus triaria, Phenacolimax annularis*, the Geometrid moth *Entephria cyanata gerennae*). In the Aggtelek Karst area, with ancient connections to the northern Carpathians, the immediate transition to the higher limestone plateaus of Slovakia is combined with the occurrence of Carpathian (land snails: *B. coerulans, Cochlodina cerata, Trichia unidentata;* ground-beetles: *Carabus obsoletus, C. zawadszkyi, A. schueppeli*), boreal and xeromontane species at surprisingly low altitudes, influenced by the meso-climatic and geomorphological features of this area. Some influences of the northern and the eastern Carpathians are to be observed at the NE marginal areas of the Pannonian lowland, i.e., along with the upper course of the river Tisza and its tributaries (e.g., the land gastropods *Vitrea diaphana, B. coerulans, Balea stabilis, Perforatella dibothrion, P. vicina*).

5.3.4 Faunal Elements of the Carpathian Basin

There are several Mediterranean-Manchurian bicentric faunal elements with disjunct range occurring in the Carpathian Basin. The distribution of this species group is connected with the Ponto-Caspian waterway-system, and displays long distance disjunctions from the vicarious eastern Asiatic taxa, which often are only subspecifically differentiated (Lepidoptera: *Apatura metis metis – A. m. substituta, Chariaspilates formosarius hungaricus – Ch.f. formosarius, Rhyparioides metelkanus metelkanus – Rh. metelkanus flavidus, Arytrura musculus* ssp. *– A. m. musculus*). These and also some other species of this group (e.g., noctuid moths *Polypogon gryphalis, Herminia tenuialis, Diachrysia nadeja*) occur at also the lower course of the Danube and the Drava as well as in swampy-boggy areas of the lowlands in Transdanubia, in the Banat and eastern Hungary. The refugia of these faunal elements were probably along the lower courses of the Danube and its tributaries. Gallery forests of the Illyrian and the Pannonian type and alluvial wetlands accompanying the large rivers of the Pannonian lowland served as corridors for the northwards expansion of these species.

Different types of long-distance disjunctions have been observed in the relict-like steppe and semidesert species. The polytypic butterfly *Melanargia russiae*, which is widespread in West and Central Asia, South Siberia and in the mountains of Italy and the Balkan Peninsula, occurred locally, as *M. russiae clotho* on tall-grass clearings of gallery forests on the sandy lowland in Kiskunság. Its extinction was partly due to the consequence of overcollecting, and mostly because of destroying the habitats (re-forestation with *Robinia pseudoacacia*). The habitats of *Chondrosoma fiduciarium* (Kasy 1965, 1981) are also tall-grass lowland and hilly steppes often mixed with slightly saline patches. Other species are confined to open dolomitic rocky swards (e.g., *Phyllometra culminaria, Lignyoptera fumidaria, Cucullia mixta lorica*) or open sandy and rupicolous grasslands (*Oxytripia orbiculosa*, vanishing).

Eremic species are restricted to semidesert-like habitats of the lowland with extreme edaphic conditions. Abundant examples can be found in strictly localized phytophagous insects, which are often connected with special halophytic plant communities. They are often represented by endemic Pannonian subspecies or allopatric sibling species of Turanian origin, e.g., the Noctuid moths *Saragossa porosa kenderesiensis* and *Hadula dianthi hungarica* or the Microlepidoptera: *Coleophora hungariae, C. klimeschiella, C. magyarica, C. peisoniella, Holcophora statices, Stenodes coenosana, Agriphila tersella hungarica*, etc. The dispersal of this species group could have taken place in the latest glacial phases, with subsequent isolation as a consequence of the postglacial expansion of the forested belts.

Mediterranean species, as sub-Mediterranean, ponto-Mediterranean elements, mostly occur as marginal isolates in the Carpathian basin. Especially the surrounding of the "Iron Gate" at the lower course of Danube, the island-like hilly parts of southern Transdanubia (Villány hills and Mecsek) and the xerothermic lanuginose oak forest and scrub-forest belts of the Hungarian Middle Range are rich in such, partly relict-like elements. The richness of the calcareous sandy area of Kiskunság in Mediterranean elements can be explained by its favorable meso-climatic character and its territorial connection with the Middle Range. The eastern Balkanic influences are significant at the western border of the W-Transsylvanian mountains and also along the great rivers of the eastern part of the Pannonian lowland. The occurrence of some southern elements in the northeastern part of the Pannonian plain (e.g., Nyírség) and in the subCarpathian lowland and hilly regions can be explained by this dispersal route. Eastern Balkanic influences reach also the warm foothill zone of the Hungarian Middle Range by relict-like occurrences of some Balkanic and Balkanic-Anatolian elements.

Influences of the Ponto-Caspian steppe belts are also characteristic for the Carpathian basin. Some of their elements are recent invaders, e.g., the butterfly *Colias erate*, dispersed during the last 25 years in the greater part of the Carpathian basin. Other members of this group threatened by retreat and fragmentation of the extended open grasslands, e.g., the great bustard, *Otis tarda* or the members of the western mole-rat (*Nannospalax leucodon* agg.) species group with three caryologically differentiated species (*N. transsylvanicus, N. hungaricus, N. montanosyrmiensis*) in the Pannonian lowland, or the steppic Pannonian birch mouse *Sicista trizona*. Numerous steppe species fluctuate at the western limit of distribution, e.g., *Sturnus roseus*. Typical inhabitants of the steppic grasslands are often restricted to isolated sites of rupicolous and loess grasslands, e.g., the grasshoppers *Stenobothrus eurasius, Arcyptera microptera* and the subendemic bush-cricket *Isophya costata*. They can be regarded as relics of the postglacial steppe period, often corroborated by geographical isolation and taxonomical differentiation, as well, e.g., *Vipera ursinii rakosiensis* or the Zephyr Blue (*Kretania sephirus*), with isolated colonies in the Carpathian basin.

Last not least, xeromontane elements are also present in the Carpathian basin. The two main groups are: the Mediterranean-xeromontane species, represented by only a few Vertebrates (e.g., *Monticola saxatilis*), but larger is the number of such species in some insect groups, e.g., Noctuidae (e.g., species of genera *Euxoa, Dichagyris* and *Chersotis*) and Orthoptera (e.g., *Paracaloptenus caloptenoides*). The continental xeromontane type is represented by widely distributed Asiatic mountain steppe species and by relics of the dolomitic rupicolous grasslands (Geometridae: *Phyllometra culminaria, Lignyoptera fumidaria*). Some genera, typical for the steppe biome, probably have a xeromontane origin (e.g., Lycaenidae: *Kretania, Polyommatus;* Satyridae: *Chazara, Pseudochazara, Hyponephele;* Noctuidae: *Euxoa, Agrotis, Dichagyris, Chersotis,* etc.). These connections give us the possibility of the continental-scale generalization of the "Ősmátra"-theory, i.e., the origin of the lowland steppic species from mountain steppic ancestors (Varga, 1989, 1995).

5.4 Most Important Habitat Types of the Pannonian Region and Their Contribution to the European Biodiversity

5.4.1 Naturalness and Conservation Status of Pannonian Habitats in Connection With Biodiversity

Between 2003 and 2006, a large-scale vegetation mapping comprising the whole of Hungary was performed, and the results were integrated in a database called 'MÉTA.' According to this survey, around 13% of the country's territory is covered by near-natural vegetation. When accounting for the naturalness of this remaining area, the study revealed that only 3.2–9.8% remains of the natural capital of former times (the actual percentage depending on whether biodiversity or other ecosystem services were given priority at weighting). Figure 5.3 shows the differences in the value of Natural Capital Index (NCI) across the microregions of the country (Czúcz et al., 2008). The most intact areas are forested mountains; the most degraded ones are agricultural lowlands. Peaks in NCI values widely overlap with high amphibian and reptile diversity values (Figure 5.4).

According to the MÉTA-database, the least endangered habitat types are the rocky habitats, certain halophytic and aquatic habitats, open acidophilous woodlands, dry shrub vegetation with *Crataegus* and *Prunus spinosa* and the beech woodlands. The most seriously endangered habitats in Hungary are the following: sand and loess steppe oak woodlands, tussock sedge communities, extensive orchards, closed lowland oak woodlands, water-fringing and fen tall herb communities, wooded pastures, vegetation of loess cliffs, rich fens and *Molinia* meadows, Cynosurion grasslands and *Nardus* swards, swamp woodlands, xeromesophilous grasslands and salt steppe oak woodlands.

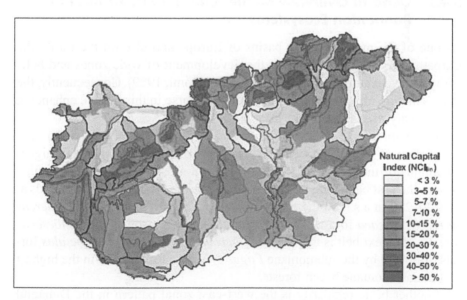

Figure 5.3 Natural capital index map of Hungarian physical geographical microregion. *Source:* Czúcz et al., (2008).

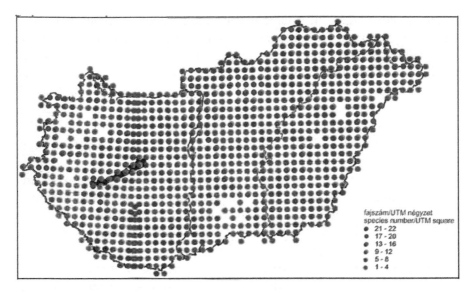

Figure 5.4   The number of Amphibia and Reptilia species present in 10 × 10 km²
UTM quadrates.

5.4.2   General Overview on the Spatial Regularities of the Pannonian Ecosystems

As one of the most complete basins of Europe and also on the Earth, the
Carpathian Basin does not favor the development of wide zones and belts
with zonal vegetation (Varga 1989, 1995; Zólyomi, 1989). Consequently, the
representation of zonal vegetation is rather diverse in this region, enhancing
the diversity of species composition.

The main coarse-scale regularity in the Pannonian region is the spatial pattern of vegetation belts in the mountains, especially in the higher
mountains of the Northern Middle Range. In most parts of the Pannonian
region, the first belt is the forest-steppe belt in the lowland. This belt ends at
ca. 220–230 m a.s.l. on the foothills, and it is followed usually by *Quercus
cerris–Q. petraea* forests or, due to mesoclimatic reasons, by *Q. pubescens*
forests. The next belt is the level of *Quercus petraea–Carpinus betulus* forests, followed by the submontane *Fagus sylvatica* forests, and in the highest
mountains, montane beech forests.

Another basic regularity is the west-east zonal pattern in the Dunántúl
region. The elsewhere altitudinal belts of forest types develop here at similar
altitudes adjacent to one another, from the sub-Atlantic and sub-Mediter-

ranean from West to the more continental East, as follows: beech forests, oak–hornbeam forests, sessile oak–turkey oak forests and partly pubescent oak forests, and the open forests of the forest steppe. The similarity of the patterns of altitudinal belts and horizontal zones may be attributed to a basin effect; that is, the strong continental influence of the Hungarian Plain on the one hand, and the effect of the surrounding high mountains, i.e., Alps in the West, Carpathians in North and East, on the other hand.

The *Quercus cerris–Q. petraea* forests were, and still are the most extensive forests in Hungary. This community belongs to both types of spatial patterns (horizontal zones and altitudinal belts). This plant community prefers deep soils, thus the greatest proportion of these forests occurs on level terrain or gentle slopes. They may form a belt of substantial size, as wide as 400 meters. The lower boundary of their belt is at 230–250 m in the mountains while the upper boundary is often at about 500 m, but this value is more variable. In the North Hungarian Middle range, this belt generally reaches as high of an altitude as 600–750 m. A reason for this is the increasing continentality towards the Northeast. The forest-steppe belt also ends at higher altitudes at this region. In the Mátra Hills, for example, forest-steppes may occur at exceptionally high altitudes, such as 300–350 m. In other hills, it occurs sometimes at 400 m just below the belt of the closed *Quercus petraea–Quercus cerris* forests.

Another major regularity in the Pannonian region is the climate gradient along the Middle Range from Southwest to Northeast. Several, mainly sub-Mediterranean plant communities (e.g., open and closed *Quercus pubescens–Fraxinus ornus* forests, dolomite grasslands, see: Jakucs 1961; Zólyomi et al., 1992) are widespread in the Southwest (in the Transdanubian Middle Range) but disappear towards Northeast. On the other hand, the frequency of several plant communities with continental character, which is mainly distributed in the Northeast (e.g., various calcareous rupicolous grasslands, *Spiraea media* thicket), diminishes towards Southwest (Zólyomi 1942).

As a consequence of the macroclimatic gradients from the periphery of the Hungarian plain towards the center of the basin, closed broad-leaved deciduous forests cover the margins of the Nagyalföld in the North, Northeast, and East, and in the valleys of the Drava and Sava Rivers, while the greatest part of the lowland belongs to the forest-steppe zone (Fekete et al., 2010). Vegetation of the landscapes of the forest-steppe zone of the Hungarian plain mostly differs due to edaphic factors, primarily geological substrate (loess, calcareous and acid sand, saline soils on loose sand or hard clay), and hydrology (azonal vegetation in stagnant or flowing waters). The species composition of the plant communities on the sand, saline soils and dry loess are clearly distinct, and are more so than the zonal plant communities of the forest zone in other parts of the Pannonian region. Nevertheless, gradients indicating climatic control may still be demonstrated in this area (see, e.g., the gradients of forest and steppe species in Fekete et al., 2010).

Deviations from the general pattern usually have singular causes that may not be formalized. Both the zonal pattern, and the extrazonal vegetation pattern can be violated thus leading to deviations. Zonal deviations are conspicuous differences between the actual potential vegetation (sensu Zólyomi, 1989) and macroclimate-based expectations (Borhidi, 1961). Extrazonal deviations are recognizable from potential vegetation maps. Extrazonal vegetation is commonly encountered in the field. Typically, it can be mapped and predicted well, since its occurrence follows clearly observable rules. Extrazonal stands usually occur nearby zonal localities of the same vegetation type. Extrazonal stands interpreted as deviations, however, are distant from their zone. The most apparent deviations, which are the most suitable for the characterization of a region, occur at the landscape level.

Perhaps the most apparent example of deviation at the landscape scale is found in Western Dunántúl. Here, in several landscapes stands of *Pinus sylvestris* intermixed with deciduous trees occur in extensive areas (Pócs, 1960). It is known that *Pinus sylvestris* was present in Dunántúl during the last glaciation and formed patches of open taiga forest (Rudner and Sümegi, 2001), which were transformed into closed and mixed taiga forests during the late glacial period (Willis et al., 1997; Juhász and Szegvári, 2007). Survival of *Pinus* to the early and middle Holocene, until 6700–7000 calibrated years in Dunántúl is almost certain (Magyari et al., 2010). Its retreat to nutrient-poor habitats took place very likely during the second half of the Middle Holocene (7,000–4,200 cal yr BP), due to the intensified expansion of *Quercus, Carpinus* and *Fagus*. Its frequency at the early Holocene may likely be explained by climatic factors. The typical plant formations in West-Dunántúl, such as transitional mires, willow swamps and fens, *Calluna* heath, etc. (see Pócs et al., 1958) are rare in other parts of the Pannonian Region. The reason why the presence of mixed *Pinus* forests is interpreted as a deviation is that transformation of the vegetation (into pure broad-leaved forests, such as acidophilous beech or oak forests) did not take place here. *Pinus sylvestris* disappeared on the lowland about 9,800–10,000 years BP (Magyari et al., 2004, 2008, 2010). There is one exception, the Kisalföld. There, in the vicinity of the cooler Bakony hills at the border of the Transdanubial range (at Fenyőfő), *Pinus* stands grow on calcareous sand. Beneath the loose forest canopy the plants of sandy grasslands cooccur with the few characteristic species of pine forests (*Monotropa hypopitys*, *Pyrola chlorantha*, *Veronica officinalis*, etc.). This forest cannot be regarded as a representative of the sub-Mediterranean forest steppe, but as that of the cool-continental forest steppe and is closely related to similar forests in the Ukrainian–Russian forest-steppe zone (Fekete et al., 2014). Its presence is regarded a deviation. According to palynological studies, in Transdanubia *Pinus sylvestris* was dominant during the early Holocene up to 6800–7000 calibrated year BP, and was less abundant and fluctuating afterwards. The high early Holocene pollen frequencies suggest the last glacial origin of the *Pinus* forests of northeastern Transdanubia (Sümegi, 2007).

Forests of the East European cool-continental forest-steppe zone are primarily composed of broad-leaved deciduous tree species instead of coniferous species. An occurrence of such forests in Hungary was hardly known until the 1960s. Fekete (1965) noticed that the usual altitudinal regularity of forest communities in the Northern Middle Range is disrupted in the Gödöllő-hills. The zonal forest stands there are formed by dry oak–hornbeam forests in which *Acer campestre* is abundant, and grows into the upper canopy layer reaching the height of the hornbeam and the oak species *Quercus petraea* and *Q. robur* and their hybrids. The deviation is caused by the mesoclimate of the Gödöllői-dombvidék, because these hills are almost surrounded by warm lowlands of the Hungarian plain. The link to the East is further emphasized by the fact that forest-steppe forests dominated by *Tilia cordata* were also identified in the Gödöllői-hills on sand soils (Fekete, 1965). Similar types of forests grow in the eastern part of the Russian forest-steppe zone between the Volga and Ural Rivers and have been also recently reported to occur in the Southern Urals (Chytrý et al., 2010). As shown by pollen diagrams, similar forests with variable composition were widely distributed around the Hungarian plain in the early Holocene (i.e., *Tilia* dominance at Bátorliget between 11,200 and 10,800 calibrated year BP, Magyari et al., 2002). The development of the forest-steppe forest with *Tilia* in the Gödöllői-dombvidék may also have commenced in the Early Holocene (Willis et al., 1997).

The absence of expected vegetation belts is also a sign of deviation. This is observed, e.g., on the hills north of the Lake Balaton (Bakony and Balaton-upland). At about 400 m a.s.l. in the Bakony hills, the plakor areas and gentle slopes are covered with beech forests. South of this plateau, towards the lake, the potential vegetation changes abruptly. The hilltops and plateaus are covered with dry oak scrub forests dominated by *Cotinus coggygria, Quercus pubescens*, and *Fraxinus ornus*, while beech forests are restricted to the northern slopes (Fekete and Zólyomi, 1966). Two vegetation belts are missing here: the subcontinental *Q. cerris–Quercus petraea* and the *Quercus petraea–Carpinus betulus* forests which regularly develop elsewhere in the Pannonian region. The close vicinity and even direct contact of the belts of mesic beech forests and dry pubescent oak forests, and their occurrence at the same altitude is unexpected in the Pannonian region. This phenomenon, however, exists also in the southern half of the Eastern Alps and along the Adriatic coast as well (Bohn et al., 2000). One reason for the lack of these vegetation belts in the Bakony–Balaton upland area is a sudden change in climatic conditions. Coming from the north, moist air is forced upward by the Bakony Mts. resulting in the moisture to precipitate on the northern part of the area. This provides humid mesoclimatic conditions for beech forests. In the southern part, on the Balaton upland on the other hand, descending warm air masses create a rain shadow, which leads to the dominance of dry oak scrub forests. This unusual spatial pattern of zonal communities has, however its background in the regional vegetation history. Migration of beech started from NW Balkanic (Slovenia) refugia during the early stages of the postglacial period between 11,500–10,200 years BP, and beech abundance rapidly increased in the areas of the Eastern Alps and Western Hungary (Magyari et al., 2002; Willis and van Andel, 2004; Magri et al., 2006). The first maximum pollen density of beech in the Balaton area was detected between 8,000 and 6,000 calibrated year BP (Magyari et al., 2002), just at the time when sub-Mediterranean oak forests were the most extended (Lang, 1994). It was concluded (Fekete et al., 2014) that the current vegetation pattern may have developed rather early. At the beginning of this period, the xerothermophilous character of oak forests is indicated by the presence of *Cotinus coggygria*. The unusual vegetation history of the Bakony–Balaton upland area also explains the fairly isolated occurrences of several West-Balkanic species (e.g., *Aethionema saxatile, Amelanchier ovalis, Carpinus orientalis, Daphne laureola, Hippocrepis emerus, Physocaulis nodosus, Scilla autumnalis, Scutellaria columnae, Stipa bromoides*, see: Fekete and Zólyomi, 1966). Thus, this area is one of the hot spots of floristic diversity in the Pannonian region.

The here considered vegetation geographical characters represent deviations at the landscape scale. Deviations, however, are also present at finer spatial scales. One of the most prominent examples is the case of scree and ravine forests. On rocky slopes, periglacial screes of the Middle range *Fraxinus excelsior* and *Tilia* species are forming typical rupicolous forest stands resulting in isolated islets in the "matrix" of the surrounding zonal oak–hornbeam and beech forests (Zólyomi, 1967). The dominant understory species of this intrazonal community are partly relict-like forest steppe species with disjunct distributions such as *Carex brevicollis, Waldsteinia geoides*, etc. The main components of the shrubs along the forest fringe are the eurosiberian–continental species *Spiraea media, Cotoneaster niger, Rosa spinosissima*, etc. According to pollen data, ash and lime appeared in the mountains of North Hungary in 9,600–9,500 calibrated year BP, whereas the first occurrence of beech is estimated at 6,000 calibrated year BP (Willis et al., 1997). The Holocene pollen diagram of a marsh on the northern slope of the highest part of the Mátra mountains suggests that lime formed mixed forests with spruce were extended on the northern slopes of the Kékes between 8,400–6,100 calibrated year BP, before the appearance of beech (Magyari et al., 2002). After the expansion of beech, lime, and ash, survived only in rocky habitats of the mountain slopes where the beech was competitively inferior (Chytrý and Sádlo, 1997). Thus the continental ash–lime rock forest is likely a remnant from vegetation historical times and represents a former phase of secular succession (Zólyomi, 1967).

Sphagnum bogs are rare representatives of the boreal vegetation zone in Hungary, and thus their presence may be qualified as a deviation. *Sphagnum* bogs with *Eriophorum* spp. (and also willow swamps with peat mosses) are preserved by the Mohos lakes at the northern foothills of the Bükk mountains, and by three oxbow lakes surviving in the northern lowland (106–124 m). It is a reasonable assumption that they originate from the Würm period, and have been continuously present since then. Indeed, pollen analytical studies found evidence of a mire community in the layers of 25,000–24,000 calibrated year BP in the Nagymohos lake (Magyari et al., 1999). Transitional *Sphagnum* mires developed twice during the Pleistocene (25,000–23,000 and 21,000–18,000 calibrated year BP). Macrofossil evidence indicates that several extant species (*Sphagnum paluster, Menyanthes trifoliata, Potentilla erecta, Drosera rotundifolia*) may be traced back to the Würm period. During the Holocene, stages of open lake, floating bog were following one another up to the current stage, and *Potentilla palustris, Sphagnum paluster*, and *Sphagnum* sect. *cuspidatum* were recorded almost in the entire 25,000 years period (Magyari et al., 1999).

5.4.3 *The Unique Features of the Pannonian Region*

Unique types of the Pannonian natural vegetation are those communities that do not occur outside this region. These endemic plant communities have also unique species composition. The vegetation of the Carpathian Basin is, however, far from being composed of endemic communities only. Based on their species composition, certain associations could fit into several landscapes of Central Europe. Others may be considered as outposts of the vegetation of neighbouring regions, such as the Illyrian beech and oak–hornbeam forests, the Carpathian calcareous rocky grasslands, and the East Alpine dolomite grasslands. The Carpathian Basin, however, is not just a simple meeting area of vegetation types from surrounding regions. Some of the unique vegetation types host endemic species, some are composed here by the combination of species typical for the neighboring regions and thus the unique species composition makes them characteristic for the region (Table 5.5).

The most important endemic plant communities of the Carpathian Basin are the following:

1   The *loess forest-steppe forest* on loess with *Acer tataricum* on the Great Hungarian plain and adjacent foothills of the Hungarian Middle range is an endemic zonal vegetation type. The sub-Mediterranean pubescent oaks (*Quercus pubescens, Q. virgiliana*) are components of the canopy of these, mostly at the margins of the lowland. *Quercus robur* and the continental *Acer tataricum* may dominate. In the forest interior, and especially along the edges and in glades, numerous tall herbs (i.e., *Nepeta pannonica, Phlomis tuberosa, Pseudolysimachion spurium*) and grasses (such as *Melica altissima*) grow. Density of the canopy is variable and frequently light-penetrated.

2   The *Quercus pubescens–Fraxinus ornus* forests and *Cerasus maha-leb–Quercus pubescens* scrub communities are extended mostly on the dry and warm southern slopes of the Middle range as a mosaic of low canopy and open forests (mainly *Quercus pubescens* with *Fraxinus ornus* in the West and with *Cerasus mahaleb* in the East) and dry rock steppes. The trees and most of the shrubs in this vegetation type have southern, often Balkanic distribution ranges, whereas the herb layer is saturated (especially in the Northern Middle range) with species having continental range in addition to sub-Mediterranean ones. This special combination of floristic elements makes these oak scrub communities endemic or subendemic.

Table 5.5 Unique Vegetation Types of the Pannonian Region and Their Occurrences (Fekete et al., 2014)

Community	Landscape	Main reasons for uniqueness
Forest-steppe forests on loess	Great Hungarian plain, Lesser Hungarian plain and margins of the Hungarian Middle range	Rare representatives of the sub-Mediterranean forest steppe with unique tree composition
Oak scrub with *Quercus pubescens* and often with *Fraxinus ornus*	Hungarian Middle range	Canopy is composed of trees with southern distribution, herb layer is rich in steppe plants
Forest-steppe forests and tall-herb meadow steppes on saline soil	Great Hungarian plain (East of river Tisza)	Rare forest type. The tall herb meadow steppe in the glades is the western outpost of vegetation type distributed from Southern Siberia and Kazakhstan to Eastern Europe
Closed oak forests and open steppic forests on sand	Great Hungarian plain, Lesser Hungarian plain	Forests close to the climatic forest limit, sources and refugia of the nemoral flora on the continental Hungarian plain
Vegetation mosaic on sand: dry perennial *Festuca vaginata* swards mosaicing with juniper–poplar forests	Great Hungarian plain, partly on Lesser Hungarian plain	Edaphic grasslands composed of many endemic species and species with Ponto-Pannonian, Pontic–sub-Mediterranean distribution
Fine-scale mosaic of *Artemisia* and *Achillea* steppes with *Puccinellia* and *Camphorosma* swards and salt lakes	Great Hungarian plain, partly on Lesser Hungarian plain	Presence of numerous Pannonian endemic species and rare species with Pontic–Pannonian, and Irano–Turanian distribution ranges
Open dolomite grassland	Transdanubial Middle range	Exclusive habitat of several Pannonian endemics and rocky grassland specialists with southern distribution
Karstic beech forests	Transdanubial Middle range	Ecotone-like community with isolated occurrences of species of distant zones (*Allium victorialis, Calamagrostis varia, Carduus glaucus, Festuca amethystina*)

3   The *alkaline forest-steppe forests* are endemic to the lowland and are found mainly east of the Tisza River on alkaline soils. In terms of their origin, some of these forests developed naturally, possibly many thousand years ago. Besides, other stands developed secondarily from hardwood gallery forests in response to lowered groundwater level that followed river regulations in the nineteenth century. The species-rich meadow steppes dominated by *Peucedanum officinale* and *Aster sedifolius* are the westernmost outposts of an intrazonal tall forb vegetation type distributed from Kazahstan and southern Siberia to Eastern Europe.

4   The *sand forest-steppe forests* are either closed pedunculate oak (*Quercus robur*) forests with nemoral character, or steppic oak forests with light-penetrated canopy on the sandy areas of the Great and Lesser Hungarian plain. They are seriously threatened due to the lowered groundwater table, which is caused partly by climate change, and partly by excessive water extraction and draining. The oak forests on sand (beside the azonal oak–ash–elm gallery forests) are the sources of and refuges for the nemoral forest flora on the Alföld. Since most dry oak forests in the Pannonian Region are characterized by a comparatively open canopy structure, they are not only surrounded by a species-rich forest margin (tall herb forest fringe communities of Trifolio-Geranietea and scrub), but are disrupted by internal fringes, where patches of clonally spreading forb species exhibit peculiar dynamics (Fekete and Varga 2006).

5   The entire *sand vegetation mosaic* of the Pannonian region is unique (Fekete, 1992). It occurs between the Danube and Tisza rivers. On coarse sand oak the forests are replaced by *Juniperus communis–Populus alba* stands forming either a loose scrub or a closed forest depending on water availability. The juniper–poplar scrubs form mosaics with dry grasslands on the dunes and wet (now mostly dried-out) grasslands in the depressions. The dominant species are *Festuca vaginata, Stipa borysthenica*, and in less extreme habitats *F. wagneri.* This community hosts numerous endemic Pannonian species (*Festuca vaginata, F. wagneri, Colchicum arenarium, Dianthus diutinus, Iris arenaria, Sedum hillebrandtii*). The rest of the species exhibit mainly Ponto-Pannonian, or Pontic–sub-Mediterranean type of distribution. In the northeastern part of the Hungarian plain, on limeless sand, another type of grasslands occurs with endemic species, *Pulsatilla flavescens,* the Ponto-Pannonian *Iris aphylla* subsp. *hungarica,*

and the south Siberian species *Pulsatilla patens* and *Pseudolysima-chion pallens*.

6   The *vegetation mosaic on alkaline soils* is also unique to the Panno-nian region. The fine-scale mosaic is determined by the microrelief, i.e., by the distance to the layer of salt accumulation, and the yearly excessive dynamics of water regime. On solonetz soils, the short-grass steppes of *Festuca pseudovina* alternate with wet *Puccinellia limosa* meadows, and ephemeric *Camphorosma annua* stands. In the late summer or autumn, annuals and succulent halophytes grow on the dried out lakebeds. The vegetation of the alkaline areas of the lowland significantly differs from the saline communities on the sea-shore. Connection to seashore communities is indicated only by a few species (i.e., *Spergularia maritima*), while vicarious taxa occur frequently (i.e., *Aster tripolium* vs. *A. tripolium* subsp. *pannonicus*, *Salicornia europaea* vs. *S. prostrata*, *Bassia hirsuta* vs. *B. sedoi-des*). The species group of *Suaeda prostrata, S. pannonica*, and *S. salsa* is complicated both taxonomically and regarding distribution history (Weising and Freitag, 2008). The characteristic species have frequently Pannonian, Pontic–Pannonian, or Turano–Iranian distri-bution ranges. Saline meadows also have endemic species (*Achillea asplenifolia, Cirsium brachycephalum, Plantago schwarzenbergi-ana*). All these facts confirm the separate vegetation history of the Pannonian saline vegetation, which may have been continuous from the Würm glaciation through the entire Holocene to present (Varga, 1989; Sümegi et al., 2000, 2005). After all, a sizeable area of the Alföld is covered with endemic communities.

7   The Pannonian dolomitic grasslands cover a much smaller area on the Transdanubian hills. The open dolomite rupicolous grassland develops on steep, eroded south-facing slopes and on ridges cov-ered with dolomite gravel. The proportion of sub-Mediterranean ele-ments and, accordingly, the proportion of semishrubs and succulents are high here. The loose vegetation hosts a rich lichen–moss layer and is dominated mainly by *Carex humilis* and *Festuca pallens*. This is the exclusive habitat of several Pannonian endemics (*Seseli leu-cospermum, Dianthus plumarius* subsp. *regis-stephani*). Many less habitat-specific endemics also occur here, as *Bromus pannonicus, Vincetoxicum pannonicum, Hieracium kossuthianum*. Furthermore, the single occurrence of the endemic *Linum dolomiticum* can also be found in an open dolomite grassland stand in Pilis hills.

8   The karstic beech forests are also very unique forest communities, which occur in the same mountainous areas, on dolomite rock. It appears mainly on steep north and northeast facing slopes, on shallow soils. Trees are rather stunted reaching only 10–12 meters in height. In the canopy beside *Fagus sylvatica,* trees of xerothermic sub-Mediterranean oak forests (*Quercus pubescens, Fraxinus ornus*), and subordinately *Sorbus* taxa (*Sorbus torminalis, Sorbus aria* and their hybrids) appear. In the shrub layer and sometimes in the ground layer *Cotinus coggygria* may be monodominant, in other cases elements of beech and oak forests grow intermingling each other. Characteristic plant is here *Carex alba,* and in numerous stands of forest unique, rare elements (*Allium victorialis, Primula auricula, Carduus glaucus, Festuca amethystina,* etc.) occur as relics from the glacial period (Zólyomi, 1958; Isépy, 1970).

5.4.4   *Vegetation Types of the Pannonian Region According to the European Habitats Directive*

In the next part the ecological and compositional characteristics of the habitat types are considered according to the Manual of the Habitat Directive, with special regard to the priority habitat types* of the Pannonian region.

The ***Pannonic salt steppes and salt marshes**** (**1530**) consist of different types of steppic or semidesert-like open grasslands, saline meadows and marshes with typical zonation, and saline bottomlands with shallow, often ephemeric lakes with annual plant communities. They cover huge surfaces along the former catchment areas of the Pannonian lowland rivers and are usually influenced by the extreme temperatures and fluctuating humidity. These habitat types are part of natural origin but strongly influenced in their extension and composition by anthropogenic drainage and traditional pastoralism (gray steppe cattle and "racka" sheep). These habitats are significant breeding places of numerous threatened bird species as the Great Bustard (*Otis tarda*), the Pratincole (*Glareola pratincola*), the Kentish Plower (*Charadrius alexandrinus*), the Lapwing (*Vanellus vanellus*) and the recently vanishing Short-toed Lark (*Calandrella brachydactyla*). The tall-forb alkali meadows show an unusual rich composition of food-plant specialized insects as the large estuarine moth (*Gortyna borelii*).

The ***saline grasslands of the Pannonian region*** can be regarded as the western-most outposts of the continental saline areas of the Eurasiatic steppe and semidesert zones. They are, at the same time, the most extended inland saline areas of Central Europe. These habitat types are shared by Hungary, Austria (Neusiedlersee area), Romania (saline areas of the Transylvanian basin and Banat) and Serbia (mostly Vojvodina). They now consist of more than 30% of the total extension of grasslands of Hungary (and about 2.9% of the whole area of the country!). Their total extension is over 270,000 ha, from which about 76% is protected. It is a recently discussed matter whether they mostly secondarily arose in their recent large extension, due to the regulation of the riverways and desiccation of the marshy areas of the bottomlands during the eighteenth to nineteenth centuries followed by extensive grazing of ancient domestic races of cattle and sheep. Primary salinization took place, however, during all seasonally arid climatic periods of the late Pleistocene and Holocene at the shallow bottomlands of the floodplains of the lowland rivers with fluctuating humidity, at the natural clearings of the hardwood gallery forests, and also in the local depressions of lousy and sandy areas without an outlet. Due to the erosion of the topsoil, a typical microrelief of the surface will be established, resulting in a small-scale mosaic-like change of szik-banks and bottomlands. On the places with eroded topsoil, the salt accumulation horizon lies near to the surface. Hence, the vegetation is strongly influenced by the high salt content and also by the extreme physical properties of the soil. The bare alkali surfaces of the shallow saline bottomlands, called "blind-szik" are covered only by scarce vegetation formed by ephemeral plants. They are covered either by a white layer of sodium-carbonate, -chloride or -sulfate (solontchak-soils) or they are covered by amorphic SiO_2 (solonetz-soils).

The plant associations of the alkali bottomlands are clearly differentiated into these both main types. The bottomlands of the solontchak-type are characterized mainly by *Juncus gerardii* and *Lepidium crassifolium*. Typical species of bottomlands are: *Aster tripolium* subsp. *pannonicus, Bassia sedoides, Camphorosma annua, Cerastium dubium, Matricaria recutita* subsp. *salina, Trifolium bessarabicum*, and in Southeast Hungary the subendemic *Plantago schwarzenbergiana*. In the depressions of the bottomlands, covered usually in the spring by shallow water a mass-production of cyanobacteria (*Nostoc* spp.) takes place. After desiccation in late spring or early summer, only a blackish layer of died cyanobacteria remains. These sites are typified by the *Puccinellia* swards and, in the "szik-rills," the endemic ephemeral association: *Pholiuro-Plantaginetum tenuiflorae*. The areas with the highest NaCl content are characterized by the presence of ephemeric halophytic associations of *Atriplex litoralis, Salicornia prostrata, Salsola sodae, Spergularia* spp., *Suaeda maritima* and *S. pannonica*.

The alkali grasslands are characterized by the short-grass swards of *Festuca pseudovina*. The main body of the species of these grasslands is the same, both on solonchak and solonetz types of soil (*Artemisia santonicum* agg., *Limonium gmelini* subsp. *hungaricum, Plantago maritima, Podospermum canum, Ranunculus pedatus, Trifolium angulatum, T. parviflorum*). On deeper topsoil, the association *Achilleo setaceae-Festucetum pseudovinae* is widely distributed, and easily can be nitrophilised. They are mostly secondary grasslands, which have replaced the forest-steppe habitats due to clear-cutting, grazing and trampling. The sites with shallow topsoil are characterized by the halophytic associations: on solonchaks *Lepidio crassifoliae–Festucetum pseudovinae*, and on solonetz soils, *Artemisio santonici–Festucetum pseudovinae*, characterized by a mixture of Pannonian, continental and maritime halophytic species (*Aster tripolium* subsp. *pannonicus, Atriplex littoralis, Limonium gmelini* subsp. *hungaricum, Suaeda maritima*) and, on the bare patches, by the appearance of ephemeric species, mosses and lichens *(Ceratodon purpureus, Rhacomitrium canescens, Polytrichum* sp. and *Cladonia foliacea, C. furcata)*. With the exception of some special patches (e.g., primary natronic pans), alkali grasslands are usually managed by grazing. The control of the number of livestock and the timing of grazing are very important factors of the optimal use.

***Natural eutrophic lakes with Magnopotamion* or *Hydrocharition vegetation* (3150)** consist of two main types of vegetation: the large pondweeds and the free-floating surface plant communities which are closely associated in the typical lacustrine zonation. They usually occur in eutrophic or slighly dystrophic lakes or ponds with turbid, mostly dirty gray to blue-green waters, rich in dissolved bases. The floating pondweed vegetation is mostly formed by *Hydrocharis morsus-ranae, Stratiotes aloides*, the *Ceratophyllum-* and *Lemna* species. The large pondweeds are characterized by extended stands of *Nymphaea alba, Nymphoides peltata*, the *Ceratophyllum, Potamogeton* and *Myriophyllum* species. The pondweed stands of the Balaton Lake are characterized by the dominance of *Ceratophyllum submersum* and the endemic subspecies of *P. pectinatus* subsp. *balatonicus*. The shallow waters are characterized by the ensembles of *Polygonum amphibium, Oenanthe aquatica, Rorippa aquatica, Alisma plantago-aquatica, Sparganium erectum* and *Butomus umbellatus*. Some thermophilous species, as *Trapa natans* and *Marsilea quadrifolia* have the northernmost strong populations in these Pannonian habitats. The pondweed vegetation is the habitat of several insects as the larvae of several Damselflies (*Erythromma* spp.) and Dragonflies (*Sympetrum* spp., *Anax imperator, Anaciaeshna isosceles,*

Aeshna viridis – the latter specialized to *Stratiotes*), of numerous water-bugs as *Nepa cinerea, Ranatra linearis, Notonecta* spp., the species of Chrysomelid beetle genera *Donacia* and *Plateumaris,* and of the larvae of *Nymphula* and *Acentropus* moths. These habitats have a crucial importance as breeding places of waterfowl, especially of the species with floating nests (*Podiceps* spp., *Chlidonias* spp.).

The Molinia meadows on calcareous, peaty or clayey-silt-laden soils (Molinion caeruleae) (6410) are humid or wet meadow types of plain to montane levels, on nutrient-poor, mostly slightly calcareous or neutral soils with often fluctuating water table. They also occur on slightly acidic soils of forest-clearings of hilly areas. The more humid calcareous type is more distributed in Western and Central Hungary and is characterized by the dominance of *Carex davalliana, Eriophorum latifolium, Schoenus nigricans* and *Sesleria uliginosa.* These stands are characterized by the occurrence of many colorful plants as *Dactylorhiza incarnata, Epipactis palustris, Veratrum album, Parnassia palustris, Pedicularis palustris, Gentiana pneumonanthe, Iris sibirica* and some glacial relict species as *Primula farinosa,* and the insect-feeding *Pinguicula.* Its dryer variation is the tall-grass *Succiso–Molinietum hungaricae* meadow dominated by *Agrostis stolonifera, Poa trivialis* and *Arrhenatherum elatius.* The acidic type of marshy meadows is mostly distributed in Northeast Hungary and is characterized by the dominance of *Juncus effusus, Festuca rubra, Thymus pulegioides, Potentilla erecta* and also by the occurrence of several species of Orchids (*Orchis pallens, Dactylorhiza majalis, D. sambucina*). These meadows often show some transitions to the next vegetation type by the occurrence of *Nardus stricta* and other acidophilous species as *Danthonia decumbens* and *Achillea ptarmica.* The *Sanguisorba officinalis* is the initial food-plant of the protected Large Blue species *Maculinea teleius* and *M. nausithous.* The larvae of these species are social parasites of some red ants (*Myrmica* spp.). The Large Copper *Lycaena dispar rutilus,* declining in most parts of Europe, still has strong populations in these habitats of the Pannonian region. These habitats should be extensively used by traditional mowing.

The species-rich Nardus grasslands* (6230) show several common features with the acidic *Molinia* meadows. They occur on siliceous substrates with shallow topsoil in mountain areas with cool-humid climate. Vegetation is varied, mostly influenced by grazing and dominated by grass species which are more or less resistant for trampling as *Nardus stricta* and *Festuca ovina.* The moderately grazed stands are often rich in typical acidophilous species as *Anthoxanthum odoratum, Alchemilla* spp., *Potentilla erecta, Polygala* *vulgaris, Antennaria dioica, Linum catharticum, Viola canina,* etc. Rare floristic elements of such meadows are the Moonwort species (*Botrychium lunaria, B. matricariifolium, B. multifidum*).

The *Nardus* grasslands usually have several common floristic elements with the **mesophilous hay meadows of montane regions** (**6520**), dominated by tall grasses as *Festuca rubra, Arrhenaterum elatius, Alopecurus pratensis,* etc. and a large number of forb species as *Betonica officinalis, Scabiosa ochroleuca, Leucanthemum vulgare, Salvia pratensis, Filipendula vulgaris,* and also numerous species of Orchids as *Gymnodenia conopsea, Orchis morio, O. ustulata,* etc. These meadows belong to the most important target areas of insect, especially of butterfly conservation due to their enormous species diversity, e.g., numerous species of Fritillaries (*Melitaea* spp., *Boloria selene, B. euphrosyne, Brenthis ino, Argynnis aglaia, A. adippe, A. niobe*), Blues (*Maculinea alcon, M. arion, Cyaniris semiargus*) and Coppers (*Lycaena virgaureae, L. alciphron*), Ringlets (*Erebia medusa, E. aethiops*), etc. The floristic and faunistic richness of these meadows should be sustained only by traditional mowing since the abandoned stands will be rapidly reforested by birch and aspen (*Populus tremula*), or they can be also degraded by grazing and trampling.

The rupicolous pannonic grasslands (*Stipo–Festucetalia pallentis* 6190) belong to the most typical habitat types of the Pannonian region, often rich in relict-like and/or protected species. They are more or less open, rock sward associations occurring on steep, dry xeric slopes in medium altitude mountains of the Pannonic basin and adjacent regions at 150–900 m a.sl. The base rock is limestone, dolomite or calcareous volcanic rock (basalt, andesite, gabbro) and the soils are shallow rendzinas. They can be subdivided into several subtypes, as the species-rich xerothermophile, circum-Pannonic *Festuca pallens* grasslands on calcareous substrate (*Bromo pannonici–Festucion pallentis*). The most valuable stands belong to the associations of dolomitic grasslands, developed in the Transdanubian Middle range. These preserve many relict and Pannonian endemic species as the *Linum dolomiticum, Ferula sadleriana,* etc. The character species of the open dolomitic swards also belong to this floristic component, as *Seseli leucospermum, Thalictrum minus* subsp. *pseudominus, Euphorbia segueriana* subsp. *minor,* and *Dianthus plumarius* subsp. *regis-stephani.* This association hosts several common elements with the calcareous sand grasslands (*Carex liparicarpos, Alyssum tortuosum, Helianthemum canum, Fumana procumbens*) and other types of rupicolous swards (*Alyssum montanum, Iris pumila, Jovibarba hirta, Scorzonera austriaca,* etc.). The closed dolomitic grasslands are rich in feathergrass species (*Stipa pulcherrima, S. eriocaulis*) and characterized by some tall forbs as *Serratula radiata*, the endemic subspecies *Jurinea mollis* subsp. *dolomitica*, or the rare *Vincetoxicum pannonicum*. They are often in contact with shrubs of lanuginose oak (*Cotino–Quercetum pubescentis*) and karstic beech woods (*Orno–Fagetum*). The rupicolous swards of the Mecsek and Villányi mountains in southern Transdanubia are forming a special group, showing a strong Mediterranean influence and with the presence of West-Balkanic, Dalmatian connections, characterized by the dominance of the Dalmatian Fescue (*Festuca dalmatica*) and the occurrence of some rare (*Galium lucidum, Trigonella gladiata, Lathyrus sphaericus, Artemisia alba* subsp. *saxatile*) and/or endemic-subendemic species as *Colchicum hungaricum, Ceterach officinarum* subsp. *javorkaeanum*. The northern calcareous swards (*Diantho lumnitzeri–Seslerion albicantis*) occur on carbonatic substrates of the Northern Middle Range, in the submontane to sub-Alpine levels. They are characterized by shallow, rocky topsoil and by the presence of montane and/or Carpathian species as *Alyssum montanum* subsp. *brymii, Dianthus praecox* subsp. *lumnitzeri, Onosma visianii, Thalictrum foetidum*, etc. The acidocline grasslands (*Asplenio septentrionalis-Festucion pallentis*) occur on volcanic rocks of the Hungarian Middle Range. It is mostly characterized by the occurrence of some northern and/or montane species as *Asplenium septentrionale, Woodsia ilvensis, Jovibarba hirta, Sempervivum marmoreum*, and *Saxifraga paniculata*.

The sub-Pannonic steppic grasslands* **(6240)** belong to the most important habitat types of the Pannonian forest-steppe. These rather diverse tall-grass steppic plant communities are developed mostly on southern xerothermic slopes of hills in the colline-submontane zone, mostly on rocky substrate and on clay-sandy sedimentation layers enriched with gravels. They are dominated by xeromorphic grasses as the feathergrass (*Stipa*) and fescue (*Festuca*) species and are rich in dwarf scrubs as *Thymus marschallianus, Th. pannonicus, Th. serpyllum*, and *Teucrium chamaedrys, T. montanum*, and herbaceous plants as *Dracocaphalum austriacum*, (rare!), *Polygala major, Pulsatilla grandis, Iris pumila*, etc. The stands of the Transdanubian vs. Northern Middle Range show essential differences in floristic composition. Several steppic species have in these habitats a western boundary of distribution, e.g., some food-plant specialist blue butterfly and owlet moth species.

Dry and semidry steppic grasslands (class: *Festuco–Brometea*) are widely distributed in hilly areas surrounding the Pannonian basin. They display a very high diversity of floristic composition. The floristic composition of the semiclosed Pannonic rupicolous swards (*Stipo pulcherrimae–Festucetalia pallentis*) is strongly influenced by the volcanic (typical species: *Asplenium septentrionale, Festuca pseudodalmatica, Poa pannonica* subsp. *scabra, Alyssum saxatile, Minuartia frutescens, Saxifraga paniculata, Sempervivum marmoreum, etc.)* or calcareous, limestone or dolomit substrate (typical species: *Festuca pallens, Melica ciliata, Sesleria heufleriana, Stipa pulcherrima, Alyssum montanum, Dianthus plumarius, Campanula sibirica* subsp. *divergentiformis, Onosma visianii, etc.)*. The closed steppic grasslands of hilly areas belong to the alliance *Festucion rupicolae*. They are floristically diverse, regionally typical extrazonal mosaic patches of the Pannonian forest-steppe, interlocking to the subcontinental, sub-Mediterranean oak forests (*Aceri–Quercion, Orno–Cotinion*). The structure is determined by sclerophyllous grasses of *Festuca* and *Stipa* spp., and by the high diversity of tall forbs, especially of Apiaceae, Asteraceae, Fabaceae and Lamiaceae. Characteristic species are: *Astragalus austriacus, Centaurea triumfettii, C. sadleriana, Dorycnium herbaceum, Inula ensifolia, I. oculus-christi, Jurinea mollis, Peucedanum alsaticum, Pulsatilla grandis, Seseli osseum, Teucrium montanum, Thymus pannonicus,* etc.

The **meadow-steppe like semidry grasslands** (*Brometalia erecti*) are distributed in Central Europe in areas with sub-Mediterranean and continental influences. In the Pannonian region they belong either to the alliance *Danthonio–Stipion tirsae* (mostly on volcanic substrates) or to *Cirsio pannonicae–Brachypodion pinnati* (mostly on calcareous substrates). They were earlier mostly seminatural, extensively used grasslands (traditionally mowed once a year), but they have been also recently developed, as more advanced stages of old field succession, especially of old vineyards (e.g., in Tokaj area), but also of former arable lands. They are phytocenologically complicated, due to their manifold origin and many transitional types. They also show an exceptional high floristic diversity, as a consequence of the mixture of the steppe grassland (e.g.: *Adonis vernalis, Cytisus procumbens, Dianthus pontederae, Genista tinctoria, Echium russicum, Hippocrepis comosa, Jurinea mollis, Lathyrus pannonicus* subsp. *collinus, Polygala maior, Prunella grandiflora, Pulsatilla grandis*) and forest-steppe and forest-fringe (*Aceri–Quercion* and *Trifolio-Geranietea*, resp., e.g.: *Aster amellus, A. linosyris, Campanula bononiensis, C. cervicaria, Centaurea scabiosa, Chamaecytisus albus, Cirsium pannonicum, Geranium sanguineum, Hypochoeris maculata, Inula ensifolia, I. hirta, Linum flavum, L. hirsutum, Libanotis pyrenaica, Peucedanum cervaria, P. alsaticum, Seseli osseum, Trifolium alpestre, T. montanum, T. rubens*) species (Borhidi and Kevey, 1996). Since their traditional use was discontinued, mostly during the "socialist" industrialization period in the 60's, different successional changes have appeared. A rapid reforestation has only started in regions with relatively more precipitation (e.g., Transdanubia, Zemplén mountain range). Under other climatic and/ or edaphic conditions (e.g., in the North Hungarian Karst area) only some typical structural changes occurred, resulting in the extension of clonal and tall-forb plant species (described as *"Versaumung"* in the German phytosociological literature).

The seminatural grasslands **on calcareous substrates** (**6210**) are completely transitional to the former habitat type and partly belong also to the priority habitats (important orchid sites). They are tall-grass, tall-forb grasslands, developed in the peri-Pannonian colline-submontane regions with transitional continental-sub-Mediterranean climate and characterized by high number of dwarf scrubs, herbaceous and sprout-colony (polycormon)-forming species, mostly from composites, papilionaceous and umbelliferous plants. They are partly natural, edaphic plant communities on limestone, dolomite or calcareous clay enriched with gravels, but more frequently they have originated by clear-cutting of xerothermic oak forests and stabilized under influence of traditional mowing. Their composition is characterized by the dominance of tall (*Bromus erectus, Stipa* spp.) and lower (*Festuca* spp.) perennial grasses and sedges, and by the presence of numerous significant steppic species as *Adonis vernalis, Echium maculatum, Chamaecytisus albus, Ch. austriacus, Ch. procumbens, Ch. ratisbonensis,* rare Orchids as *Anacamptis pyramidalis, Orchis tridentata, O. purpurea,* etc. Abandonment results in the development of thermophile scrubs (see: 40A0) with an intermediate stage of thermophile fringe vegetation with *Geranium sanguineum, Trifolium alpestre, T. montanum, T. rubens, Iris aphylla* subsp. *hungarica, I. variegata,* etc. These associations are already transitional to the next, *priority** habitat type. These grasslands, together with the former and next type of vegetation, belong to the most important target areas of insect, especially of butterfly conservation due to their enormous species diversity. Connected with the tall-grass, tall forb patches of vegetation, the Praying Mantis *(Mantis religiosa)* and numerous steppic bush-crickets occur in these habitats, as *Ephippigera ephippiger, Pachytrachis gracilis, Platycleis grisea, Pterolepis germanica, Saga pedo, Pholidoptera transsylvanica* (in NE Hungary). Some grasshoppers require short-grass patches as *Arcyptera microptera, Euchorthippus pulvinatus, Paracaloptenus caloptenoides, Stenobothrus eurasius.* Among the characteristic beetles of this habitat type, there are numerous food plant specialists, as the Chrysomelid species *Entomoscelis adonidis, E. sacra (Adonis verna-*

Global Biodiversity, Volume 2

lis), the Flower Chafers *Netocia hungarica, Potosia fieberi* (*Jurinea mollis, Carduus collinus*), the longhorn beetles of the genus *Agapanthia, Phytoecia,* etc. Many Zygaenidae, and Polyommatinae are connected to Fabaceae, as *Zygaena angelicae, Z. carniolica, Z. lonicerae,* or *Cupido osiris, Polyommatus admetus, P. thersites* (*Onobrychis arenaria*), *Polyommatus bellargus, P. coridon* (*Coronilla varia, Hippocrepis comosa*). Some Fritillary species are also food-plant specialists, as *Brenthis hecate* (*Filipendula vulgaris*), *Melitaea britomartis* (*Veronica teucrium*), *M. ornata* (*Cirsium pannonicum*). There are some food-plant specialists also among the Noctuid moths, as some Cuculliinae species (*Cucullia campanulae, C. dracunculi, C. gnaphalii, C. lucifuga, C. tanaceti, C. xeranthemi,* and *Shargacucullia gozmanyi, Sh. lychnidis,. Sh. thapsiphaga*), specialized to Campanulaceae, Asteraceae, and Scrophulariaceae. Some Noctuinae species prefer the rupicolous steppic habitats, as the *Euxoa, Dichagyris* és *Chersotis*-species.

The subcontinental peri-Pannonic scrub* (**40A0**) communities are also mostly distributed in the peripheries of the Carpathian basin, on both calcareous and siliceous substrates forming mosaic-like vegetation with steppe grasslands, rupicolous Pannonic grasslands and often along the fringes of woodlands. They can be subdivided into three main types. The habitats of the **subcontinental rupicolous scrubs of Spiraea** are mostly rocky ridges and screes on calcareous or basic volcanic substrates. They occur in isolated patches of the Hungarian Middle ranges, in the upland North of Lake Balaton, but more extended in Northeast, in the Mátra, Bükk and Zempléni hills, and in karstic areas of North Hungary and Slovakia. They are mostly characterized by the occurrence of continental shrubby (*Cotoneaster matrensis, C. niger, Prunus spinosa, Rosa spinosissima*), clonal (*Carex brevicollis, Waldsteinia geoides*) and tall grass-tall forb species as *Melica altissima, Aconitum anthora, Carduus collinus, Doronicum hungaricum,* etc. The *forest-steppe scrubs* of the lowland or loessy pediments consist mostly in rosaceous species as *Amygdalus nana, Prunus fruticosus,* several *Rosa* spp., etc. They are penetrated by loess steppic plants, as *Salvia nemorosa, S. austriaca, Thalictrum simplex, Bupleurum falcatum,* in more nature-like stands also *Ajuga laxmannii, Phlomis tuberosa, Vinca herbacea,* and on the famous Tokaj hill also *Aster oleifolius.* **The forest-fringe scrubs** are tall thorny thickets, partly composed by the shrub species of forests (e.g., *Acer tataricum, Crataegus* spp., *Corylus avellana*), partly by some Southern, heliophilous species as *Cotinus coggygria, Cerasus mahaleb,* etc.

The Pannonic loess steppic grasslands* (**6250**) originally covered large areas of the lowland and hillfoots, nowadays restricted to specific landforms like loess screens and ridges formed by fluviatile erosion and accumulation. The *loess screen vegetation* (*Artemisio-Kochion*) is developed on nearly vertical surfaces under semidesert-like extreme conditions of aridity and permanent erosion. The widely distributed but highly fragmented *loess steppe grasslands* (*Salvio-Festucetum rupicolae*) have several regional subtypes according to the edaphic and climatic conditions. These communities show a high structural and compositional diversity like the true zonal meadow steppes of Ukrania and southern Russia. They are rich in tall and lower perennial grasses, tall forbs as Lamiaceae (*Phlomis tuberosa, Nepeta pannonica*) composites (*Aster linosyris, Centaurea* spp., *Crepis pannonica, Hypochoeris maculata*) and umbelliferous plants (*Peucedanum alsaticum, P. cervaria*), polycormon-forming herbs (*Inula* spp.) and dwarf scrubs (*Thymus* spp., *Teucrium chamaedrys*, etc.). Overgrazed stands have a simplified structure and a selected set of more tolerant species.

The **loess grasslands** are usually tall-grass and tall-forb, floristically rich communities which can be found either on the lower slopes of hilly areas (e.g., on the famous Tokaj hills), on the escarpments of the loess plateaux (e.g., at the northern shore of Balaton, the western border of Danube) or on the loess ridges of the plain (e.g., in the county Békés). These are tall-grass/tall-forb continental steppe grasslands, displaying a peculiar mixture of cold-continental (southern Siberian-Mongolian) and xerothermic (Ponto-Pannonian) steppe elements (e.g., *Agropyron pectinatum, Bassia laniflora, Galium verum, Phlomis tuberosa, Salvia nemorosa* **vs.** *Adonis vernalis, Ajuga laxmannii, Anchusa barrelieri, Aster oleifolius, Astragalus dasyanthus, Crambe tataria, Echium maculatum, Nepeta parviflora, Nonnea pulla, Salvia nutans, Sternbergia colchiciflora, Vinca herbacea, Viola ambigua*) resulting in a rather high species diversity and high number of constant-subconstant species (Zólyomi, 1967).

As a consequence of the high species diversity, the loess grasslands are, as a rule, richly structured by tall-forbs (many Apiaceae:, e.g., *Peucedanum alsaticum, P. cervaria, Seseli osseum* and Asteraceae:, e.g., *Carduus, Centaurea* and *Cirsium* spp., *Hypochoeris maculata, Jurinea mollis*) and display a patchy structure of dicotyledonous polycormons (e.g., *Ajuga genevensis, Phlomis tuberosa, Salvia nemorosa, Thalictrum minus, Thymus* spp.) which can be considered as a compartmentalization under strong competition pressure. The fragmented stands of the northeastern part of the lowland (e.g., Hortobágy area) are, however, more degraded and impoverished, and display a decrease in the number of constant species. Loess grasslands are rather stable under natural conditions. They are, however, sensitive against nitrophilization and many of their valuable plant species cannot tolerate over-grazing, dunging and trampling. Thus, their sustainable use must be only a very extensive one. The floristically less outstanding areas may be tradition-ally grazed, but the most valuable stands should be carefully mowed, spar-ing the tall-forb (e.g., *Phlomis*) and dwarf-scrub (*Amygdalus nana, Prunus fruticosa*) patches and avoiding the use of heavy machinery.

Based on the homogeneity of the floristic composition, the Hungarian phytocenologists did not subdivide the loess grasslands into more associa-tions (Soó, 1968, 1973; Borhidi, 1996). The geographically slightly differen-tiated zonal loess grassland was described as *Salvio nemorosae–Festucetum rupicolae* Zólyomi (in Soó, 1964), which is widely distributed on the whole Pannonian lowland and adjacent hilly areas from the western "*Porta Hun-garica*" of the Carpathian basin to the Transsylvanian basin in the East and to the loess ridges of the Banat in the South. The successional connections of this associations are also simple (Fekete 1992). A single "linear" possibil-ity of successional change was described, only, from the pioneer-like tall-grass–dwarf scrub association of loess escarpments (*Agropyro–Kochietum prostratae*), through the tall-grass–tall-forb loess grassland (*Salvio–Festuce-tum rupicolae*) and loess dwarf-scrub associations (*Prunetum fruticosae* and *Amygdaletum nanae*) to the loess steppe forest (*Aceri tatarici–Quercetum*).

The Pannonic sand steppes* (**6260**) are distributed in calcareous and siliceous sandy regions between the Danube and Tisza, and in the northeast-ern part of the lowland (Nyírség). They are more or less closed formations dominated by medium or tall perennial tuft-forming grasses (e.g., the steppic feather-grass *Stipa borysthenica*), together with the associated therophyte communities developed on mobile or fixed sands. The sandy grasslands are characterized by Pannonian endemic grass (*Festuca wagneri*) and herba-ceous species (*Dianthus serotinus, D. diutinus, Pulsatilla flavescens*, etc.). The plant associations of calcareous sand show significant connections with the steppic and dolomitic grasslands of hilly regions. Extended sandy grass-lands of the plain were traditionally used as pastures, dominated by the false sheep fescue.

Sandy grasslands have been developed mostly on the alluvial fans of the Pannonian lowland rivers which were transformed by wind erosion dur-ing the cold-continental arid climatic periods of the last glaciation and post-glacial times. The two most important areas are: the calcareous sandy area between the Danube and Tisza (Kiskunság) and the mostly limeless (but in patches loessy) sandy area of Nyírség, on the northeastern part of the low-land. Due to the different edaphic and climatic conditions, they display also several differences in the composition of their vegetation. In the calcareous sandy area of Kiskunság meadow-steppe-like grasslands are hardly ever found. Their characteristic species are: *Festuca rupicola, F. wagneri*, Carex liparicarpos*, Iris humilis* subsp. *arenaria, Astragalus exscapus**. The "generalist" steppe grassland species (*Festucion rupicolae*) are the following: *Stipa capillata, Astragalus onobrychis*, Inula oculus-christi*, Jurinea mollis*, Melandrium viscosum, Onobrychis arenaria, Ranunculus illyricus* and *Vinca herbacea** (the species marked with * do not occur in the limeless sandy areas). The permanent, semiclosed grasslands of *Festucion vaginatae* are the most widely distributed grassland types of the Kiskunság area. They consist of several biogeographically significant floral elements, as the sub-endemic pannonian *Festuca vaginata, Stipa borysthenica* and *Holoschoenus vulgaris*. Further characteristic species are: *Achillea kitaibeliana, Alkanna tinctoria, Alyssum tortuosum, Astragalus varius, Colchicum arenarium, Dianthus serotinus, Euphorbia pannonica, Fumana procumbens, Gypsophila paniculata, Linum hirsutum* subsp. *glabrescens, Sedum hillebrandtii*, etc. Recently, a rapid spreading of *Cleistogenes serotina* was observed from the xerothermic slopes of the Hungarian Middle Range to the northern part of Kiskunság, which resulted in a structural change due to its competition pressure (Fekete, 1992). The pioneer therophytic swards (*Brometum tectorum*) are also important components of the vegetation. They consist of therophytic grasses (*Bromus squarrosus, B. tectorum, Secale sylvestre*) and dicots (*Anthemis ruthenica, Kochia laniflora, Polygonum arenarium, Tribulus terrestris*). The soil surface is covered by xerophytic mosses and lichens, e.g., *Syntrichia ruralis* and *Cladonia foliacea, Cl. convoluta, C. magyarica*. The extension of the bare wind-blown sand surfaces has been reduced since the discontinuing of grazing. As a consequence of the sinking water-table, the extension of the *Molinieta*-marshy meadows, mixed with *Salix rosmarinifolia* dwarf scrubs, is also in decrease. Some adventive weeds are increasing, e.g., *Asclepias syriaca, Ambrosia elatior*, mostly on the disturbed places near to the pine (*Pinus sylvestris, P. nigra*) plantations.

The **sandy grasslands of the Nyírség** (NE Hungary) could only survive in some protected areas. Their floristically richest stands often consist of typical species of steppe forests and forest-fringes (*Aceri–Quercion* and *Trifolio–Geranietea*, respectively), e.g., *Chamaecytisus ratisbonensis, Geranium sanguineum, Peucedanum oreoselinum, Teucrium chamaedrys, Trifolium alpestre*; with populations of the endemic *Iris aphylla* subsp. *hungarica, Pulsatilla flavescens*, and with isolated occurrence of the Baltic-continental *Pulsatilla patens*. There are two associations of sandy steppic grasslands: *Pul-*

satillo hungaricae-Festucetum rupicolae and on the sand-dunes: *Pseudolysimachio pallentis-Chrysopogonetum.* Characteristic species are: *Iris humilis* subsp. *arenaria, Dianthus arenarius* subsp. *borussicus, D. giganteiformis* subsp. *pontederae, Euphorbia segueriana, Onosma arenarium, Peucedanum arenarium, Pseudolysimachia pallens* (endangered!), *Pulsatilla grandis, P. flavescens.* The greatest part of sandy grasslands was transformed into the pasture *Potentillo arenariae–Festucetum pseudovinae* due to persistent grazing. The healthy management of sandy grasslands is quite difficult. Grazing must be stopped in some periods (March–June and September) to protect the vulnerable plant species (e.g., *Pulsatilla* spp.). Control of livestock is important, especially in dry years, to avoid overgrazing and excessive trampling. Otherwise, bare spots arise, which will be overgrown by invading weeds, mostly by *Conyza canadensis* and recently, also by *Ambrosia elatior.* The effect of drought recently has enhanced the dispersal of these species.

These types of sandy grasslands show hardly any common features with the zonal steppe grasslands. The grasslands on calcareous substrates are characterized by the presence of the Ponto–Pannonian and sub-Mediterranean–Ponto-Mediterranean elements of flora, while the grasslands on the limeless sand by Baltic and South-Siberian species. As a consequence of the floristic richness and due to the structural diversity of the sandy grassland associations, the sandy grasslands are subdivided into many associations, which are network-like interconnected by the possible dicho- or polytomic successional changes (Fekete, 1992).

The acidophilous Luzulo-Fagetum beech forests (9110) and **neutrophilous Asperulo–Fagetum beech forests (9130)** represent the dominant woody associations of the montane zone in Central Europe and also in the Carpathian basin. The forests developed on acidic soils usually have a poor scrub and herbaceous layer, but relatively rich in moss (*Polytrichum*) and fern (*Pteridium aquilinum, Dryopteris* spp.) species. Sub-Alpine stands near to the Austro-Hungarian border are often mixed with fir (*Abies alba*) and have a more varied floristic composition. The beech stands developed on near-neutral soils are characterized by a strong representation of geophytic species (*Anemone nemorosa, Corydalis* spp., *Galanthus nivalis, Isopyrum thalictroides, Scilla* spp.) and are rich in forest-type character species as *Aegopodium podagraria, Carex pilosa, Dentaria* spp., *Galium odoratum* and *Melica uniflora,* etc. The soil invertebrate communities are strongly influenced by the slowly decomposing, humide litter of beech.

The **Medio-European limestone beech forests of the alliance Cephalanthero–Fagion (9150)** are xerophile stands developed on calcareous, often superficial soils, usually of steep slopes or stony ridges. They have a mixed, somewhat light-penetrated canopy with the regular presence of *Fraxinus excelsior, Acer pseudoplatanus* and *Ulmus scabra*, thus they are rich in relict species as *Sesleria* spp., *Taxus baccata, Atragene alpina, Arabis alpina, Aquilegia vulgaris, Valeriana tripteris* and numerous Orchids (e.g., *Epipactis* spp.). The more light-penetrated stands are rich in tall forbs as *Aruncus sylvestris, Hesperis matronalis, Lunaria rediviva, Scrophularia vernalis*, and *Stachys sylvatica*. In the upper beech forests zone of the Northern Middle range some further Carpathian floristic elements occur, as *Aconitum moldavicum, Aconitum variegatum* subsp. *gracile, Dentaria glandulosa, Geum aleppicum, Petasites albus, Daphne mezereum, Polygonatum verticillatum, Scopolia carniolica, Senecio nemorensis* subsp. *fuchsii*. The sub-Mediterranean karstic beech forest type (*Fago–Ornetum*) of Transdanubia is characterized by the dominance of *Fraxinus ornus*, the occurrence of *Sorbus* species and on rocky patches of screes by the occurrence of dealpine species as *Carduus glaucus, Primula auricula, Rubus saxatilis*.

The Tilio–Acerion forests of slopes, screes and ravines* (9180) are mixed forests of ash, lime (*Tilia cordata, T. platyphyllos*) and maple (*Acer platanoides, A. pseudoplatanus*) species, developed in steep gorges, on screes, abrupt rocky slopes, particularly on calcareous, but also on siliceous substrates. The richest stands occur in the montane beech zone, where they are dominated by the Sycamore Maple and the Excelsior Ash. Their herbaceous layer is rich in tall forbs (*Anthriscus nitida, Hesperis matronalis, Lunaria rediviva, Scrophularia vernalis, Scutellaria columnae*) and nitrophilous species (*Geranium phaeum, Impatiens noli-tangere, Parietaria officinalis, Scopolia carniolica*) due to their humous soil covered by deep litter layer. The more thermophilous submontane stands are characterized by the presence of *Quercus petraea, Q. cerris, Carpinus betulus, Acer campestre, Sorbus torminalis* and by the rich shrub level (*Cornus mas, Corylus avellana, Rhamnus cathartica, Viburnum lantana*, etc.). These forests are rich in rare and/or protected herbaceous species as *Erythronium dens-canis, Asyneuma canescens, Carduus collinus, Scutellaria altissima* and *Waldsteinia geoides*.

These forests have an important soil protecting function. The invertebrate assemblages are characterized by the occurrence of several Carpathian land snails (*Vestia turgida, V. gulo*) and ground beetles (e.g., *Carabus glabratus, C. intricatus, Abax carinatus, A. schueppeli*), and also by high diversity of xylophagous buprestid (*Lampra rutilans, Agrilus angustulus, A. auricollis, Anthaxia manca*) and cerambycid (*Alosterna tabacicolor, Exocentrus lusitanus, Rhopalopus insubricus, Stenostola ferrea*) species.

The alluvial forests with *Alnus glutinosa* and *Fraxinus excelsior* (*Alno-Padion, Alnion incanae, Salicion albae*)* (91E0) are riparian forests of lowland and hill watercourses, woods of *Alnus incanae* of montane and submontane rivers, galleries of tall willow and poplar species (*Salix alba, S. fragilis* and *Populus nigra*), along lowland, hilly or submontane rivers. All types occur on alluvial deposits periodically inundated by the annual rise of the river level, but otherwise well-drained during low-water. This habitat includes several subtypes: ash-alder woods of springs and fast-flowing rivers; ash-alder woods of slow-flowing rivers; white willow gallery forests (*Salicion albae*). The most diverse species composition can be observed in the woods of pre-Alpine western Transdanubia and in the Zempléni Mts. They are characterized by numerous hygrophilous tall forb species and ferns. The alder-ash gallery forests have a mixed canopy layer (*Alnus glutinosa, Fraxinus excelsior, F. angustifolia* subsp. *pannonica, Prunus padus*) a dense shrub layer (*Rubus caesius, Viburnum opulus, Frangula alnus, Ribes rubrum, Sambucus nigra*), and are rich in horsetails (*Equisetum sylvaticum, E. telmateia*), ferns (*Matteuccia struthiopteris, Thelypteris palustris, Dryopteris robertiana*) and tall forbs (*Aconitum variegatum* subsp. *gracile, Petasites hybridus, Ranunculus lanuginosus, etc.*). Many meso-and hygrophilous forb species (*Asarum europaeum, Dentaria bulbifera, D. enneaphyllos, Galanthus nivalis, Lathraea squamaria, Leucojum vernum, Majanthemum bifolium, Oxalis acetosella*) also occur in this habitat type.

The willow-poplar gallery forests of the Tisza lowland consist of several different types. The most widely distributed type is dominated by *Salix alba* and *S. fragilis, Populus alba* and *P. nigra*, and they are characterized by numerous creeping plants as *Humulus lupulus, Clematis vitalba* and *Vitis sylvestris*. The typical association of the shallow depressions is the willow gallery forest (*Leucojo aestivo–Salicetum albae*), which is rich in marshy species as numerous sedge species (*Carex gracilis, C. riparia, C. vesicaria*), *Galium paluster, Myosotis palustris, Polygonum hydropiper, Rorippa amphibia* and *Stachys palustris*.

The old gallery forest stands are important habitats of numerous sapro and xylophagous beetles as *Osmoderma eremita, Liocola lugubris*, of jewel beetles (*Dicerca aenea, Agrilus viridis, Anthaxia salicis*) and longhorn beetles (*Aromia moschata, Saperda* spp.). The softwood gallery forests have an outstanding importance as sites of heron–colonies and breeding places of the Black stork (*Ciconia nigra*) and White-tailed Sea-eagle (*Haliaetus albicilla*).

Riparian mixed forests of *Quercus robur*, *Ulmus laevis* and *Ulmus minor*, *Fraxinus excelsior* or *Fraxinus angustifolia*, along the great rivers

(Ulmenion minoris) **(91F0)** are mixed forests of hardwood trees (*Quercus robur, Fraxinus excelsior, F. angustifolia* subsp. *pannonica* and *Ulmus laevis, U. campestris*), with high quality timber production on the uppermost level of the floodplains, and developed mostly on recent alluvial deposits. The soil may be well drained between inundations or remain wet. The stands of the Pannonian lowland are characterized by the dominance of *Fraxinus angustifolia* subsp. *pannonica*. The dense shrub layer is formed by numerous species (*Acer tataricum, Cornus sanguinea, Frangula alnus, Viburnum opulus*) and is regionally diverse. The undergrowth is also generally well developed and contains many nemoral species which are restricted in the Pannonian region to hilly or montane areas (*Aegopodium podagraria, Allium ursinum, Asarum europaeum, Dentaria bulbifera, Galanthus nivalis, Galeobdolon luteum, Lathraea squamaria, Majanthemum bifolium,* etc.). This association has numerous common elements with the closed Pedunculate oak forests of the lowland (*Convallaria majalis, Corydalis cava, Galium odoratum, Polygonatum latifolium, Pulmonaria officinalis, Salvia glutinosa, Vinca minor,* etc.).

The invertebrate assemblages of the soil and litter are rich in hygro- and mesophilous land snails (e.g., *Bradybaena fruticum, Cochlodina laminata, Helicigona arbustorum, Laciniaria plicata, Monachoides vicinus, Ruthenica filograna*). Some species are restricted to the northeastern–eastern part of the Pannonian lowland, as the Carpathian blue slug (*Bielzia coerulans*) or the endemic relict species *Drobacia banatica* and *Kovacsia kovacsi*. Due to the richness in prey, these forests are populated by numerous large ground beetle species as *Cychrus attenuatus, Carabus coriaceus, C. violaceus, C. nemoralis, C. intricatus,* etc. The occurrence of the Transylvanian species *Carabus hampei* is restricted to the northeastern part of the lowland. Due to the high diversity of woody species, the assemblage of phytophagous insects is exceptionally rich, as the Notodontidae species *Cerura erminea, Clostera curtula, C. pigra, C. anastomosis, Drymonia* spp., *Furcula bifida, F. furcula, Gluphisia crenata, Harpya milhauseri, Peridea anceps, Pheosia tremula, Spatalia argentina, Tritophia tritophus* or the Drepanidae species *Drepana falcataria, D. curvatula, Falcaria lacertinaria, Sabra harpagula, Watsonalla binaria*. These forests are important habitats of the Black stork, the White-tailed-eagle and many passerine bird species.

The Pannonic woods with *Quercus petraea* and *Carpinus betulus** (91I0) belong to the most widely distributed forest types of the region. They occur in zonal position in some more cool-humid marginal areas of the lowland (Dráva and Upper Tisza plain) and in the colline-submontane zone of the Middle Range. The lowland subtype is characterized by the prevalence of *Quercus robur*, while the colline-montane stands are dominated by species of the *Quercus petraea* complex. These forests have several differentials, mostly Balkanic or sub-Mediterranean, ponto-Mediterranean species (e.g., *Q. cerris, Acer tataricum, Sorbus* spp., *Staphylea pinnata, Helleborus* spp., etc.) opposed to the Central European oak-hornbeam woods. They have several regional variations in composition. The highly diverse stands in southern Transdanubia show sub-Atlantic and sub-Mediterranen, Balkanic influences with the occurrence of *Castanea sativa* and of numerous western Balkanic (Illyrian) floristic elements, as *Aruncus sylvestris, Cyclamen purpurascens, Galium sylvaticum, Knautia drymeia, Lamium orvala, Primula vulgaris*, etc. The mostly differentiated stands are the karstic oak-hornbeam forests (*Waldsteinio-Carpinetum*) in the northern Hungarian (and southern Slovakian) karstic region, with transitions to the ravine forests, and characterized by forest-steppe species. The semiopen canopy consist of the mixture of ravine forest (*Fraxinus excelsior, Tilia cordata, Acer pseudoplatanus, Ulmus scabra*) and xerophilous calcareous oak forest species (*Sorbus torminalis, S. aria* agg., *Acer campestre, Ulmus minor*). The shrub (*Cornus mas, Lonicera xylosteum, Viburnum lantana*) and the herbaceous layers are also rather diverse, the latter with the presence of forest-steppe elements (*Carex brevicollis, Lathyrus niger, Waldsteinia geoides*), and colorful geophytic species in the spring aspect (*Corydalis cava, C. solida, Erythronium dens-canis, Scilla drunensis*).

Some natural stands of the oak-hornbeam forest have rather diverse insect assembleges, e.g., in the Drava and Upper Tisza lowland consisting of numerous protected butterflies, as *Parnassius mnemosyne, Euphydryas maturna, Neptis rivularis*, etc. The karstic stands are characterized by the occurrence Carpathian land snails (*Bielzia coerulans, Cochlodina cerata, Isognomostoma isognomostoma, Oxychilus orientalis, Vestia turgida*) and ground beetle (*Abax schueppeli, Carabus zawadskyi, C. obsoletus*) species. These stands and the adjacent forest fringes are important habitats of *Parnassius mnemosyne, Leptidia morsei, Limenitis populi, Neptis sappho, N. rivularis* and *Lopinga achine*.

The Pannonian-Balkanic turkey oak–sessile oak forests (91M0) are representing the other most extended zonal forest type of the region. These are the zonal forests of the colline-submontane level between 250–500 m, with double, mixed canopy layer consisting of 3–4 oak species, broad-leaved lime, ash and maple species, etc. They also show several regional subtypes, characterized by differential species as the *Fraxinus ornus, Tilia tomentosa* and *Lonicera caprifolium* in southern Transdanubia. The character species of the undergrowth are *Poa nemoralis, Festuca heterophylla, Melica uniflora, Vicia cassubica, Tanacetum corymbosum, Potentilla alba, Euphorbia polychroma*, etc. The composition is mostly influenced by the calcarous vs. siliceous substrate. Due to its distribution, this community has several regional types. The stands of the Northern Middle range mostly show the nearly uniform Central European character while the Transdanubian stands are mostly influenced by the acidic substrate with the presence of sub-Atlantic elements (*Asphodelus albus, Genista pilosa, G. germanica, Veronica officinalis, Veratrum nigrum*). The mostly differentiated type is distributed in South Transdanubia, in Mecsek and Villányi mountains, and is characterized by the occurrence of numerous sub-Mediterranean and Balkanic elements. Some of these species are widely distributed as *Carex strigosa, Luzula forsteri, Genista ovata, Helleborus odorus, Potentilla micrantha, Rosa arvensis, Ruscus aculeatus*, while *Paeonia banatica* and *Doronicum orientale* only occur on some isolated spots of Mecsek Mts.

These forest exhibit a very rich insect assemblage, e.g., the large ground-beetles: *Calosoma inquisitor, Carabus cancellatus, C. hortensis, C. ullrichi, C. scheidleri*, the sub-Mediterranean Giant Cicada *Tibicina haematodes*, numerous notodontid (*Drymonia ruficornis, D. querna, D. dodonaea, Harpya milhauseri, Peridea anceps, Spatalia argentina*) and noctuid moths (*Catephia alchymista, Catocala nymphagoga, C. promissa, C. sponsa, Minucia lunaris, Dicycla oo*) specialized on *Quercus* spp. but also some defoliating pest species as the leafrollers *Aleimma loeflingianum, Tortrix viridana*, the Gipsy moth (*Lymantria dispar*) and Procession moth (*Thaumetopoea processionea*).

The Pannonian woods with *Quercus pubescens** (91H0) are xero-thermophile oak woods of the colline and submontane levels of hilly regions. They are mostly developed in extrazonal position, on more or less steep southern slopes with shallow topsoil. One group of them consist of thermophilous mixed oak forests dominated by the Lanuginose (*Quercus pubescens, Q. virgiliana*) and Sessile oak (*Q. petraea, Q. dalechampii*) species, and mostly in Transdanubia, as sub-Mediterranean influence, with *Fraxinus ornus, Sorbus domestica, Coronilla emerus, C. coronata, Cotinus coggygria*, etc. These communities show a regional differentiation within the Carpathian basin. The western, Transdanubian stands show a more significant sub-Mediterranean influence with the presence of the following typical species: *Campanula bononiensis, Dictamnus albus, Erysimum odoratum, Euphorbia polychroma, Iris graminea, I. variegata, Muscari tenuiflorum,* etc. Oppositely, the East Hungarian stands are affected by a marked steppic, continental impact, as signalized by such species as *Acer tataricum, Aconitum anthora, Adonis vernalis, Anemone sylvestris, Chamaecytisus albus, Melica picta, Phlomis tuberosa, Vinca herbacea, Waldsteinia geoides.*

The other group contains light-penetrated scrub-forest stands with numerous rupicolous and steppic grassland species and is subdivided into a southwestern, sub-Mediterranean and a northeastern, continental type. The most important structural trait of the former type is the growing of a dense scrubby "collar" of *Cotinus coggygria* around the low *Quercus* trees. The Balkanic character species *Carpinus orientalis* occur in the Vértes Mts. only. Typical species of this type are: *Anthericum ramosum, Buglossoides purpureo-coerulea, Colutea arborescens, Dictamnus albus, Geranium sanguineum, Serratula lycopifolia,* and *Vicia sparsiflora.* However, the presence of colorful Irises (*Iris graminea, I. variegata*) and Orchids, as *Himantoglossum adriaticum, Orchis simia, Anacamptis pyramidalis,* etc. also characterize this habitat type. The typical tall shrub of the northeastern stands is *Cerasus mahaleb,* and this association is characterized by the presence of continental, Ponto-Pannonian forest-steppe or even steppic species as *Aconitum anthora, Asyneuma canescens, Carduus collinus, Coronilla varia, Dracocephalum austriacum* (in the North Hungarian and Slovakian karst only), *Ferula sadleriana* (in the Pilis and Bükk Mts. only), *Lathyrus pannonicus* subsp. *collinus,* etc.

These forests and shrub forests belong to the most important components of the Pannonian forest-steppe association complex and often represent relict-like habitats of threatened and protected plant and insect species, e.g., the jewel beetle *Anthaxia hungarica* living in old *Quercus pubescens* trees, the butterflies Balkanic small white (*Pieris ergane*) living on *Aethionema saxatile,* the large Hungarian blue (*Jolana jolas*) living on *Colutea arborescens,* the Winter Geometrid moth *Erannis ankeraria* living on *Q. pubescens,* the Purple Owlet moth (*Pyrrhia purpura*) living on *Dictamnus albus,* the Owlet moths *Dioszeghyana schmidtii* living on *Q. cerris,* and *Asteroscopus syriacus* living on *Fraxinus ornus,* etc.

The Euro-Siberian steppic woods with Quercus spp.* 91I0 represent a widely distributed but mostly fragmented type of xero-thermophile oak woods of the plains and lower hills of southeastern Europe. The climate is very continental, with a large temperature range. These woods can be subdivided according to the substrate which often consists of loess, alkali soil or sand. The loess steppe tartar maple-oak forests (*Aceri tatarici–Quercetum*) have a mixed, two-layer canopy with *Quercus robur, Q. cerris, Acer camp-*

estre, A. tataricum, and *Ulmus minor.* Their undergrowth is rich in nemoral, geophytic species. They are usually structured by some natural gaps and scrubby, tall-forb fringes, rich in typical forest steppe plant species as *Amygdalus nana, Prunus fruticosus, Iris variegata, Doronicum hungaricum,* etc.

The alkali steppe forests have a similar composition of forest canopy, but these are mostly light-penetrated stands with strongly fluctuating humidity and with the presence of numerous steppic and alkali grassland species as *Festuca pseudovina, Limonium gmelini* subsp. *hungaricum, Aster sedifolius, Serratula tinctoria,* and *Peucedanum officinale,* etc. Other important forest-steppic components are: *Melica altissima, Aster linosyris, Buglossoides purpureo-coerulea, Doronicum hungaricum, Lychnis viscaria, Melampyrum cristatum, Phlomis tuberosa.* The shadowed stands are, however, habitats of typical nemoral species as *Brachypodium sylvaticum, Poa nemoralis, Geum urbanum,* also including some geophytic species as *Anemone ranunculoides, Corydalis bulbosa, Scilla drunensis.*

The sandy oak forests are subdivided into regional subtypes on calcareous (Central Hungary) vs siliceous (Northeast Hungary) substrate. These are partly closed, nemoral stands, rich in ferns (*Athyrium filix-femina, Dryopteris carthusiana, Polypodium vulgare*) and umbrophile herbaceous plants as *Geranium phaeum, Salvia glutinosa, Stachys sylvatica, Viola mirabilis, V. riviniana* and also in some geophytic species and Orchids (*Cephalanthera longifolia, C. damasonium, Neottia nidus-avis, Platanthera bifolia*). The xerothermic type is rich in Ponto–Pannonian forest-steppe species as *Iris variegata,* and *I. aphylla* subsp. *hungarica, Trifolium alpestre,* and *T. pannonicum, Lychnis coronaria,* and *L. viscaria,* etc.

The Illyrian Fagus sylvatica (Aremonio–Fagion) forests (91K0) are the beech forests of the western Balkanic ranges and hills, with irradiations in the southeastern Alps and also to the southwestern Pannonic hills (County Zala). Species diversity is here much higher than in the Central European beech woods, thus the alliance Aremonio-Fagion constitutes an important center of species diversity. The forest canopy is strongly mixed (*Acer pseudoplatanus, Tilia cordata, T. platyphyllos*) and is also characterized by the presence of thermophilous species as *Quercus cerris,* locally also *Q. farnetto, Tilia tomentosa,* and on siliceous substrate, the Chestnut *Castanea sativa.* The presence of some lianes is also significant as *Tamus communis* and *Hedera helix.* The undergrowth is rich in western Balkanic (Illyrian) and eastern Alpine floristic elements, as, e.g., *Anemone trifoliata, Cardamine trifoliata, Cyclamen purpurascens, Doronicum austriacum, Lamium orvala, Vicia oroboides, Scilla drunensis* subsp. *drunensis,* etc. A special type of this association occurs in the Mecsek mountains, characterized by several sub-Mediterranean–West Balkanic species as *Helleborus odorus, Lathyrus venetus, Lonicera caprifolium, Ruscus aculeatus, R. hypoglossus,* etc.

The Illyrian oak –hornbeam forests (Erythronio–Carpinion) (91L0) are woods of *Carpinus betulus, Quercus robur, Q. petraea*-aggr., often mixed with thermophilous species as *Q. cerris, Fraxinus ornus* and *Tilia tomentosa,* mostly on deep neutral to slightly acidic brown forest soils, in the SE-Alpine-Dinaric region, extending northwards to Lake Balaton mostly in hilly and submontane regions, river valleys and the plains of the Drava and Sava. The riverine forests of the Drava and Sava lowland are characterized by the dominance of *Fraxinus angustifolia* subsp. *pannonica.* They have a much higher species richness than the Central European oak woods. The humid stands are characterized by some ferns, as *Polystichum setiferum,* by the mass occurrence of *Allium ursinum* and *Adoxa moschatellina.* The undergrowth is very similar to the Illyrian beech forests and is characterized by the diversity of western Balkanic (Illyrian), eastern Alpine and sub-Mediterranean elements, e.g., *Anemone trifoliata, Cardamine trifoliata, Cyclamen purpurascens, Doronicum austriacum, Knautia drymaea, Lamium orvala, Lathyrus venetus, Primula acaulis, Ruscus aculeatus, Senecio ovirensis, Tamus communis, Vicia oroboides,* etc.

The Pannonic inland sand dune thickets (Junipero–Populetum albae)* (91N0) represent a very characteristic endemic plant association of the central and western Pannonian sandy plains. They are more or less light-penetrated assemblages of *Juniperus communis, Populus alba, P. nigra,* locally mixed with *Acer campestre* and *Ligustrum vulgare.* The groups of trees are bordered by a fringe of polycormon-forming forbs, as *Fragaria vesca, Geranium sanguineum, Polygonatum odoratum, Vincetoxicum hirundinaria,* etc. The slopes of dunes are covered by pioneer-like swards with a nearly semidesert character and they are rich in steppic species as *Festuca vaginata, Stipa borysthenica, Euphorbia seguierana* subsp. *pannonica, Linum hirsutum* subsp. *glabrescens, Fumana procumbens,* etc. In the shallow depressions a high diversity of scrubs and forest-steppe species occur, as *Berberis vulgaris, Asparagus officinalis, Cynoglossum officinale, Lithospermum officinale, Hieracium umbellatum, Cephalanthera rubra,* and the endemic *Epipactis bugacensis,* etc. These softwood forests have a crucial importance in the conservation of several bird species, as *Coracias garrulus, Dryocopus martius, Picus canus, Jynx torquilla, Oriolus oriolus* and *Lanius minor.* The structure of the vegetation is locally strongly influenced by the introduced rabbit (*Oryctolagus cuniculus*).

The caves not open to the public **(8310),** including their water bodies and streams, are hosting specialized or high endemic species, and are of outstanding importance for the conservation of Annex II species (e.g., bats, amphibians). They exhibit a very specialized and highly endemic cavernico-lous fauna. These species often exclusively occur in a single cave or subter-ranean water system of adjacent cave systems, e.g., the tiny Pannonian water snail *Bythiospeum hungaricum* (Mecsek Mts.), the ancient relict Syncarida (*Bathynella hungarica*), the white Amphipod *Niphargus aggtelekiensis* which occurs in water systems of the Aggtelek and Slovakian karstic areas. The cavernicolous terrestrial invertebrates are mainly coleoptera, which are carnivorous and have a very limited distribution, e.g., *Duvalius hungaricus* which occurs in several caves of the Aggtelek and Slovakian karstic area.

5.5 Main Drivers and Threats to Biodiversity

5.5.1 General Trends of Economic Development and Pressures on Ecosystems

Most of the threats to biodiversity are dependent on human activities. An ecological footprint is a measure of the overall human pressure on nature as it shows how much land and water area are necessary for a given human population to produce the resources and to absorb its wastes, using prevail-ing technology. According to The Ecological Footprint Atlas, Hungary's per capita ecological footprint was 3.5 global hectares in 2005. As the corre-sponding biocapacity was only 2.8 hectares, the country had an ecological deficit of 0.7 hectares. (In comparison: the world total ecological footprint per capita was 2.7, while the biocapacity was 2.1). This means that the popu-lation of the country is putting more pressure on the ecosystems than the sus-tainable level, which seriously affects biodiversity. Between 1989 and 1993, with the collapse of socialist regime and economy, Hungary's footprint has substantially decreased (from 4.5–5 ha to 3–3.5 ha), and since it is oscil-lating around 3.5. This is happening in spite of a marked overall economic growth during the last years. During the Landscape Ecological Vegetation Mapping of Hungary (MÉTA) numerous factors threatening plant biodiver-sity in Hungary were unraveled. The most important among them are the fol-lowing: invasive alien species, the overpopulated big game stands, drainage, development of secondary shrub habitats after disturbance or abandonment, trampling, large-scale intensive forest use by clear-cutting and plantation, and abandoning the mowing and/or grazing of grasslands.

The Hungarian Biodiversity Monitoring System (HBMS) monitors five invasive alien plant species (*Ailanthus altissima, Amorpha fruticosa, Asclepias syriaca, Solidago gigantea, Solidago canadensis*) at the landscape, community and population levels since 1998. The distribution and impact of invasive alien species on natural vegetation of the whole area of Hungary were evaluated during the MÉTA vegetation survey conducted between 2003–2006. The results indicate that 5.5% of the country is covered by perennial alien species. This number is only slightly lower than the half of the total area of natural vegetation. With inclusion of alien annuals in the arable fields and alien plantations, this number would be much higher. The outcome of the survey suggests that the most invaded region is the Lesser Hungarian plain, wherein floodplains of the Danube and Rába rivers, and in the large dried out marsh and fen areas of the Hanság region huge stands of certain alien species have developed. The Hungarian Middle range with high percentage of (semi)natural forest cover represent the less invaded belt. The most important alien plant species are *Acer negundo, Ailanthus altissima, Amorpha fruticosa, Asclepias syriaca, Elaeagnus angustifolia, Fraxinus pennsylvanica, Robinia pseudoacacia* and *Solidago* spp.

5.5.2 Fragmentation

The economic model and path prevailing in Hungary since the mid-1990s has resulted in a rapidly growing demand for land for economic development. Lands have been withdrawn from cultivation for the purposes of residential construction, green-field investment for shopping centers, highways, and industrial parks. Further development of infrastructure (especially linear facilities) may increase the fragmentation of the spatial structure of the landscape and may lead to a decline of natural areas, fragmentation of habitats, and the isolation of natural populations. The rapid adaptation to western style consumerism had wide-ranging pressures on biodiversity (Mihók et al., 2017).

5.5.3 Climate Change

The effects of climate change are already present in Hungary. According to the calendar kept by beekeepers since 150 years, Black Locust trees are flowering 3–8 days earlier than in the nineteenth century. Bird migration times are changing and some previously migrant birds stay in the country in mild winters. New, thermophile plant species coming from South have appeared in the country, some of them are invasive. According to data provided by the forestry light trap network, new moth species are arriving from South, among them also some pest species, and the abundance of meso-hygrophilous species is decreasing. In forests, the accumulation of years of drought increases tree-dying and damages caused by insects. Abundances of previously insignificant species may undergo a dramatic increase so that these insects become strong pests. This is due to two reasons: the decrease in resistance potential of trees because of drought and the advantageous temperature favoring the gradation of pests. Effects of climate change interact with those of forest management.

5.5.4 Natural Vulnerability of Hungary

Hungary is situated in the center of the Carpathian basin, which has dual consequences for its conservation potentials. Firstly, the whole territory of Hungary is situated in the Pannonian biogeographical region, which is characterized by a wide range of special biological conditions, such as the presence of a number of endemic species and habitat types. The deliberate or accidental introduction of non-native species in Hungary therefore carries a particular hazard. Secondly, Hungary is particularly exposed to enhanced climatic fluctuations and negative transboundary environmental impacts due to the geographical characteristics of the basin, for example, it receives 95% of its surface water from abroad.

5.5.5 Biodiversity of Agricultural Ecosystems

Agriculture is one of the most important sectors of the Hungarian national economy. The unique natural endowments of the country's topography, climatic conditions and highly fertile soils make it possible to achieve yields of outstanding quality and quantity in most crops. Hungary has a total area of 9.3 million hectares. 83% (7,721,000 ha) of Hungary is cultivable land, including forests, reed-beds and fishponds. The total agricultural area is 5,807,000 hectares (62.4%), which represents an outstandingly high proportion in Europe. 78% of this agricultural area is made up of arable land and 17% is grassland, while kitchen gardens, orchards and vineyards account for 5%. More than half of the production involves plants, 34% animals and 8.5% services, and others.

The area of organic farming more than doubled between 2000 and 2004, but since then numbers are stagnating. In 2007, ecological production covered 122,270 hectares (1.3% of Hungary's territory). Besides the declining trends of the area occupied by natural vegetation, some attention should be directed towards plow-lands abandoned from cultivation, where vegetation can start to regenerate. In 2006, there were about 350,000 ha of uncultivated land of 2–50 years of age in Hungary. In 2007 and 2008, this number has diminished because of the increase of food prices and changes in European Union regulation.

5.6   Overview of Status of Agro-Biodiversity

Hungary is rich in native plant genetic resources as its territory belongs to a secondary center of crop diversity, where a number of local types and landraces developed even in relatively recently introduced New World crops (green and red peppers, tomato, maize, etc.). The natural flora is a rich source of wild fruits, medicinal plants (including diverse chemotaxa), forage grasses and legumes, and some crop wild relatives (*Aegilops, Lactuca, Daucus, Secale, Vitis, Prunus, Pyrus*, etc.). A great variation of local types of temperate fruits and grapes are still grown in so-called "restricted garden areas," and backyards. In the Eastern part of the country, seminatural fruit forests (walnut, plum) still exist and maintained in protected areas. In situ conservation of crop wild relatives and landraces is closely associated with nature conservation. Populations of several crop wild relatives live in protected natural habitats, and such areas can also play an important role in "*in situ, on farm*" conservation of locally developed landraces. Although the main aim of the Ramsar Convention is the protection of wetlands (as birds-habitats) a number of protected plants including crop wild relatives live in these areas. The ex situ collection of seed-propagated plant genetic resources contains 86,756 accessions of 1,877 taxa maintained under medium and long-term chambers at the Research Centre for Agrobotany at Tápiószele (RCAT). In Hungary, the law on animal breeding regulates the conservation of native and endangered domestic animal breeds and considers this task as a state responsibility. At present, 46 breeds are found in the list of native domestic animal breeds of national importance and the list must be updated every five years.

5.6.1   Trends in Agro-Biodiversity

The farmland diversity in Hungary is declining, calculated by applying the Shannon-Wiener diversity measure to landscape composition, using the area of the 16 most important land cover types (grassland, lands withdrawn from cultivation) and crop types (e.g., wheat, maize, barley, oat, potato, sunflower, sugar beet, alfalfa, fruits, vegetables, abandoned land).

The change of regime and joining the EU has resulted in drastic changes in land use. Many arable lands and other agricultural areas were abandoned, the overgrazing stopped, and the intensity of agriculture, e.g., the use of chemicals decreased. These changes favored the biodiversity of agricultural lands; however traditional land-use forms have been disappearing. An agri-environmental program was started, which subsidies extra performance associated with environmentally aware farming, sustainable landscape management, and reimburse income losses resulting from these activities. The program allows only a limited level of chemical use in participating sites. However, the impact on biodiversity of the program is not assessed yet. Considerable changes occurred in the state of diversity of the country's agricultural production system between 1996 and 2007. Comparing the relevant issues of the national list for cultivars published in 1996 and 2007 it turns out that in spite of a slight decrease of taxa the number of registered cultivars has almost doubled. Although the number of taxa has slightly declined in the cases of ornamentals and vegetables, the number of registered varieties has increased in these crop groups as well. The remarkable change in case of vegetables is representing the highest increase among the crop groups (312%).

5.6.2 Main Threats to Agro-Biodiversity

Cheap international transportation of agricultural products affects national production as the prices of agricultural products coming from far-away places are often lower than the prices of nationally cultivated products. The diversity of fruit and vegetables accessible at the markets and supermarkets has decreased, less landraces and farmer varieties can be found on the market. Several traditional varieties are not cultivated due to economic reasons and lack of knowledge. After a sharp decline beginning in 1990, the use of chemicals is growing since year 2000. However, management intensity in Hungary is lower than in Western Europe, e.g., herbicides were being applied to 25%, insecticides and fungicides to 11% of the total agricultural land in 2002. Meanwhile, the target of reducing by 20% the 1999 level of use of chemicals dangerous for the environment and human health has been achieved for 2008.

5.7 Overview of Status of Forests and Their Biodiversity

The forest area has grown in Hungary mostly with plantations. By this way 20.3% of the country is forested. The proportion of state-owned forests is

56%, while 43% is private and 1% is community-owned. Forest areas under management of national parks have increased from 27,458 ha in 2003 to 33,504 hectares in 2008. Considering the land use type of protected areas, 46.7% of nationally protected areas are covered by forests. According to a scientific estimate 37% of the forests are considered seminatural. The ratio of indigenous tree stocks exceeds 57%, while non-native species (black locust, red oak, pine) trees grow on 23%, and poplar clones on 6.9%. 63.5% of forests have primarily an economic function, while 35.2% have protective functions. The role of the remaining 1.3% is health-care, tourism, education and research.

Forest reserves are protected forest areas, where human activities are permanently prohibited in order to let natural processes dominate in the long-term and to provide sites for research. The national network of forest-reserves currently comprises 63 reserves. Therefore, the total area of the network increased from 9,730 hectares in 2000 to 13,000 hectares in 2008. Researches in several reserves were conducted on treestand structure, dynamics and forest-ecology, soil science, botany and methodology. According to a recent study comparing protected and nonprotected forests with natural tree composition in mountain areas of Hungary, the naturalness of protected forest is slightly higher than that of unprotected ones. The national average of naturalness in 'forest stands with native tree composition' is 58.5%; that of 'forest stands with site-alien species' is 53.5% and in case of 'forest stands with nonindigenous species' it is 40.4%. Threats to forest biodiversity are the high proportion of clear-cutting in private and protected stands, dead wood removal from the forests, overpopulation of game animals at certain areas and spread of alien tree species, conservation measures are difficult to implement at certain privately owned forests.

5.7.1 Grasslands, Overview of Status of Biodiversity and Trends

Hungarian grasslands have outstanding nature conservation importance in European scale. Most of them are secondary grasslands, which means that they were formed in the past few thousand years due to the certain land use methods. Their maintenance and the preservation of their rich biodiversity depend on the human use in the future too. Approximately 10.8% of the country's territory, i.e., 1,000,000 ha is covered by grasslands. This includes: (i) colline and montane hay meadows, acid grasslands and heaths, (ii) halophytic habitats, (iii) dry open grasslands, (iv) dry and semidry closed grasslands, (v) nonruderal pioneer habitats, and (vi) other nonwoody habitats.

A strictly protected grassland species with high conservation value is the Hungarian meadow viper (*Vipera ursinii rakosiensis*) which occurs only in Hungary. This small viper disappeared from most of its known range during the last decades. Most of the species' grassland habitats were plowed, and the remaining most ones were mowed intensively. The survived small and isolated populations became vulnerable and local damages can fully destroy them. Recently, it is the most endangered member of the Hungarian fauna, as estimations put its numbers under 500 individuals. A special conservation program is running to save it since 1993.

Populations of European ground squirrels (*Citellus citellus*), living on grasslands and airports are monitored by the Hungarian Biodiversity Monitoring System. No significant temporal changes have been detected in the period 2000–2008, although at some sites squirrels have disappeared in the last years. Hungarian great bustard (*Otis tarda*) is another typical species of agricultural areas and grasslands, its stand was slightly growing during the last period.

Botanical surveys of seven rocky grassland communities in the Transdanubian and Northern Mountains were recorded in the 1930–60s and reinvestigated in 1991–94. Significant changes in species composition have been detected: the frequency of rocky grassland specialists and generalists has decreased, probably due to the overpopulated game stand, especially of wild boar. Species adapted to extreme water and nitrogen limitation have become less frequent, while the frequency of ephemeral plant species has increased. On the other hand, the communities have conserved their characteristic species composition despite the observed changes.

Grasslands are often abandoned, which leads to the spontaneous spread of shrubs and/or invasive alien species, spontaneous afforestation and accumulation of household wastes. Improper management methods, as well as human-induced damages (e.g., offroad driving, motocross and quadruped riding, trampling), may also negatively affect biodiversity.

5.8 Inland Waters

5.8.1 Overview of Status of Biodiversity and Trends

Hungary, situated in the heart of the Danube Basin, has considerable surface water resources 95% of which comes from abroad. The country's borders are crossed by 24 incoming rivers, bringing in 112 billion m³ of water per year. As the difference between domestic waterfall and evaporation is 6 billion m³/

year, the total stock is 118 billion m³/year. The Hungarian Danube traverses 417 km, forming the border with Slovakia in the northwest and thereafter flowing south. In the east, also flowing southwards is the Tisza, covering 595 km before reaching Serbia and Montenegro where it later flows into the Danube. About 25% of the country is comprised of floodplains and 25% of the population is living in reclaimed floodplains.

Of 876 natural and 150 artificial water bodies identified in Hungary, 579 freshwater surface bodies (56%) have been classified as being 'at risk' from organic, nutrient or priority hazardous substances (according to the EU Water Framework Directive definitions).

Approximately, 70% of artificial lakes (mainly fishponds) are 'at risk' from organic and nutrient loads. None of the 108 groundwater bodies identified are considered to be 'at risk' due to human intervention, but 46 sites are listed as 'possibly at risk' (mostly from nitrate pollution from diffuse sources).

Twenty-eight Hungarian wetland habitats are listed on the Ramsar List with a total area of over 233,000 ha. The Ramsar sites include all characteristic types of wetland areas in the Carpathian Basin: lakes, marshes, alkaline lakes, bogs, backwaters, river stretches, wet meadows, man-made fish farms and reservoirs. Certain sites were qualified of international importance (e.g., Hortobágy, Kardoskúti Fehértó, Lake Fertő and Gemenc), while other Ramsar sites in Hungary also meet at least two criteria. Since 2005, five new Ramsar sites and two extensions have been declared, with a total area of over 53,000 ha and an already existing Ramsar site was declared a transboundary site with Slovakia (Ipel/Ipoly).

The ecological condition of the big lakes (Balaton, Velencei, and Fertő) has been improved due to smaller nutrient loads as a result of the drastic drop in fertilizers' use, as well as governmental measures and environmental investment programs. For other lakes, the eutrophication rate has generally been decreasing, but the threat still remains.

Pannonian salt steppes and salt marshes occur only in a few countries of the European Union; the largest surface area and the center of distribution of this habitat type are in Hungary. The enrichment of salt in the soil of these habitats is due to high evaporation of groundwater during summer. There is a characteristic zonation of vegetation, based on inland flooding regime, with dominant salt-tolerant grasses and herbs that tolerate or even demand salt concentrations in the soil water. Compared with other salt lakes and marshes of the world, the alkaline lakes of the Carpathian Basin are characterized by lower salt content but higher alkalinity. Due to limited geographical distribution, they belong to the most threatened European communities. Many Pannonic salt steppes and salt marshes were totally destroyed for agricultural purposes. Those remained are threatened by the impact of eutrophication and lack of management, as well as by water management. The sinking of water table connected with river regulations and building of canals had a negative impact on those ecosystems. Primary alkali steppes do not need any active management. Grasslands are relatively fragile and can only stand extensive grazing.

One of the threats to inland biodiversity is habitat loss due to the conversion of waterbed and shoreline, which frequently happened in the past. Other threats to inland waters include inadequate water quantity and nonnatural water dynamics due to uncontrolled use of surface and underground water resources, problems with certain water power plants (Bős-Nagymaros Water Power Plant), etc. Pannonian salt steppes and salt marshes are threatened by agriculture (impact of eutrophication and lack of management) as well as by water management. Other difficulties include removal of perverse incentives and legislation supporting the cultivation of regularly flooded areas.

Keywords

- agro-biodiversity
- habitat types
- inland waters
- Pannonian region

References

Báldi, A., & Batáry, P., (2011). The past and future of farmland birds in Hungary. *Bird Study*, *58*, 365–377.

Báldi, A., & Faragó, S., (2007). Long-term changes of farmland game populations in a post-socialist country (Hungary). *Agriculture, Ecosystems and Environment*, *118*, 307–311.

Bennett, K. D., Tzedakis, P. C., & Willis, K. J., (1991). Quaternary refugia of north European trees. *J. Biogeography*, *18*, 103–115.

Berkes, F., (2016). Sustainability policy considerations for ecosystem management in Central and Eastern Europe. *Ecosystem Health and Sustainability*, *2*(8), e01234. http://dx.doi.org/10.1002/ehs2.1234.

Bohn, U., Gollub, G., & Hettwer, C., (2000). *Karte Der Natürlichen Vegetation Europas*, *1*(2), 500, 000. Bundesamt für Naturschutz, Bonn-Bad Godesberg.

Borhidi, A., & Kevey, B., (1996). An annotated checklist of the Hungarian plant communities. In: *Critical Revision of the Hungarian Plant Communities*, Borhidi, A., (ed.). Janus Pannonius Univ. Pécs, pp. 138.

Borhidi, A., (1961). Klimadiagram und klimazonale Karte Ungarns. *Annales Universitatis Budapest, Ser. Biol.*, *4*, 21–50.

Bunje, P. M. E., & Lindbergh, D. R., (2007). Lineage divergence of a freshwater snail clade associated with post-Tethys marine basin development. *Molecular Phylogenetics and Evolution*, *42*, 373–387.

Burger, A., (2001). Agricultural development and land concentration in a central European country: A case study of Hungary. *Land Use Policy*, *18*, 259–268.

Chytrý, M., & Sádlo, J., (1997). *Tilia*-dominated calcicolous forests in the Czech Republic from a Central European perspective. *Annali di Botanica*, *55*, 105–126.

Chytrý, M., Danihelka, J., Horsák, M., Kočí, M., Kubešová, S., Lososová, Z., et al., (2010). Modern analogs from the Southern Urals provide insight into biodiversity changes in the early Holocene forests of Central Europe. *J. Biogeography*, *37*, 767–780.

Csuzdi, C., & Pop, V. V., (2007). The Lumbricids of the Carpathian basin. In: A. *Kárpát-medence Faunájának Eredete. The Origin of the Fauna of the Carpathian Basin in Hung*, Forró, L., (ed.). Magyar Természettudományi Múzeum, Budapest, pp. 13–20.

Czúcz, B., Molnár Z., Horváth, F., Nagy, G. G., Botta-Dukát, Z., & Török, K., (2008). Using the natural capital index framework as a scalable aggregation methodology for local and regional biodiversity indicators. *J. Nature Conservation*, *20*(3), 144–152.

Dányi, L., & Traser, G., (2007). Magyarország ugróvillásai. (The springtails of Hungary) In: *A Kárpát-Medence Faunájának Eredete*, Forró, L., (ed.). (The Origin of the Fauna of the Carpathian Basin, in Hung.). Magyar Természettudományi Múzeum, Budapest. pp. 21–28.

Deli, T., Sümegi, P., & Kiss, J., (1997). Biogeographical characterisation of the Mollusc fauna on Szatmár-Bereg Plain. In: *ANP Füzetek Aggtelek, vol. I*, Tóth, E., & Horváth, R., (eds.). Aggtelek, Hungary, pp. 123–129.

Feher, Z., Varga, A., Deli, T., Domokos, T., Szabó, K., Bozsó, M., & Pénzes Z., (2007). Védett puhatestűek filogenetikai vizsgálata. (Phylogenetic survey of some protected Molluscs) In: *A Kárpát-Medence Faunájának Eredete*, Forro, L., (ed.). (The Origin of the Fauna of the Carpathian Basin, in Hung.). Magyar Természettudományi Múzeum, Budapest, pp. 183–200.

Fekete, G., & Varga, Z., (2006). *Magyarország Tájainak Növényzete és Állatvilága* (Vegetation and fauna of landscapes in Hungary, in Hung.). *MTA Társadalomkutató Központ*, Budapest.

Fekete, G., & Zólyomi, B., (1966). Über die Vegetationszonen und pflanzengeographische Characteristik des Bakony-Gebirges. *Annales Historiconaturalis Musei Nationalis Hungariae*, *58*, 197–205.

Fekete, G., (1965). *Die Vegetation im Gödöllőer Hügelland*. Akadémiai Kiadó, Budapest.

Fekete, G., (1992). The holistic view of succession reconsidered. *Coenoses*, *7*, 21–29.

Fekete, G., Király, G., & Molnár, Z., (2016). Delineation of the Pannonian vegetation region. *Community Ecology*, *17*, 114–124. http://dx.doi.org/10.1556/168.2016.17.1.14.

Fekete, G., Molnár, Z., Magyari, E., Somodi, I., & Varga, Z., (2014). A new framework for understanding Pannonian vegetation patterns: Regularities, deviations and uniqueness. *Community Ecology*, *15*, 12–26.

Fekete, G., Somodi, I., & Molnár Z., (2010). Is chorological symmetry observable within the forest steppe biome in Hungary? A demonstrative analysis of floristic data. *Community Ecology*, *11*, 140–147.

Haraszthy, L., (2014). *Natura 2000 fajok és élőhelyek Magyarországon*. (Natura 2000 species and habitats in Hungary, in Hung.) – *Pro Vértes Közalapítvány*, Csákvár.

Henle, K., Alard, D., Clitherow, J., Cobb, P., Firbank, L., Kull, T., McCracken, D., Moritz, R. F., Niemelä, J., Rebane, M., & Wascher, D., (2008). Identifying and managing the conflicts between agriculture and biodiversity conservation in Europe–A review, *Agriculture, Ecosystems and Environment, 124*, 60–71. http://dx. doi.org/10.1016/j. agee.2007.09.005.

Horvat, I., Glavac, V., & Ellenberg, H., (1974). *Die Vegetation Südosteuropas*. G. Fischer Verlag.

Isépy, I., (1970). Zönologische Verhältnisse der *Primula auricula* L. ssp. *hungarica* (Borbás) Soó in Ungarn. Annales Universitatis Scientiarum Budapest. *Rolando Eötvös Nom.*, Sect. Biol., *12*, 133–141.

Jakucs, P., (1961). *Die Phytozönologischen Verhältnisse der Flaumeichen-Buschwälder Südöst-Mitteleuropas*. Akadémiai Kiadó, Budapest.

Juhász, I., & Szegvári, G., (2007). The pollen sequence from Mezőlak. In: *Environmental Archaeology in Transdanubia*, Zatykó, C., Juhász, I., & Sümegi, P., (eds.). Varia Archaeologica Hungarica, Budapest, pp. 316–325.

Kasy, F., (1965). Zur Kenntnis der Schmetterlingsfauna des östlichen Neusiedlersee-Gebietes. – *Wissenschaftliche Arbeiten von Burgenland* (Eisenstadt), *34*, 75–211.

Kasy, F., (1981). Naturschutzgebiete im östlichen Ö sterreich als Refugien bemerkenswerter Lepidopteren-Arten. Beih. *Veröff. Natursch., Landsch. Baden-Württ., 21*, 109–120.

Király, G., (2007). *Vörös Lista. A Magyarországi Edényes Flóra Veszélyeztetett Fajai.* (Red list of the vascular flora of Hungary). Private edition, Sopron.

Kis, B., (1965). *Zubovskia Banatica*, eine neue Orthoptera-Art aus Rumänien. *Reichenbachia, Abhandlungen des Museums für Tierkunde Dresden, 5*, 5–8.

Kis, B., (1980). Die endemischen Orthopteren in der Fauna von Rumänien. *Muzeul Brukenthal, Studii si Comunicare, 24*, 421–431.

Korsós, Z., (2007). A magyarországi hüllőfauna története a jégkorszak után (The history of the Reptile fauna in Hungary after the Last Glaciation, in Hung.). In: *A Kárpát-Medence Faunájának Eredete,* Forró, L., (ed.). (The Origin of the Fauna of the Carpathian Basin, in Hung.). Magyar Természettudományi Múzeum, Budapest, pp. 283–296.

Kovács, E., Bela, G., & Kiss, D., (2014). *A Nemzeti Biodiverzitás Stratégia Hatásvizsgálata.* (Impact assessment of the National Biodiversity strategy, in Hung.). Environmental Social Science Research Group, Budapest.

Krolopp, E., & Sümegi, P., (1995). Palaeoecological reconstruction of the late Pleistocene, based on loess malacofauna in Hungary. – *GeoJournal, 36*(2–3), 213–222.

Lang, G., (1994). *Quartäre Vegetationsgeschichte Europas*. Methoden und ergebnisse. Gustav Fischer Verlag, Jena.

Liira, J., Aavik, T., Parrest, O., & Zobel, M., (2008). Agricultural sector, rural environment and biodiversity in the Central and Eastern European EU member states. In: *AGD Landscape & Environment 2. Landscapes and Landforms of Hungary,* Lóczy, D., (ed.). Springer, pp. 46–64.

Lóczy, D., (2015). *Landscapes and Landforms of Hungary*. Springer, Berlin-New York.

Magri, D., Vendramin, G. G., Comps, B., Dupanloup, I., Geburek, T., Gömöry, D., Latalowa, M., Litt, T., Paule, L., Roure, J. M., Tantau, I., Van der Knaap, W. O., Petit, R. J., & Beaulieu, J. L., (2006). A new scenario for the Quaternary history of European beech populations: Palaeobotanical evidence and genetic consequences. *New Phytologist, 171*, 199–221.

Magyari, E. K., Chapman, J. C., Passmore, D. G., Allen, J. R. M., Huntley, J. P., & Huntley, B., (2010). Holocene persistence of wooded steppe in the northern Great Hungarian Plain. *Journal of Biogeography*, *37*, 915–935.

Magyari, E. K., Jakab, G., Sümegi, P., & Szöőr, G., (2008). Holocene vegetation dynamics in the Bereg Plain, NE Hungary – The Bábtava pollen and plant macrofossil record. *Acta GGM Debrecina, Geology, Geomorphology, Physical Geography Series*, *3*, 33–50.

Magyari, E. K., Kuneš, P., Jakab, G., Sümegi, P., Pelánková, B., Schäbitz, F., et al., (2014). Last glacial maximum vegetation in East Central Europe: Are there true analogs in Siberia? *Quaternary Science Reviews*, *95*, 60–79. doi:10.1016/j.quascirev.2014.04.020.

Magyari, E., (2002). The Holocene expansion of beech (*Fagus sylvatica* L.) and hornbeam (*Carpinus betulus* L.) in the eastern Carpathian basin. *Folia HistoricoNaturalia Musei Matraensis*, *26*, 15–35.

Magyari, E., Jakab, G., Rudner, E., & Sümegi, P., (1999). Palynological and plant macrofossil data on Late Pleistocene short-term climatic oscillations in NE-Hungary. *Acta Palaeobotanica*, *2*, 491–502.

Mahunka, S., (1993). *Hungaromotrichus baloghi* gen. et sp. n. (Acari: Oribatida), and some suggestions to the faunagenesis of the Carpathian basin. *Folia Entomologica Hungarica*, *54*, 75–83.

Mahunka, S., (2007). A Kárpát-medence páncélosatkái. In: *A Kárpát-Medence Faunájának Eredete*. (The Origin of the Fauna of the Carpathian Basin, in Hung.), Forró, L., (szerk.). Magyar Természettudományi Múzeum, Budapest, pp. 37–44.

Mihók, B., Biró, M., Molnár, Z., Kovács, E., Bölöni, J., Erős, T., Standovár, T., Török, P., Csorba, G., Margóczi, K., & Báldi, A., (2017). Biodiversity on the waves of history: Conservation in a changing social and institutional environment in Hungary, a postsoviet EU member state. *Biological Conservation*, *211*, 67–75.

Németh, F., (1989). *Száras növények*. (Vascular plants). In: *Vörös Könyv*, Rakonczay, Zoltán, (ed.). *A Magyarországon kipusztult és veszélyeztetett állat- és növényfajok*. [Red Data Book. Extinct and threatened animal and plant species of Hungary]. Akadémiai Kiadó, Budapest, pp. 265–325.

Orci, K. M., Nagy, B., Szövényi, G., Rácz, I. A., & Varga, Z., (2005). A comparative study on the song and morphology of *Isophya stysi* Cejchan, (1958), *I. modestior* Brunner von Wattenwyl, 1882. *Zoologische Anzeiger*, *244*, 31–42.

Pócs, T., (1960). Die zonalen Waldgesellschaften Südwestungarns. *Acta Botanica Acad. Sc. Hung.*, *6*, 75–105.

Pócs, T., Domokos-Nagy, É., Pócs-Gelencsér, I., & Vida, G., (1958). *Vegetationsstudien im Őrség*. Akadémiai Kiadó, Budapest.

Potori, N., Kovács, M., & Vásáry, V., (2013). The common agricultural policy 2014–2020: An impact assessment of the new system of direct payments in Hungary. *Studia in Agricultural Economy*, *115*, 118–123.

Rudner, E., & Sümegi, P., (2001). Recurring tajga forest steppe habitats in the Carpathian Basin in the Upper Weichselian. *Quaternary International*, *76–77*, 177–189.

Schmitt, T., & Varga, Z., (2012). Extra-Mediterranean refugia: The rule and not the exception? *Frontiers in Zoology*, *9*, 22 doi: 10.1186/1742–9994-9-22.

Simon, T., (2000). A magyarországi edényes flóra határozója. [Taxonomic handbook of the Hungarian vascular flora.] – Tankönyvkiadó, Budapest.

Soó, R., *A magyar flóra és vegetáció rendszertani-növényföldrajzi kézikönyve* (*Manual of the Hungarian Flora and Vegetation, in Hung.*) I. 1964, 590 pp, II. 1966, 655 pp., III. 1968, 506 pp., IV. 1970, 614 pp., V. 1973, 724 pp., VI. 1980, 557 pp.

Soós, L., (1943). *A Kárpát-medence Mollusca-faunája.* (The Mollusc funa of the Carpathian basin, in Hung,) Magyar Királyi Természettudományi Társulat, Budapest.

Standovár, T., & Primack, R. B., (2001). A természetvédelmi biológia alapjai. (*Foundations of Conservation Biology, in Hung.*) Nemzeti Tankönyvkiadó, Budapest, pp. 17–34.

Stoate, C., Báldi, A., Beja, P., Boatman, N. D., Herzon, I., Van Doorn, A., et al., (2009). Ecological impacts of early twenty-first century agricultural change in Europe – a review. *J. Environmental Management, 91*, 22–46. http://dx. doi. org/10.1016/j.jenvman.2009.07.005.

Sümegi, P., (2007). The vegetation history of the Mezőlak area. In: *Environmental Archeology in Transdanubia,* Zatykó, C., Juhász, I., & Sümegi, P. Varia Archeologica Hungarica, *20,* 329–330.

Sümegi, P., Bodor, E., & Törőcsik, T., (2005). The origins of alkalisation in the Hortobágy region in the light of the palaeoenvironmental studies at Zám-Halasfenék. In: *Environmental Archaeology in Northeastern Hungary,* Gál, E., Juhász, I., & Sümegi, P., (eds.). Varia Archaeologica Hungarica XIX. Archaeological Institute of the Hungarian Academy of Sciences, Budapest, pp. 115–126.

Sümegi, P., Hertelendi, E., Magyari, E., & Molnár, M., (1998). Evolution of the environment in the Carpathian Basin during the last **30,000** BE years and its effects on the ancient habits of the different cultures. In: *Archimetrical Research in Hungary,* Költő, L., & Bartosewicz, L., (eds.). Budapest, pp. 183–197.

Sümegi, P., Molnár, A., & Szilágyi, G., (2000). Szikesedés a Hortobágyon. (Salinisation in Hortobágy, in Hung.) *Természet Világa, 131,* 213–216.

Sutcliffe, L. M. E., Batáry, P., et al., (2015). Harnessing the biodiversity value of Central and Eastern European farmland. *Diversity and Distributions, 21,* 722–730.

Varga, Z., & Rákosy, L., (2007). Biodiversität der Karstgebiete im Karpatenbecken am Beispiel der Gross-Schmetterlingsfauna der Turzii-Schlucht bzw. des Aggteleker Karstgebietes. *Entomologica Romanica, 12,* 15–29.

Varga, Z., (1989). Die Waldsteppen des pannonischen Raumes aus biogeographischer Sicht. *Düsseldorfer Geobotanische Kolloquien, 6,* 35–50.

Varga, Z., (1995). Geographical patterns of biological diversity in the Palaearctic region and the Carpathian Basin. *Acta Zoologica Academiae Scientiarum Hungariae, 41,* 71–92.

Varga, Z., (2003). A Kárpát-medence állatföldrajza. (The zoogeography of the Carpathian basin, in Hung.) In: *Flóra, Fauna, Élőhelyek,* Láng, I., Bedő, Z., & Csete, L., (eds.). (Flora, Fauna, Habitats) Magyar Tudománytár III, MTA Társadalomkutató Központ, Budapest, pp. 89–119.

Varga, Z., (2011). Extra-Mediterranean refugia, postglacial vegetation history and area dynamics in Eastern Central Europe. In: *Survival on Changing Climate – Phylogeography and Conservation of Relict Species,* Habel, J. C., & Assmann, T., (eds.). Springer, Heidelberg, pp. 51–87.

Varga, Z., (2014). Biogeography of the high mountain Lepidoptera in the Balkan Peninsula. *Ecologia Montenegrina, 1*(3), 140–168.

Weising, K., & Freitag, H., (2007). Phylogeography of halophytes from European coastal and inland habitats. *Zoologischer Anzeiger, 246,* 279–292.

Wesche, K., Ambarli, D., Kamp, J., Török, P., Treiber, J., & Dengler, J., (2016). The Palaearctic steppe biome: A new synthesis. *Biodiversity and Conservation*, 25, 2197–2231.

Williams, P., Humphries, C., & Araujo, M., (1999). Mapping Europe's Biodiversity. In: *Facts and Figures on Europe's Biodiversity. State and Trends 1989–1999*, Delbaere, (ed.). European Centre for Nature Conservation, Tilburg, pp. 12–20.

Willis, K. J., & Van Andel, T. H., (2004). Trees or no trees? The environments of central and eastern Europe during the Last Glaciation. *Quaternary Science Reviews*, 23, 2369–2387.

Willis, K. J., Braun, M., Sümegi, P., & Tóth, A., (1997). Does soil change cause vegetation change or vice versa? A temporal perspective from Hungary. *Ecology*, 78, 740–750.

Willis, K. J., Sümegi, P., Braun, M., & Tóth, A., (1995). The late quaternary environmental history of Bátorliget, N. E. Hungary. *Palaeography, Palaeoclimatology, Paleoecology*, 118, 25–47.

Young, J., Richards, C., Fischer, A., Halada, L., Kull, T., Kuzniar, A., Tartes, U., Uzunov, Y., & Watt, A., (2007). Conflicts between biodiversity conservation and human activities in the central and eastern European countries. *AMBIO*, 36, 545–550. http://dx.doi.org/10.1579/0044-447(2007)36%5b545:CBBCAH%5d2.0.CO.2.

Zólyomi, B., (1942). A középdunai flóraválasztó és a dolomitjelenség. (The central Danube floristic gap and the dolomit-phenomenon, in Hung.) *Botanikai Közlemények, 39,* 209–231.

Zólyomi, B., (1958). Budapest és környékének természetes növénytakarója. (Natural vegetation of the environment of Budapest, in Hung.) In: *Budapest Természeti Képe (The Natural Landscape of Budapest, in Hung.)*, Pécsi, M., Marosi, S., & Szilárd, J., (eds.). Akadémiai Kiadó, Budapest. pp. 509–642.

Zólyomi, B., (1967). Tilio-Fraxinetum excelsioris. In: *Guide der Exkursionen des Internationalen Geobotanischen Symposiums*, Zólyomi, B., (ed.). pp. 36–38. Eger-Vácrátót.

Zólyomi, B., (1989). Természetes növénytakaró (Natural vegetation, in Hung.) 1: 1,500,000. In: *Magyarország Nemzeti Atlasza*, Pécsi, M., (ed.). Kartográfiai Vállalat, Budapest, *89.*

Zólyomi, B., Kéri, M., & Horváth, F., (1992). A szubmediterrán éghajlati hatások jelentősége a Kárpát-medence klímazonális növénytársulásainak összetételére. (The importance of the sub-Mediterranean climatic influences on the composition of the climate-zonal plant communities of the Carpathian basin, in Hung.) In: *Hegyfoki Kabos Klimatológus Születésének 145. évfordulója alkalmából rendezett tudományos emlékülés előadásai, (Lectures on honor of the 145th anniversary of birth of the climatologist Hegyfoki Kabos)* Debrecen-Túrkeve., pp. 60–74.

Landscapes, Habitats

Photo L1   Dolomitic hills of Vértes Mts. (Transdanubia) – open dolomitic grasslands and xeric scrub forests of *Quercus pubescens* and *Fraxinus ornus*.

Photo L2   Mosaic of *Festuca dalmatica* calcareous grassland and scarce *Fraxinus ornus* stand on the Szársomlyó hill in Southern Transdanubia. Important species: *Dictamnus albus, Vincetoxicum pannonicum* (endemic).

Photo L3 Typical Pannonian forest-steppe mosaic of *Stipa joannis–Festuca rupicola* grassland and *Quercus pubescens* shrub forest, with forest-steppe fringe of *Dracocephalum austriacum* and *Euphorbia polychroma* (Aggteleki National Park, North Hungary).

Photo L4 Alkaline grassland with short-grass turf of *Festuca pseudovina* and bottomland with ephemeric *Camphorosma annua* in the Hortobágy National Park (Eastern Hungary).

Photo L5 Junipero-Populetum shrub forest and open calcareous sand grassland with dominant *Festuca vaginata* in the Kiskunság National Park (Central Hungary).

Photo L6 Tall-grass *Molinia* meadow with orchids *Gymnadenia conopsea* in the Kiskunság National Park (Central Hungary).

Photo L7 Peucedano-Asteretum sedifolii, endemic tall grass-tall forb association of the Pannonian lowland (East Hungary, Bihari lowland, protected landscape area).

Photo L8 Mosaic of marshy meadow (Molinietum), willow shrub (*Salicetum cinereae*) and hardwood gallery forest in West Hungary (Őrség National Park).

Photo L9 Wetland associations of *Trapa natans, Scirpus lacustris,* Magnocaricion, Willow shrub and softwood gallery forest in West Transdanubia.

Photo L10 Inundated oak-ash-elm (*Quercus robur, Fraxinus angustifolia, Ulmus minor*) hardwood gallery forest in Northeast Hungary (Bátorliget nature reserve).

Photo L11   Calcareous ravine forest of *Fagus sylvatica, Acer pseudoplatanus, Fraxinus excelsior*, etc. in the Bükk National Park (North Hungary).

Photo L12   Floristic diversity of a traditionally managed semidry grassland with *Salvia pratensis, Anthyllis vulneraria, Onobrychis arenaria, Trifolium* spp., etc.

Plant Species

Photo P1   *Diphasium complanatum*, rare northern species of Lycopsida in Luzulo-Fagetum forests.

Photo P2   Glacial relict species *Allium victorialis* in beech forest in Bakony Mts. (Transdanubia).

Photo P3   *Iris aphylla* subsp. *hungarica*, endemic Pannonian subspecies of the Ponto-Pannonian species, typical species of forest fringe habitats (Bátorliget nature reserve, Northeast Hungary).

Photo P4   *Himantoglossum jankae*, endemic orchid species of the Pannonian region, typical for Transdanubian forest-steppe.

Photo P5 *Pulsatilla patens*, rare continental steppic species of limeless sandy grasslands.

Photo P6 *Astragalus dasyanthus*, Ponto-Pannonian steppic species with local occurrences in loess and sand grasslands, species of Community Interest in the European Union (Tokaj hill protected landscape region).

Photo P7 *Dracocephalum austriacum*, Ponto-Pannonian steppic species with local occurrences in North Hungary, southern Slovakia and in the **Transylvanian** Basin (Romania), species of Community Interest in the European Union (Aggteleki National Park, North Hungary).

Photo P8 *Echium maculatum,* Ponto-Pannonian steppic species with local occurrences in loess and sand grasslands, species of Community Interest in the European Union (Aggteleki National Park, North Hungary).

Animal Species

Photo A1   *Pholidoptera transsylvanica*, endemic bush-cricket of the eastern part of the Carpathian basin, species of Community Interest in the European Union (Aggteleki National Park, North Hungary).

Photo A2   *Saga pedo*, the largest, predatory bush-cricket of Central Europe, typical steppic species, species of Community Interest in the European Union (Aggteleki National Park, North Hungary).

Photo A3 *Parnassius mnemosyne*, the Clouded Apollo, flagship species of Pannonian hardwood forests, species of Community Interest in the European Union (Bükk National Park).

Photo A4 *Araneus grossus*, one of the largest spiders in Hungary, inhabitant of xerothermic *Quercus pubescens* forests.

Photo A5 *Zootoca vivipara pannonica,* endemic subspecies of the boreo-montane species.

Photo A6 *Glareola pratincola,* typical shore-bird of the alkaline bottomland habitats.

Photo A7   *Otis tarda,* the great bustard, the largest bird of Hungary, one of typical steppic species.

Photo A8   *Nannospalax hungaricus,* endemic subterranean rodent of the Pannonian lowland.

CHAPTER 6

Biodiversity in Italy

GIANNIANTONIO DOMINA[1] and MARZIO ZAPPAROLI[2]

[1]Department of Agricultural, Food and Forest Sciences, University of Palermo, viale delle Scienze, Bldg. 4, 90128, Italy, E-mail: giannantonio.domina@unipa.it

[2]Department for Innovation in Biological, Agrofood and Forest systems, University of Tuscia, Via San Camillo de Lellis s.n.c. – 01100 Viterbo, Italy, E-mail: zapparol@unitus.it

6.1 Introduction

Italy is a country with a long-standing tradition in studies of Natural Sciences. In this country, the first botanical gardens of the whole world have been founded in Pisa and Padua; Genoa, Florence, Naples host natural history museums and research institutions among the oldest in Europe. These institutions are today still active in biodiversity research, together with many University departments. Italy boasts of having hosted hundreds of native and foreign scholars, who, with a peak of activity in the nineteenth century, left a heritage of great value for future generations. After a break due to the war events of the beginning of the twentieth century, the research on the territory has been resumed with increasing dynamism. In this way, the whole national territory is well known. Every year, new species of plants and insects are described. This testifies the continuity of fieldwork carried out by professionals and amateurs who often turn to research facilities to benefit of the support the modern genetic investigation methodologies.

Italy, with an area of 301,338 km², is a peninsula, about 1,000 km long and 170 km wide, with northwest-southeast orientation, at the center of the Mediterranean Basin, between 47° 05' and 35° 29' North latitude, separated from the bordering states by the mountain range of the Alps. It includes two large islands, Sardinia and Sicily and more than 800 smaller islets. Geographically the Italian territory is predominantly hilly (41.6%), partly mountainous (35.2%) and, to a lesser extent, flat (23.2%). The extension of coastline is of 7,458 km. The highest mountains reach an altitude of 4,810 m a.s.l. in the Alps, which stretch across northern Italy and form Italy's northern border, and 2,912 m a.s.l. in the Apennines, a chain of mountains that runs south along the whole peninsula.

Italy is administratively divided into 20 regions and has a population of almost 61 million inhabitants predominantly distributed along the coasts and flat areas.

At present, the level of knowledge of biodiversity in Italy is very high, with more detail for the areas with higher biological richness, islands, and mountains. This base overview comes from the collation made, in the last two decades, by the Ministry for the Environment and the Sea Protection in collaboration with the Italian Botanical Society, together with many University departments, and the Italian universities that started new data collection campaigns.

6.2 National Biodiversity Strategy

The approaches and methods that have been adopted in Italy, as well as the ongoing results, provide a clear example of how the EU biodiversity strategy can be implemented at the national level (MATTM, 2010) and have promoted a constructive scientific and technical debate with other EU member states (http://biodiversity.europa.eu/maes/maescatalog-of-case-studies).

The main base studies, prepared in the last years, for the assessment of Biodiversity in Italy are:

- The Map of Marine coastal biocenosis in scale 1:250,000 (http://www. va.minambiente.it/);
- Italic 5.0: The information system on Italian lichens (http://dryades. units.it/italic/index.php);
- The Checklist of the species of the Italian fauna (Minelli et al., 1993–1995);
- The annotated Checklist of the Italian vascular flora (Conti et al., 2005). This is being followed by a substantial update that includes all the taxonomic, nomenclatural and distributive novelties published during the last 13 years (Bartolucci et al., 2018; Galasso et al., 2018);
- The Map CORINE Land Cover, level III, in scale 1:100,000 (APAT, 2005);
- The Checklist and distribution of the Italian fauna (Ruffo and Stoch, 2006a);
- The Phytoclimatic Map of Italy (1:250,000) (see Blasi and Michetti, 2007);
- The Second National Inventory of Forest and Forest Carbon Pools (IFN, 2007). This is being followed by the third national inventory, done in 2015, but the data are not yet available;
- The checklist of the flora and fauna of the Italian seas (Relini, 2008, 2010);
- The Map of Vegetation Series of Italy in scale 1:500,000 (Blasi, 2010);

- The report about the species and habitat of community interest in Italy (Genovesi et al., 2014).

Some of these data are freely available in digital format, some other have limitations in the analysis that can be done online. A meticulous work of data mining was needed for the preparation of this contribution to homogenize data through new exemplifications.

Also worth mentioning is the "Fauna d' Italia" series, published under the auspices of the "Italian National Academy of Entomology" and the "Italian Zoological Union" (Ruffo and Vigna Taglianti, 2004a). The first volume was published in 1956 and so far (2018) 51 volumes have been issued and many are in preparations.

Recently, the assessments of extinction risk based on the IUCN Red List Categories and Criteria of the Italian vascular flora and some animal groups have been published by the Italian Ministry of Environment, IUCN Italian Committee and Federparchi (http://www.iucn.it/liste-rosse-italiane.php).

6.3 Biogeographic Zones

The great altitudinal variation of the Italian peninsula and the complex orographic system generate a great variety of climatic regions and bioclimates (Blasi et al., 2007). The specialization of thermopluviometric data from 400 metereological stations resulted in two climatic regions (temperate and Mediterranean), subdivided into nine bioclimates (Figure 6.1A). These bioclimates differentiate distinctly large regions that are homogeneous in physical features and climatic parameters: (a) the high mountains in the Alpine arc and Apennines; (b) the Po Valley; (c) the hill areas in the peninsula; (d) the hinterland of Sardinia and the Tyrrhenian coast of the peninsula; and (e) the coastal areas of Sardinia, coastal and inland Sicily and coastal and the flat inland areas in the southern peninsula. These areas are connected by transitional zones. Lower hierarchical levels show the complex pattern of continentality and thermo-pluviometric belts (Blasi and Michetti, 2007).

On the basis of the Bioclimate, Biogeography, Geomorphology, Vegetation and Land Cover, in accordance to what already done in several areas of the world, Italy, was classified in ecoregions in order to support environmental, social and economic strategies with a holistic approach (Blasi et al., 2014). This classification identified 2 divisions, 13 provinces, 33 sections, and about 80 subsections. It distinguishes distinctly the alpine range, the Po Plain valley, the Apennines, the Tyrrhenian coast of the peninsula, the Adriatic coast of the peninsula, Sardinia and Sicily (Figure 6.1B).

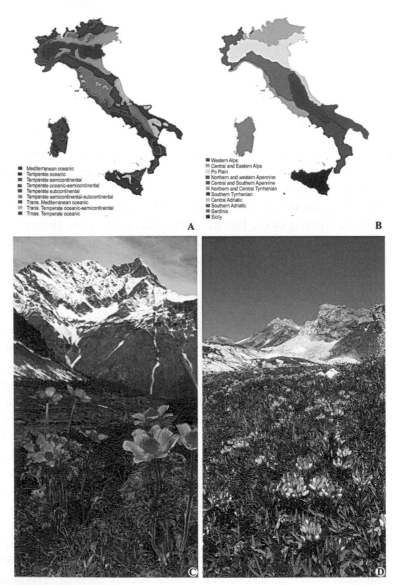

Figure 6.1   (A) Bioclimate map of Italy (redesigned with data from Blasi & al.
2007: 63); (B) Ecoregions of Italy (redesigned with data from Blasi &
al. 2014); (C) Valle d'Aosta, National Park "Gran Paradiso", *Pulsatilla
alpina* subsp. *alpiifolia* in the foreground and the "Grande Rousse" Mt.
in the background (photo by Maurizio Broglio); (D) Valle d'Aosta, Saint-
Rhémy, Alpine prairie with *Trifolium alpinum* (photo by Maurizio Broglio).

Figure 6.2 (A) Natural stand of *Pinus sylvestris* in Val di Cembra, Trentino-Alto Adige (photo by G. Domina); (B) The "Giara di Gesturi" one of the largest temporary wetland areas in the inner Sardinia (photo by G. Domina); (C) *Pancratium illyricum* in Sardinia where it occurs from the sea level to 1,350 m (photo by G. Domina); (D) Pine forest of *Pinus halepensis*, along the Gargano coast in Apulia (Photo by Robert P. Wagensommer); (E) *Erica forskalii* in Salento peninsula, Apulia (photo by R. P. Wagensommer); (F) Cushion-shaped vegetation dominated by *Astracantha sicula* on the Etna Mt. (photo by G. Domina); (G) The monotypic genus *Siculosciadium* including *Siculosciadium nebrodense*, endemic to the Madonie Mts. in Sicily (photo by Beppe di Gregorio); (H) *Agave sisalana* invasive alien species on the island of Marettimo in Sicily.

Figure 6.3 (A) Adult *Perlodes* sp., a stonefly genus (Insecta Plecoptera Perlodidae) inhabiting Italian Alpine ad Apennine rivers (Photo by Roberto Messori); (B) Adult male of *Oryctes (Oryctes) nasicornis* (Linnaeus), a rhinoceros beetle (Insecta Coleoptera Dynastidae) common in Italy; larva develops in soil under rotting woods (Photo by Massimo Ceci); (C) *Rana italica* Dubois, Italian stream frog (Amphibia Anura Ranidae), endemic to the Italian peninsula (Photo by Massimo Ceci); (D) *Podarcis muralis* (Laurenti), common wall lizard (Reptilia Squamata Lacertidae), a species largely spread in southern Europe; the subspecies *nigriventris* Bonaparte is endemic to peninsular Italy (Photo by Massimo Ceci); (E) *Austropotamobius italicus* (Chiesa et al.), Italian white-clawed crayfish (Crustacea Decapoda Astacidae); taxonomy of *A. pallipes* complex is still confusing, at present there is some consensus in considering the taxon as represented by a complex of two species, *A. pallipes* (European, in Italy only in Liguria) and *A. italicus* (endemic to the remaining areas of the Italian peninsula) (Photo by Massimo Ceci); (F) Male of *Oedemera nobilis* (Scopoli), false oil beetle (Insecta Coleoptera Oedemeridae), spread in Western and Southern Europe, very common in Italy and in other Mediterranean countries (Photo by Massimo Ceci); (G) *Hipparchias bordonii* Kudrna, the Ponza grayling (Insecta Lepidoptera Nymphalidae), a butterfly species with a very restricted range, endemic to Ponza Island, Tyrrhenian Sea, Italy (Photo by Valerio Sbordoni); (H) *Oxidus gracilis* C.L. Koch, the greenhouse millipede (Diplopoda Polydesmida Paradoxomatidae), an alien species to Italy (Photo by Massimiliano Di Giovanni).

According to Ruffo and Vigna Taglianti (2004b), 6 zoogeographic provinces should be identified in Italy (see also Minelli et al., 2006b) (Figure 6.3): the Alpine Province, including the Alps; the Po Province, represented by the Po Valley; the Apennine province, represented by the Italian peninsula with the Apennines as its central axis; the Apulian Province, including the Gargano, Murge and Salento areas; the Sicilian Province, Sicily and circum-Sicilian islands; the Sardinian Province, Sardinia and circum-Sardinian islands.

6.4 Biodiversity in Italy

Biodiversity hotspots are areas of exceptionally high biodiversity and rates of endemism. According to Médail and Quézel (1999), three areas of Italy fall within the 10 Mediterranean Basin hotspots. These are the Maritime and Ligurian Alps, Sardinia and Sicily. In addition to these hotspots that fall within the limits of the Mediterranean area, the whole Alpine range is another biodiversity hotspot. In this case, the endemic elements have not to

Figure 6.4 (A) *Sardaphenops supramontanus* Cerruti & Henrot, a polytypic species of highly specialized troglobitic ground beetles (Insecta Coleoptera Carabidae), belonging to a genus and a species endemic to Sardinia (Photo by Enrico Lana); (B) *Sabella spallanzanii* (Gmelin), the Mediterranean fanworm (Annelida Polychaeta Sabellidae), a best known and widely distributed species native to the northeastern Atlantic Ocean, the North Sea and the Mediterranean Sea, also introduced to various other parts of the world (photo by Caterina Longo); (C) *Sittilong longipes* (Canestrini), a jumping spider (Arachnida Aranea Salticidae) native to Alps, ranging in France, Switzerland, Austria and Northern Italy (Photo by Paolo Pantini); (D) *Lophophanes cristatus* (Linnaeus), European crested tit (Aves Passeriformes Paridae), European bird common resident in Italian Alpine and North-Apennine coniferous forests (Photo by Roberto Bianconi); (E) *Paramuricea clavata* (Risso), the violescent sea-whip (Anthozoa Alcyonacea Plexauridae), a typical gorgonian species of sublittoral communities of rocky substrates, endemic to Mediterranean Sea (photo by Frine Cardone); (F) *Rupicapra pyrenaica ornata* Neumann, the Apennine chamois (Mammalia Arctiodactyla Bovidae), a subspecies endemic to Italy, restricted to the Abruzzo, Majella, and Gran Sasso-Monti della Laga National Parks (Photo by Roberto Bianconi); (G) *Epinephelus marginatus* (Lowe), the Dusky grouper (Pisces Perciformes Serranidae), a well-known species largely spread from the English Channel to southern Africa also present in the Mediterranean, in western Indian Ocean and in the east coast of South America, distributed along all the sea coasts of Italy (Photo by Maria Mercurio); (H) *Pinna nobilis* (Linnaeus), the noble pen shell (Mollusca Bivalvia Pinnidae), in meadow of *Posidonia oceanica* (Linnaeus) Delile, the Neptune grass (Liliopsida Najadales Posidoniaceae), both endangered species endemic to Mediterranean Sea (Photo by Carlo Cerrano).

be calculated on the basis of country limits but with the range of the whole Alpine arch that embraces more countries.

Since 1995 the nongovernmental organization Planta Europa developed the program IPA (Important Plant Areas) in order to identify the most important areas for floristic and vegetational diversity. An Important Plant Area is a natural or seminatural site exhibiting exceptional botanical richness and/or supporting an outstanding assemblage of rare, threatened and/or endemic plant species and/or vegetation of high botanic value (Anderson, 2002). In Italy 312 IPAs that cover a surface of 4,476,830 hectares, equal to about 15% of the national territory have been identified. The designation of a site as an IPA is not associated with any management prescription. Many of the IPAs identified are already contained either fully or partially within other protected areas systems: 307 out of 312 mapped IPAs fall into protected areas and/or into the Natura 2000 network (Blasi et al., 2010a). Using the same methods applied for IPAs, also Important Faunal Areas (IFAs) have been established in Italy (Stoch, 2008). IFAs classification is based on invertebrates, freshwater fish, amphibians, reptiles and small mammals. As these animal groups represent near the 99% of the freshwater and terrestrial Italian fauna, the contribution of IFAs to the protection of national biodiversity is very important. IFAs complement the IBAs (Important Bird and Biodiversity Areas, as defined by BirdLife International) and the areas for the protection of large mammals (i.e., carnivores and ungulates). The IFAs cover near the 46% of the country; 172 IBAs have been classified, for a total area of 4,987 hectares. In Italy, the framework law on protected areas dates back to 1991. It provides a definition of natural heritage and sets out guidelines for conservation, application of forest management and restoration methods, promotion of naturalistic education activities, defense and reconstruction of hydraulic and hydrogeological equilibria and valorization of compatible production activities. Protected areas are considered no longer a static place where only naturalistic aspects are privileged, but environmental, landscaping and socioeconomic valorization tools vital to reestablish a rational man-territory relationship in the context of integrated and ecofriendly development.

The protected natural areas in Italy can be divided according to their extent and purpose in national and regional parks, national and regional natural reserves, and other regional natural protected areas and marine protected areas.

To these can be added the areas defined by the EU Community directives then transposed by national laws as the Network NATURA 2000 (SIC/ZSC and ZPS) and the Wetlands of International importance (see Table 6.1).

The 871 protected areas listed in the National official list (MATTM 2010) occupy an area of 3,163,590.71 hectares (about 9.5% of the entire national territory) on land and 2,853,033.93 hectares on the sea. In this way, almost all of the habitats and species present are protected throughout the national territory. All these areas constitute a network that allows not only the conservation of individual species and habitats, but also the creation of interconnected areas through corridors for the diffusion of plants and animals at national and international level. The crucial point for the proper functioning of these tools is to guarantee an adequate financial and staff allocation to carry out the purposes for which they have been instituted.

Following an agreement between Italian Ministry for the Environment, Land and Sea (MATTM) and Italian Botanical Society (SBI), between 2013 and 2016, the extinction risk assessment, according to IUCN criteria (IUCN,

Table 6.1 The Protected Areas in Italy Divided per Typology

Typology	Number	Terrestrial surface ha	Marine surface ha
Areas defined by national legislation			
National parks	24	1,465,681.01	71,812.00
Regional and Interregional Natural parks	134	1,294,655.87	0
National natural reserves	150	122,775.90	2,557,477.00
Regional Natural reserves	365	230,240.21	1,284.00
Other Regional Natural Protected Areas	171	50,237.72	18.40
Marine protected area	27	0	222,442.53
Total	871	3,163,590.71	2,853,033.93
Areas defined by EU Community directives			
Special Protection Areas ZPS	227	2,821,875.00	200,246.00
SIC/ZSC	1997	3,101,652.00	375,110.00
Wetlands of International importance	53	60,759.00	0

Sources: http://www.minambiente.it; http://www.ramsar.org/.

2012), of all the vascular plants endemic to Italy was compiled. This assessment should provide a roadmap to implement a future national strategy for the conservation of plant biodiversity prioritizing the efforts towards the most threatened taxa. As stressed by Rossi et al. (2014) the implementation of strategies, the establishment of protected areas or the implementation of laws alone are not sufficient to prevent the loss of biodiversity. Further active measures (e.g., conservation of germplasm in seed banks, long-term monitoring programs) have to be studied and applied based on a systematic conservation planning approach.

6.5 Ecosystem of Italy

The Ecosystem Map of Italy (1:100,000) was based on a reinterpretation of Corine Land Cover on the basis of the Italian Potential vegetation map (279 types), Bioclimatic map (28 classes, 8 bioclimates, 3 regions), Biogeographic regionalisation (Blasi et al., 2007) and Land units map (149 types). Like several other countries, especially those in the Mediterranean Area, Italy is characterized by a marked natural heterogeneity and a significant human-induced fragmentation of its land (Smiraglia et al., 2013). Around 77 natural and seminatural and 13 artificial ecosystem types have been identified and mapped on a nationwide scale. This clearly illustrating the outstanding national heterogeneity at the species, community, and landscape levels. Landscapes include 230 different habitats that can be subdivided into 7 categories (Angelini et al., 2009):

1   Coastal and halophilous communities
2   Non-marine waters
3   Bushes and prairies
4   Forests
5   Bogs and Marshes
6   Cliffs, Screes and Sands
7   Cultivated and Built areas

A medium landscape conservation status has emerged at the national level, though the ecosystem conservation status considerably varies from bad to medium and high depending on the regional and ecoregional sectors. The areas with a higher conservation degree of the ecosystems at national level are the Alps and small parts at the top of the mountains of central Apennines, Sicily and Sardinia. They occupy 12.17% of the total surface of the Country. A medium conservation status characterizes the higher parts of the Apennines, the Sardinia's hinterland and the mountain chains of Sicily (14.37% of the total). The hilly part of the Apennines has low level of conservation (14.98% of the total). The rest 58.48% of the national territory including hills and flat areas used as cultivated fields or used as inhabited and industrial areas have a bad conservation degree. In this part of the territory the degree of naturalness is very low and only in some points survive species of biogeographical interest.

Main restoration priorities that have been identified in medium and low level of conservation areas include several forest ecosystem types, especially deciduous oak forests in the plain and hilly sectors, as well as ecosystem types of sandy coasts, salt waters, and hygrophilous environments (Capotorti et al., 2015).

6.5.1   Forests

Forests in Italy and in the rest of the Mediterranean have been heavily exploited by humans since the dawn of the agriculture. It is realistically to say that in this area there are not anymore natural forests but only old-growth forests can be found. These are characterized by the presence of large old trees and the lack of human disturbance (Blasi et al., 2010b). In terms of number of species, Mediterranean forests have lower values than open areas that represent their degradation stages but they and above all the old-growth forests have an impact on biological diversity of invertebrates, lichens, mosses, fungi, birds and vascular plants. At present, forests are predominantly spread in mountain and hilly areas, from the alpine range to the islands (Table 6.2). Oak forests and Beech forest are the most common formations from the sea level to 1,900 m. In several areas, these have been replaced with Chestnut forests that give higher income for the fruits and wood. The twentieth-century war events have greatly reduced the wooded areas for the need for fuels and for the search for areas to be cultivated. Since the 1960s to mitigate the damage caused by the abandonment of mountain and less productive agricultural areas, intensive reforestation programs have been carried out to increase the forested areas. Such interventions have often been carried out with monospecific stands of native or nonnative species, but they are nevertheless used to prevent hydrogeological breakdowns and to avoid building speculation. The forestry policies of the last 30 years are aimed at converting these low natural systems into stable formations, with different native species.

6.6   Floral Diversity in Italy

6.6.1   Vascular Plants

The Italian Peninsula and the major islands are the second highest area, after the Iberian Peninsula and the Balearic Islands, of plant species richness in Europe (Bilz et al., 2011). Their total surface is less than 3% of the whole Europe, but they host about half of the plant species found throughout Europe. Following the climatic events that occurred in the various geological eras, the Italian flora includes species of Eurasian, Middle Eastern and North African origins. The overall biodiversity of the flora is enriched by the presence of taxa coming from very different areas of the world that have reached Italy in different geological eras and, finished the climatic event that have allowed the settlement, survived in particular bio-refugia. The higher Apennine for boreal flora and vegetation and the Tyrrhenian coastal valleys for many tropical and subtropical species are examples of such refugia.

Several projects such as the second edition of the "Flora d'Italia" by S. Pignatti, and its digital tool (Pignatti et al., 2017–2018), the "Flora critica d'Italia" by Peruzzi et al. (http://www.floraditalia.it), the "Index of names of plants occurring in Italy" by Acta Plantarum (http://www.actaplantarum.org/flora/flora.php), the "Archive taxonomic Checklist" (Lucarini et al., 2015) and the new "The updated checklist of the vascular flora of Italy" (Barto-

Table 6.2 Forest Types and Their Distribution in Italy

Forest Type	Dominant species	Area ha	Area %	Altitude m a.s.l.	Distribution
Oak forests	*Quercus petraea, Q. pubescens, Q. robur*	1,084,247	13	100–1300	Alps, Apennines, Sicily, Sardinia
Beech forests	*Fagus sylvatica*	1,035,103	12	400–1900	Alps, Apennines, Sicily
Turkey oak, Hungarian oak, Macedonian Oak and Valonia oak forest	*Q. cerris, Q. crenata, Q. congesta, Q. frainetto, Q. macrolepis, Q. trojana*, etc.	1,010,986	12	0–1100	Po Valley, Apennines, Salentine peninsula
Mixed deciduous broadleaves forests	*Fraxinus excelsior, Betula pendula, Alnus cordata, Robinia pseudoacacia*, etc.	994,777	12	0–1600	Alps, Apennines, Sicily, Sardinia
Hornbeam forests	*Carpinus* spp.*, Ostrya carpinifolia*	852,202	10	0–3600	Alps, Peninsular Italy, Sicily, Sardinia
Chestnut forests	*Castanea sativa*	788,408	9	200–1000	Alps, Apennines, Sicily
Holm oak forest	*Quercus ilex*	620,318	7	0–1900	Peninsular Italy, Sicily, Sardinia
Norway spruce forests	*Picea abies*	586,082	7	1600–1900	Alps, N Apennines
Larch and stone pine forests	*Larix decidua, Pinus cembra*	382,372	4	1300–2000	Alps
Black pines forests	*Pinus nigra, P. laricio, P. leucodermis*	236,467	3	0–2100	E Alps, Apennines, Etna
Hygrophilous forests	*Populus* spp.*, Salix* spp.	229,054	3	0–1500	Alps, Po valley, Apennines, Sicily, Sardinia
Mediterranean pines forests	*Pinus domestica, P. halepensis, P. pinaster*	226,101	3	0–1000	Peninsular Italy, Sicily
Cork oak forests	*Quercus suber*	168,602	2	0–700	C Italy, Sicily, Sardinia

Table 6.2   (Continued)

Forest Type	Dominant species	Area ha	Area %	Altitude m a.s.l.	Distribution
Scots pine and mountain pine forests	*Pinus sylvestris, P. mugo*	151,671	2	250– 2100	Alps, Emilia Romagna
Silver fir forests	*Abies alba*	68,460	1	800– 1900	Alps, Apennines
Other evergreen broadleaves forests	*Quercus coccifera, Ceratonia siliqua, Laurus nobilis, Ilex aquifolium*	84,712	1	0–1500	Peninsular Italy, Sicily, Sardinia
Other coniferous forests	*Cupressus sempervirens, Abies nebrodensis,* etc.	63,407	1	0–1400	Peninsular Italy, Sicily, Sardinia
Total		8,582,969			

Source: INFC (2007).

lucci et al., 2018; Galasso et al., 2018) are currently carried on. The updated checklist of the vascular flora of Italy has been continuously updated on the basis of the most updated taxonomic studies of the Italian flora with the data of the hundreds of floristic records published in the last 12 years. The vascular native flora of Italy (Bartolucci et al., 2018) includes 8195 specific and subspecific taxa. This increase is due not only to the floristic and taxonomic research but also to the choice of including the subspecies of *Hieracium* and *Pilosella* deliberately excluded in the previous contribution. Adding the 1597 taxa not-native to Italy (Galasso et al., 2018) the whole checklist of spontaneous plants of Italy rises to 9792 taxa. Despite these numbers are still subject to variations (Table 6.3). 30 specific and subspecific taxa of gymnosperms are included in the checklist of native species. Further 33 alien taxa, most of which are cultivated and naturalized or casual, can be added, for a total of 63 taxa. The pteridophytic native flora of Italy (lycopodiophytes and monilophytes) includes 20 families, 36 genera and 131 species including 3 species of *Isoetes* endemic of the Italian territory. Further 13 alien species occur in the Italian territory rising the number of species to 144. The most represented families in terms of number of genera and specific and subspecific taxa are Asteraceae, Poaceae, Fabaceae, and Caryophyllaceae (Table 6.4). The genera *Hieracium, Taraxacum, Centaurea, Ranunculus,* and *Limo-*

Table 6.3   Number of Taxa in the Different Groups of the Native Italian Flora
Including Doubtfull Taxa (from Bartolucci et al., 2018)

Lycopodiophytes		Pinidae	
Lycopodiidae		**Pinidae**	
Families	3	Families	3
Genera	5	Genera	6
Species	23	Species	20
Species & Subspecies	23	Species & Subspecies	24
Monilophytes		**Angiosperms**	
Equisetidae		**Magnoliidae**	
Families	1	Families	97
Genera	1	Genera	783
Species	9	Species	5353
Species & Subspecies	9	Species & Subspecies	6885
Ophioglossidae		**Liliidae**	
Families	1	Families	25
Genera	2	Genera	260
Species	9	Species	1325
Species & Subspecies	9	Species & Subspecies	1460
Polypodiidae		**Paleoherbs**	
Families	15	Families	3
Genera	28	Genera	4
Species	99	Species	13
Species & Subspecies	.109	Species & Subspecies	14
Gymnosperms			
Gnetidae			
Families	1		
Genera	1		
Species	5		
Species & Subspecies	6		

nium, in which is known apomixis, the reproduction via asexually formed seed, are the richest of taxa (Table 6.4).

6.6.2 *Endemicity*

The taxa endemic to Italy (Peruzzi et al., 2014) includes 1462 species and subspecies, grouped in 320 genera, 69 families. Three taxa belong to

Table 6.4 Most Represented Families in the Italian Flora

Family	Genera	Sp. & Subsp.	Endemics	%	Genus*	Sp. & Subsp.	Endemics	%
Asteraceae	126	2202	582	26.43	Hieracium*	1167	337	28.87
Poaceae	123	550	65	11.82	Taraxacum	154	33	21.43
Fabaceae	41	495	80	16.16	Centaurea	116	73	62.93
Brassicaceae	74	333	78	23.42	Limonium	111	98	88.29
Caryophyllaceae	37	359	97	27.02	Ranunculus	110	37	33.64
Rosaceae	28	312	40	12.82	Ophrys	103	61	59,22
Apiaceae	89	263	39	14.83	Silene	87	29	33.33
Lamiaceae	33	240	38	16.10	Festuca	83	17	20.48
Orchidaceae	27	236	87	36.86	Allium	69	25	36.23
Ranunculaceae	24	231	60	25.97	Campanula	58	19	32.76
Plumbaginaceae	6	135	117	86.66	Viola	56	20	35.71
Boraginaceae	25	115	34	29.57	Dianthus	53	28	52.83
Amaryllidaceae	7	93	29	31.18	Genista	45	29	64.44

It is reported the number of genera, species and subspecies, the number of endemics and their rate in comparison with the whole flora.

* Species only.

Lycopodiidae, one to Polypodiidae, two to Pinidae and 1456 to Magnoliidae. This list also includes the taxa whose distribution embraces Corsica or Malta in the light of the strong biogeographical links connecting these islands with Italy.

The Italian endemic taxa correspond to 18.9% of the whole flora (Bartolucci et al., 2018). The most represented families and genera are highlighted in Table 6.5. Seven genera are endemic to Italy: Castroviejoa Galbany et al., Eokochia Freitag & G. Kadereit (Amaranthaceae), Morisia J. Gay (Brassicaceae), Nananthea DC. (Asteraceae), Petagnaea Caruel, Rhizobotrya Tausch and Siculosciadium C. Brullo et al. (Apiaceae). 965 taxa (66% of the

Table 6.5 Number of the Italian and Regional Endemic Species and Subspecies Per Single Administrative Region (From Peruzzi et al., 2014)

Region	No. Italian endemics	No. Regional endemics	% of regional endemics
Sicily	443	305	68.85
Sardinia	329	256	77.81
Calabria	298	64	21.48
Abruzzo	252	45	17.86
Tuscany	217	70	32.26
Basilicata	217	9	4.15
Lazio	212	13	6.13
Campania	195	24	12.31
Puglia	157	35	22.29
Marche	146	11	7.53
Molise	137	1	0.73
Umbria	119	0	0.00
Emilia-Romagna	89	5	5.62
Trentino-Alto Adige	80	19	23.75
Veneto	78	23	29.49
Lombardia	78	23	29.49
Liguria	70	12	17.14
Piedmont	58	24	41.38
Friuli-Venezia Giulia	34	13	38.24
Valle d'Aosta	10	5	50.00
Total	1,462	965	66.01

total endemics) are narrow endemics occurring in a single region and 569 of them (39% of the total) occur only in Sicily or Sardinia. The analysis of the distribution of endemics confirms the phytogeographic isolation of Sardinia and Sicily and the distinctiveness of the peninsular regions compared to those set in the Po Valley or Alpine area (Peruzzi et al., 2014).

6.6.3 Crop Wild Relatives and Wild Harvested Plants

Crop Wild relatives are those wild plant taxa more or less closely related to species of direct socioeconomic importance including food, fodder and forage crops, medicinal plants, condiments, ornamental and forestry species, as well as those related to crops used for industrial purposes. The interest in this group of plants grew in the last 20 years because they possess many beneficial traits that can be bred into crops.

Landucci et al. (2014) list 5,072 native species of Crop Wild Relatives for peninsular Italy, 2,254 for Sicily and 2,031 for Sardinia. The most represented genera are: *Hieracium* (232 species), *Taraxacum* (137), *Pilosella* (97), *Ranunculus* (95), *Centaurea* (77), *Festuca* (56), and *Euphorbia* (52). Most of the Crop Wild relatives are not yet included in conservation programs. Parks and Natural reserves provide indirect protection for some of them and the Italian seed banks preserve 6029 accessions of 229 CWR taxa, belonging to 11 families and 57 genera (Magrini et al., 2016). However, practical conservation actions are still lacking for the large part of this heritage.

According to the data in Landucci et al. (2014), the wild harvested native plants used as food, medicine, forestry, fodder and forage are 1363 in peninsular Italy, 783 in Sardinia, and 737 in Sicily. Traditional plant uses is often confined to oral tradition of older people in rural communities. This poses a serious danger of loss of this information. Several projects are in place at national and regional scale to limit the loss of knowledge of the Intangible Cultural Heritage related to plants.

6.6.4 Bryophytes

Bryophytes occur in every environment on the earth with the lone exception of the marine one. These organisms include about 24,000 species in the world. In the last years, bryological research in Italy has made remarkable progresses. The intensification of bryological investigation has been matched by a significant increase in scientific publications that brought to the identification of new taxa for the Country (Aleffi, 2007). As regards mosses the number of species known has risen from 818 in 1992 (Cortini Pedrotti

1992) to the actual 864 (Martellos et al., 2013). The checklist of Bryophytes of Italy (Martellos et al., 2013) consists of 1,214 specific and infraspecific taxa subdivided in 302 hornworts and liverworts and 912 mosses. The lists of hornworts and liverworts include 292 species, 5 subspecies and 5 varieties, for 87 genera and 42 families. The list of mosses includes 864 species, 8 subspecies and 40 varieties, included in 222 genera and 59 families. The Bryological flora of Italy is characterized by taxa with a boreal, subartic-sub-Alpine and suboceanic distribution. These taxa occur in the Alps, the higher Apennines and on the top of the highest mountains of Sardinia and Sicily. An important part of the flora has Oceanic-Mediterranean distribution. The taxa with this chorotype occur in areas of transition between the Mediterranean and the Atlantic climate and includes several species which show a disjunct distribution. In Sardinia and Sicily there is the highest percentage of Mediterranean and sub-Mediterranean taxa (Aleffi, 2007).

6.6.5 *Lichens*

The checklist by Nimis (2016) includes 2704 accepted specific and infraspecific taxa, 2,565 of which are lichenised. The increase in the number of taxa from 1993 to 2016 is of 26%. This was due to the efforts of a new generation of Italian lichenologists that grew up, and several nonItalian lichenologists that published many new records from Italy: more than 1200 papers appeared after 1993, which contain at least one lichen record from the country.

The high diversity of lichens in Italy is mainly due to the presence of some biogeographic groups of species with different ecological requirements. These groups reflect fairly well the climatic diversity of the country, from cold-alpine to warm suboceanic climates, with a prevalence of warm-temperate, moderately humid climates, and, despite the summer drought period of some regions in the south, with an overall scarcity of truly arid climate-types. According to Nimis and Tretiach (1995) the main groups occurring in Italy are:

- A "northern" group with arctic-boreal affinities, restricted to the highest mountains, most frequent on the Alps on acid substrata and becoming progressively nearer southwards.
- A rather poorly defined set of species occurring the southern European mountains, which in Italy is mostly found on calcareous substrata in the alpine belt of the Alps.
- Another poorly defined element restricted to the lowlands and lower mountains of the Mediterranean Region, which in Italy has a mainly Tyrrhenian distribution in the Mediterranean belt.

- A mainly temperate group of lichens which is well represented in all regions and is most frequent in the deciduous forest belt.
- A suboceanic to oceanic group with subtropical affinities, bound to humid climates, which is most frequent along the western side of the Peninsula and Sardinia.
- A small set of widespread xerophytic species, occurring in the most arid parts of Sardinia and Sicily and in the driest parts of C and W Alps.

6.6.6 *Algae*

A complete list of freshwater algae in Italy does not yet exist. Recently, studies on Chlorophyta and Rhodophyta have been done but the accounts for the entire peninsula collect heterogeneous reports from administrative regions with different levels of base knowledge (Abdelahad and Bazzichelli, 2007). There is a great difference on the data collected from N Italy, from which over 1,600 taxa are reported and those collected from S Italy from which little more than 500 taxa are known. These data derive more from a higher concentration of research in the north of the country than from a real greater richness of alga biodiversity.

The research done on marine algae and vascular plants in Italy includes a considerable number of contributions. Nevertheless, knowledge of the macro-benthic flora of the Italian coasts is still patchy and incomplete. In fact, there are some areas particularly well studied (e.g., the north Adriatic, the Gulf of Taranto, the Gulf of Naples, Sicily), and others with a still incomplete knowledge (Liguria, Lazio, Sardinia, and Calabria). The account by Furnari et al. (2010) includes 944 taxa of algae at specific and infraspecific level (46 Cyanophyta, 534 Rhodophyta, 214 Ochrophyta, 150 Chlorophyta) giving for each of them the occurrence in the 9 biogeographic sectors in which was divided the sea around Italy (Bianchi, 2004).

More limited knowledge is available for planktonic and benthic organisms. The phytoplankton catalog includes 1,484 identified species of microalgae belonging to 356 genera subdivided into 13 classes, and others classified as *incertae sedis* (Cabrini, 2010). This list includes autotrophic, heterotrophic and mixotrophic species belonging to classes Dinoflagellates, diatoms and other classes commonly named phytoflagellates. Microphytobenthos includes the microscopic aquatic algae (diatoms, dinoflagellates, phytoflagellates, and cyanobacteria) living on or close to the bottom. The checklist of microphytobenthos by Cibic and Facca (2010) includes 524 taxa in the following phyla: Bacillariophyta (506), Myzozoa (9), Chlorophyta (2), Cyanobacteria (4) and Heterokontophyta (3).

6.6.7 *Fungi*

The floristic mycology research saw a period of flourishing activity between the end of the nineteenth century and the beginning of the twentieth century. After this period mycological studies in the universities were addressed to morphology, genetics and physiology. Only in the last decades, also thanks to the contribution of amateurs, the floristic mycology is living an encouraging revival. The number of taxa recorded in Italy for Basidiomycota is 4,296 (3,973 species, 6 subspecies, 263 varieties and 54 forms) (Onofri et al., 2007). This represent about the 20% of the known taxa in the whole earth. Data for the Ascomycetes are still incomplete but a preliminary study (Onofri et al., 2007) collected a thesaurus of 532 accepted names. Given the low availability of data and their heterogeneity is unreliable a regional comparison or estimation of rarity of species. The data available indicates that 56 species, up to now, are known only from Italy.

6.7 Animal Diversity in Italy

The recent *Fauna Europaea Project* – a database of all living, currently known, multicellular, European land and freshwater animal species (de Jong et al., 2014) – stated that Italy have the largest number of animal species in Europe. In the *Checklist of the species of the Italian Fauna* (Minelli et al., 1993–1995) near 55,600 metazoans and 1,800 protozoans are listed. This number is continuously increasing as many new species are described or recorded for the first time in Italy each year, so that near 60,000 species are estimated to exist today in the area and the census needs to be updated. Vertebrates account for near 2% of the total. More than 82% are represented by arthropods. Among these, insects dominates with over 37,000 species, near 67% of the whole Italian fauna. Species living in freshwater habitats (excluding protozoans) are nearly 5,500, 10% of the total Italian fauna. More than 9,000 species belong to the marine fauna.

Faunistic hotspot in Italy are represented by the Alpine and Pre-Alpine areas, the Ligurian Apennines, the Tuscan-Emilian Apennine, the Central Apennine (Abruzzo and Latium). Lower species richness has been recorded in southernmost regions and the larger island, Sicily and Sardinia. This may be due to an insularity and peninsular effect, as well as reduced faunistic field research. Hotspots in these regions are in Gargano, Calabrian Apennine, Northeast Sicily and central Sardinia. All of these areas include the most extensive endemic animal species hotspots in Italy (Stoch, 2008).

6.7.1 Endemic Species

According to the *Checklist of the species of the Italian fauna* (Minelli et al., 1993–95), 4,777 species of metazoan are strictly endemic to Italy (not trespassing the Italian political borders). It means 8.6% of the overall species richness, although Stoch (2008) estimated this value should be of 10%. This remarkable number and rate of endemic species are mainly due to terrestrial and freshwater invertebrates in which the percentage of endemic species may be up to 25%. Minelli et al. (2006) stated that the endemic rate is significant in lower groups of invertebrates of the soil fauna, such as Annelida, Oligochaeta, Arachnida, Pseudoscorpionida (more than 50% of the species), Chilopoda (e.g., in genus *Lithobius*), Coleoptera Carabidae (in *Carabus, Trechus, Pterostichus, Lessinodytes, Italaphaenops, Percus*) and Staphylidnidae Aleocharinae (e.g., *Leptusa* spp.). In groundwater habitats, many endemics species are in Gastropoda Hydrobiidae as well as in many groups of Crustacea, such as Copepoda, Bathynellacea, Thermosbenacea, Amphipoda and Isopoda. In surface waters, several endemic species are in groups inhabiting lotic habitats (e.g., in Plecoptera, Coleoptera Hydraenidae, Trichoptera). In lentic habitats, although the number of endemic species is lower the Crustacea Anostraca of the genus *Chirocephalus* should be mentioned (three species in the northern Apennines). Less than 3% of the vertebrates and 2% of the marine species are endemic. Peninsular Italy (Table 6.6) hosts the higher number of endemics (ca. 1,825 species recorded in the *Checklist*), equal to 7.60% of the total macroregional fauna (ca 24,300

Table 6.6 Species Numbers of Total (Nt) and Endemic (Ne) Species of Vertebrates (Vert) and Invertebrates (Inv), Excluding 'Protozoans,' Compared with Area of: Northern Italy, Emilia Romagna Excluded (N); Central and Southern Italy (S); Sicily (Si); Sardinia (Sa); Western Tyrrhenian Sea (3); Northern and Central Adriatic Sea (4); Southern Adriatic, Jonic Sea and Other Seas (5)

	Nt_Vert	Ne_Vert	Nt_Inv	Ne_Inv	%Ne/Nt	Area (sq. km)
N	629	22	33,414	1,720	5.12	97,741
S	545	22	24,297	1,825	7.60	153,710
Si	399	5	12,988	776	5.83	25,708
Sa	365	6	9,841	676	6.68	24,090
3	455	1	6,529	159	2.29	-
4	317	2	3,958	79	1.89	-
5	394	0	3,717	44	1.07	-

From Stoch and Minelli (2004); See Minelli et al. (1993–95) for a Map of Sectors.

species), whereas in Northern Italy the endemic species (ca. 1,720) represent a percentage of about 5% of the total (ca. 33,400 species). The number of endemics are limited to Sardinia (ca. 700 species, 7% of the 9,800 species in the island) or Sicily (ca. 780, almost 6% of the ca. 13,000 species listed). All these values are, however, only indicative as the faunal knowledge of Italy is still incomplete, and species/subspecies ranks are often attributed subjectively to a population or to a group of populations, affecting the estimate of numbers of endemics (Stoch and Minelli, 2004; Minelli et al., 2006a).

A synthetic overview of the metazoan taxa represented in the Italian fauna is presented below. Figures on the Italian fauna has been gathered mostly from Minelli et al. (1993–1995), Ruffo and Stoch (2006a) and Blasi et al. (2007), considering the updates by Relini (2008, 2010) on marine fauna. Other updated references will be provided case by case. General figures from Zhang (2011, 2013).

6.7.2 Protozoa

Protozoans represent a large and heterogeneous group of unicellular eukaryotes, including divergent lineages of solitary or colonial, free-living, symbiont or parasitic organisms belonging to the Kingdom Protista. Free-living protozoans colonize mostly aquatic habitats, such as seas, fresh- and brackish waters, where they mostly live in plankton as well as in benthos communities, some colonize soils. According to the classification scheme in Minelli et al. (1993–1995), the Italian fauna is represented by near 1,800 species, in ca. 630 genera, 320 families, 80 orders, in 6 classes. About 1,000 species have been recorded in the Italian seas.

6.7.3 Porifera

Sponges are among the oldest pluricellular organisms, as fossil records date back to more than 700,000 years ago. Species of this group colonize a wide range of aquatic habitats, from shallow waters to oceanic trench as well as freshwaters of all continents but Antarctica. Nearly 8,350 species are globally recognized. The checklist of Italian sponges includes 180 genera and 492 marine species, arranged in three classes (Hexactinellida, Calcarea, Demospongiae), 18 orders and 76 families, this figure representing about 76% of the near 560 species of Mediterranean Porifera. Demospongiae is the richer and articulated class, with 450 species, in 161 genera, 65 families and 13 orders. The knowledge of the Italian marine Porifera is generally good, although some shortcomings still exist, especially on the deep fauna.

Four genera and seven species, all belonging to the Demospongiae family Spongillidae, are included in the Italian freshwater fauna.

6.7.4 Cnidaria and Ctenophora

Cnidarians represent a mostly marine group, including a number of freshwater species. They are characterized by a complex life-cycle in which two basic forms, planktonic medusa and benthic polyp, alternate sometimes omitting one of the stages. Actually, more than 11,000 species are recognized all over the world, arranged in 6 classes: Anthozoa, exclusively polypoid, colonial, clonal, or solitary, skeleton-less or with a mineralic and/or proteinaceous skeleton (approx. 7,500 species), Cubozoa (36 valid species), Hydroidomedusa and Automedusa (included in the superclass Hydrozoa, with approx. 3,500 valid species altogether), Scyphozoa (approx. 200 species), Staurozoa (50 species) (Daly et al., 2007). The checklist of Italian Cnidaria includes 301 genera and 493 species, arranged in 6 classes, at least 23 orders and 142 families. Among them Anthozoa includes 129 species, 11 orders, 53 families, 98 genera; Cubozoa are represented by 1 species; Hydroidomedusa 313 species, 7 orders, 70 families, 163 genera; one genus, *Hydra*, with 4 species in freshwater habitats; *Craspedacusta sowerbyi*, an alien species of tropical origin, also in freshwater habitats); 33 species, 9 families, 20 genera of Automedusa are listed; Scyphozoa account for 13 species, in 4 orders, 8 families and 15 genera; 4 species, 1 order, 1 family, 4 genera of Staurozoa are known. The present knowledge on taxonomy and distribution of Italian cnidarians is however still limited, especially on freshwater species, although many studies on single species or groups has been carried out especially in marine habitats.

Ctenophora are exclusively marine metazoans, mostly planktonic, few species are benthonic. About 110 species are presently known all over the world. The checklist of Italian Ctenophora includes 32 planctonic species, arranged in 5 orders, 13 families and 23 genera.

6.7.5 Platyhelminthes

Flatworms, or platyhelminths, include near 22,500 known species, most of them are parasitic and some are of high interest from sanitary and veterinary point of view. Free-living species mostly inhabit marine and freshwater habitats. A few are terrestrial. Four classes are traditionally recognized, "Turbellaria," Digenea, Monogenea and Cestoda. Turbellarians is presently considered a polyphiletic and therefore disregarded as natural group.

6.7.5.1 Turbellaria

About 4,000 species of "Turbellaria" are presently known, taxonomic and faunistic knowledge of this group is very poor. Near 520 species, have been recorded in Italy, 338 of them inhabit marine habitats, arranged in 156 genera, 48 families and 10 orders; 180 species belonging to 65 genera, 21 families, 7 orders are in freshwater habitats, whereas 6 species in four genera, two families, one order are terrestrial.

6.7.5.2 Digenea

Among the parasite flatworms, Digenea is the most diverse taxon. Near 7,200 species are known worldwide. The life cycle of the species generally alternates between two hosts, a gastropod (intermediate host) and a vertebrate (definitive host). In Italy two orders are represented. In marine habitats, 198 species, in 120 genera and 38 families, are parasites mostly of bony fishes, whereas 213 species, in 125 genera and 36 families, has been recorded as parasites of vertebrates in freshwater and terrestrial habitats.

6.7.5.3 Monogenea

Monogenea are flatworms mostly external parasites of fishes, amphibians and other aquatic vertebrates, some species parasitize invertebrates. About 2,000 species are known. In Italy, 88 species of parasites of marine fishes (both Chondrichthyes and Osteichthyes), in 50 genera and 14 families have been recorded. One species is known as parasite of freshwater fishes but knowledge in this subject is very lacunose. Other two species parasitize amphibians (Anura) and freshwater reptiles (Testudines), respectively.

6.7.5.4 Cestoda

Cestoda are a class of flatworms exclusively endoparasites of terrestrial or aquatic vertebrates, with one or more intermediate hosts. About 3,500 species are known. In Italy 89 species parasites of marine fishes, mostly cartilagineous, in 42 genera, 20 families and 8 orders; 241 species, in 108 genera, 22 families and 8 orders, are known as parasites of vertebrates in freshwater and terrestrial habitats have been recorded.

6.7.6   Aceola

Traditionally considered as an order of Platyhelminthes, acoels is now interpreted as a lineage of basal bilaterians with phylum rank. Acoela are free-living and some 380 species generally related to sediments in marine habitats has been described. Knowledge of the Mediterranean Acoela is still insufficient and a preliminary list of the Italian fauna includes 54 species, arranged in 16 families and 30 genera.

6.7.7   Nemertodermatida

A lineage of marine metazoa recently separated from Platyhelminthes. Only eight marine species are known. Most occur in sediments, one is known as parasite of holothurians. Five species belonging to two families and four genera have been recorded to date in Italy.

6.7.8   Gnathostomulida

Gnathostomulida are a group of acoelomata microscopic marine worms. These animals belong to interstitial marine fauna, only one species is known in shallow brackish coastal waters. Near 80 species are known. Nine species belonging to two orders, six families and seven genera have been to date recorded in Italy.

6.7.9   Nemerteea

A group of mostly marine worms, few species colonize freshwaters, generally benthic, some pelagic, or commensal. Nemertea are difficult to study and therefore poorly known. Near 1,100 species are known. The Italian list includes 96 marine species arranged in two classes, two superorders, at least 20 families and 53 genera. One species in fresh waters, *Prostoma graecense* (Bohmig, 1892).

6.7.10   Gastrotricha

Gastrotricha are microscopic, free-living, aquatic worms. Mainly interstitial in marine habitats, they are a common component of the benthos (including periphyton) and, in a limited extent, plankton assemblages in freshwaters. About 750 species are globally known, subdivided in two orders, Macrodasyida, almost exclusively marine, and Chaetonotida, with marine and freshwater taxa. The Italian marine fauna is well known, 154 species have been identified so far, 91 Macrodasyida (in 7 families and 24 genera) and 63 Chaetonotida (in 3 families and 10 genera). Freshwater Gastrotricha of the Italian fauna include only members of the order Chaetonotida, with 90 species, in 16 genera and four families.

6.7.11   Rotifera

Rotifers are microscopic, aquatic, semiaquatic, endo- and ecto-parasite organisms, occurring in all habitats where water is present. Near 2,000 species are known in the phylum (Segers 2007) and about 400 species have been reported from saltwater habitats worldwide. The Italian fauna is still poorly known. A provisional list of 246 species has been provided (3 orders, 26 families, 61 genera). More than 54 species/subspecies have been found in the Italian seas (see De Smet et al., 2015 for details on the Italian marine fauna), the other are in freshwaters.

6.7.12   Nematoda

A highly diversified group of free-living and parasite roundworms occurring in all types of habitat, generally dominant in sediments of aquatic habitats. Globally, near 25,000 species are recognized but estimates suggest about 1 million species exists. The Italian representatives includes near 1,570 species in 648 genera, 164 families, 14 orders. About 430 species of the Italian fauna are parasites of vertebrates, mostly mammals (250), 14 species are related to invertebrates, mostly insects. In marine habitats near 580 species have been listed, belonging to near 280 genera, 58 families and 7 orders. Near 200 species colonize freshwater habitats.

6.7.13   Nematomorpha

A group including two classes, Nectonematoidea, with marine species parasites of crustacean decapods during juvenile phases, and Gordioida, parasites of terrestrial or freshwater animals in juvenile phases. About 230 species are known. Nectonematoidea are represented in Italy by only one species, whereas Gordioida are represented by 22 species, in 2 families and 5 genera.

6.7.14 Acanthocephala

A phylum of specialized parasites, with complex life cycles (adults in vertebrates, mostly fishes, larvae in invertebrates). The group includes about 1,000 species. The Italian fauna is poorly known and 27 species in 8 families in 15 genera has been listed. Twenty species parasitize freshwater bony fishes, three parasitize mammals, three parasitize birds and one amphibians.

6.7.15 Kinorhynncha (=Echinodera)

Kinorhyncha represent a group of poorly known meiobentonic, microscopic, marine animals including near 140 described species. In Italy 48 species belonging to six genera, five families and two orders, has been recorded.

6.7.16 Loricifera

Exclusive marine organisms, loricifers are representatives of the permanent benthonic meiofauna. Generally very poorly known, this phylum was established in 1983; only 21 species have been to date described. In the Italian fauna, five species belonging to one order, two families and three genera are to date known.

6.7.17 Priapulida

Rare macro or meiobenthonic marine invertebrates, 19 species are known all over the world. Four species belonging to two orders and two families has been collected in the Mediterranean Sea, including Italy.

6.7.18 Kamptozoa (=Entoprocta)

A phylum of about 150 known species of small solitary or colonial aquatic filter-feeding animals, mostly inhabiting marine waters, only one species is known in freshwater habitats. Records of this group in the Italian waters are sporadic and refer to 17 species, belonging to 4 genera and 3 families in two orders.

6.7.19 Mollusca

This phylum has great diversity with nearly 117,000 known species that account for about 8% of the world fauna. Eight classes are traditionally recognized. The small class Caudofoveata includes marine benthic vermiform molluscs, accounting in Italy for six species under three genera, three families and one order. Solenogastres, another small group of worm-like, shell-less marine molluscs, include in the Italian seas 16 species under 15 genera, 10 families, 3 orders. Polyplacophora, also called chitons, are represented in Italy by 29 marine species under 11 genera and 6 families. Very few species are known in Monoplacophora, a group of bentic marine molluscs represented in Italy by only one species.

Gastropoda (snails and slugs) is the largest group of the phylum (60,000–100,000 extant species worldwide have been estimated) and can be found in almost all types of habitats, marine, terrestrial and freshwaters. According to Minelli et al. (1993–1995), Relini (2010), Ruffo and Stoch (2006a) the Italian fauna includes about 1,700 species in near 700 genera, 200 families and 20 orders. Stylommatophora and Neotaenioglossa are the most richest orders, including near 400 species each. Although the classification scheme of the group is still unstable two subclasses are recognized, Prosobranchia (seven orders, mainly marine but also terrestrial and freshwater) and Heterobranchia (three orders, marine, terrestrial and freshwater). About 140 species arranged in 14 families colonize Italian freshwater habitats.

The class Bivalvia is also highly diversified. This group include in Italy 341 marine species under 184 genera, 66 families and 9 orders in 2 subclasses. Conversely, the Italian bivalve fauna of freshwater habitats includes five families, 9 genera, 25 species. The class Scaphopoda, a small class of bentonic marine molluscs known as tusk shells, includes in Italian seas 2 orders, 6 families, 7 genera and 13 species, whereas Cephalopoda, squid, octopuses and their relatives, includes 58 species, falling under 40 genera, 22 families, and 4 orders.

6.7.20 Annelida

The known world annelid fauna is estimated to be of 17,210 species. Two main groups (classes) are traditionally recognized, Polychaeta and Clitellata, the latest in turn divided in Oligochaeta and Hirudinea. Close to annelids are other two phyla, Sipunculida and Echiura, whose relationships with annelids are still debated.

6.7.20.1 Polychaeta

Polychaetes are a group of segmented worms very common in marine habitats and very diverse. At least 9,000 species are known in the world. They mostly inhabit sediments, mud or sands, where burrow or build tubes; others are free swimming; commensal and parasitic species are also known. Some species colonize fresh and brackish water and very few are terrestrial, in the soil. A total of 876 species belonging to 384 genera, 71 families in 21 orders are listed in Italy. Taxonomy and distribution of the Italian polychaete fauna is relatively well known.

6.7.20.2   Oligochaeta

Oligochaeta is a subclass of clitellate annelids mostly inhabiting freshwater and terrestrial environments, from where about 3,500 species are known worldwide; about 650 species inhabit marine habitats, in the intertidal and subtidal benthic communities, at all latitudes and depths. In Italy, about 300 species has been recorded in all, belonging to 13 families and 92 genera. In freshwater habitats 127 species and about 45 genera are known, mostly represented by the "microdrila" families Lombriculidae, Haplotaxidae, Tubificidae (mostly), Naiadidae (mostly), Propappidae, together with some species of Enchytraeidae, Lumbricidae and Criodrilidae, all three also including terrestrial representatives. Terrestrial habitats are colonized by 79 species and 20 genera of the "megadrila" families Hormogastridae, Lumbricidae (mostly), Megascolecidae, Ocnerodrilidae, Acanthodrilidae, and Octochaetidae. About 88 species and 14 genera of Enchytraeidae has been also recorded in Italian terrestrial habitats. Along the Italian coasts, 34 species in 16 genera of tubificids, and three species of Enchytraeidae, all belonging to genus *Grania*, has been to date recorded.

6.7.20.3   Hirudinea

Hirudinea, leeches, is a subclass of clitellate annelids including 680 known species worldwide. Predators, they mostly inhabiting freshwater; about 100 species colonize marine habitats and few are terrestrial. Of the 28 species of Hirudinea to date recorded in Italy, 7, belonging to 2 families and 5 genera, are marine and 21, belonging to 6 families and 14 genera, are related to freshwater. The freshwater-related groups Branchiobdellae and Aphanoneura include 6 species in 2 genera and 1 family and 8 species in 2 genera and 1 family, respectively.

6.7.21   *Pogonophora*

A group of vermiform benthic, tube-dwelling, marine animal, living also at great deep. About 150 species are known. Previously considered at phylum rank, Pogonophora has been recently assigned to the anellid polychaete family Siboglinidae. The only Pogonophora species known in the Mediterranean Sea has been recorded along the western coast of Corsica. No records from Italian waters.

6.7.22 Echinura

A phylum of benthic marine worms close to annelids, including 236 species known. The present knowledge of Echiura in the Mediterranean area is scanty. In Italy, five of the six species are known in the Mediterranean.

6.7.23 Sipunculida

A group of marine worms, close to anellids. About 1500 species has been described. The phylum is represented in the Italian seas by 25 species, belonging to four orders, five families and nine genera, accounting about 76% of the 33 sipunculid species in the Mediterranean.

6.7.24 Arthropoda

Arthropoda is the largest animal phylum in terms of known species. Over 1.5 million species have been described, accounting for about 80% of the total number of species in the animal kingdom (Zhang, 2011, 2013). Insects are the most successful group as it accounts for over 80% of all known species of arthropods. Another important group is Arachnida (112,400 species). Other highly diverse arthropod groups include Crustacea (67,600 species) and Myriapoda (>15,000 species).

6.7.24.1 Arachnida

Arachnida is a class of arthropodans including spiders, scorpions, ticks, mites, harvestmen, falsescorpions and other groups. Globally 43,700 species of spiders are known. Besides, about 54,400 species of acari, 3,500 species of pseudoscorpions, 1,980 species of scorpions and 6,500 species of harvestmen have been recorded, as well as about 500 species belonging to few other small groups not represented in Italy (Zhang, 2013).

In Italy, the dominant groups of this class are represented by Acari, commonly known as mites and ticks, and Araneae, the spiders. The Italian fauna of Acari includes 2,863 species arranged in six orders, 280 families and 964 genera. The most richer order of mites is Actinedida, with 1,160 species in

355 genera and 88 families, followed by Oribatida, including 721 species, 249 genera, 93 families, Gamasida, with 580 species in 150 genera and 51 families, Acaridida, with 365 species in 187 genera and 45 families, Opilio-acarida, with only one species, one genus and one family. Ixodida, ticks, are represented by 36 species, in 22 genera and two families.

According to Pantini and Isaia (2016), spiders are represented in the Italian fauna by 1,620 species under 426 genera and 54 families, the most representative are Linyphiidae (470 species), Gnaphosidae (156), Salticidae (139), and Theridiidae (108).

Other rich arachnid orders are Pseudoscorpionida, the false scorpions, including in Italy 209 species arranged in 43 genera and 12 families, and Opiliones, the harvestmen, with 120 species, 43 genera, 11 families. A smaller number of the representative are in Scorpiones (scorpions), with one family, one genus, and 10 species, Palpigradi (microwhip scorpions), with one family, one genus, nine species; Solifugae (solifuges), represented by only one family, one genus, two species. As essentially tropical-subtropical groups, these three taxa are only just represented in more temperate regions.

6.7.24.2   Pycnogonida

Pycnogonida (sea spiders) are a group of benthic marine arthropods of uncertain systematic position, including 700 species. The present knowledge of the Italian fauna is far to be complete. A not definitive list of the species includes 44 of the 56 taxa recorded in the Mediterranean, belonging to 8 families and 14 genera.

6.7.24.3   Crustacea

The global diversity of crustacean has been estimated around 67,700 species, about 5% of the known extant arthropods. About 3,230 species have been recorded in Italy, approximately belonging to 1,219 genera, 390 families and 30 orders, mostly inhabiting marine and freshwater habitats, although also terrestrial species are known. They account for the 7% of the whole Italian arthropod fauna. Six class are represented in the study area, Branchiopoda, Ostracoda, Maxillopoda, Mystacocarida, Cirripedia, Malacostraca.

Branchiopoda includes in Italy 132 species, arranged in four orders (Anostraca, Notostraca, Spinicaudata, Cladocera), 20 families, 59 genera. They inhabit fresh- and hyperaline waters. Cladocera, commonly known as water fleas, is the largest group including 450 species worldwide, mostly inhabiting freshwater habitats; 110 species has been recorded in Italy and only six are marine. Ostracoda, ostracodes or seed shrimps, are small crustaceans mostly bentonic or planktonic, living in a wide variety of aquatic habitats; about 8,000 species are known worldwide; in Italy 377 species, belonging to 102 genera in 31 families has been recorded in marine habitats, whereas 157 species, belonging to 57 genera and 12 families are known in non marine habitats (Pieri et al., 2015). The Italian fauna of Maxillopoda includes three subclass: Copepoda, Branchiura and Pentastomida. Copepoda is a very large group of planktonic and bentonic crustaceans living in marine and continental waters, the class includes also parasitic species of fishes; about 11,500 species are known worldwide. According to Minelli et al. (1993–1995), the Italian fauna consist of 970 species in 8 orders, 117 families, 377 genera (see also Relini 2010 for updating). Branchiura (fish lices) is a small group of fish ectoparasites including only four species in Italy. Pentastomida is an enigmatic group of parasitic crustaceans related to respiratory organs of vertebrates; two species have been recorded in Italy. Mystacocarida is a group of microscopic interstitial crustaceans including dozen species worldwide. Only one species is known in the Italian seas. Cirripedia (barnacles) is also considered an infraclass of the crustacean subclass Thecostraca; they are among the most modified crustaceans, mostly bentonic, one order (Rhizocephala) includes species parasites of crabs; the Italian fauna includes 49 species in 4 orders, 13 families and 29 genera.

With near 1,600 species assigned to about 200 families and more than 600 genera, Malacostraca is the most biodiverse group of crustaceans of the Italian fauna (Minelli 1993–1995; Relini 2010). Twelve orders are represented in Italy. Bathynellacea and Thermosbenacea are entirely stygobiont (subterranean waters) and include five to four species respectively. Five species are included in Leptostraca and eight in Stomatopoda, the mantis shrimps, both groups living in benthic or bathypelagic marine habitats. Only three marine species are included in Lophogastrida, whereas Mysida includes 76 marine and three freshwater species. Cumacea, living in marine mobile sediment, includes in Italy near 40 species and Tanaidacea, mostly inhabiting soft bottoms and vegetated marine systems, are in Italy with 54 species. Euphausiacea, shrimp-like exclusive marine crustaceans includes 13 species. Isopoda, Amphipoda and Decapoda are the richest and the most diversified orders of Malacostraca. They include marine, freshwater and terrestrial species, adapted to a wide range of habitats. Isopoda diversity consists of about 580 species in near 50 families and 179 genera, marine species are 194 in 90 genera and 28 families, about 44 species belonging to 11 genera and seven families inhabits freshwater habitats and ca 360 species in 21 families and 79 genera are terrestrial (Oniscida). Amphipods of the Italian seas are 458, mostly belonging to suborder Gammaridea (365 species), whereas about 104 species in 17 genera and 11 families are known in freshwater habitats (Ruffo and Stoch, 2006b). Decapoda are mainly marine, a number of species are restricted to continental surface and subterranean waters. 295 species, arranged in 152 genera and 66 families are known in Italian seas. Twelve species belonging to 10 genera and six families are currently recorded in freshwater habitats.

6.7.24.4   Myriapoda (Chilopoda, Diplopoda, Symphyla, Pauropoda)

Our knowledge of myriapods, a group including only terrestrial forms, mostly inhabiting soil, is highly incomplete. Approximately, 3,300 species are known in the class Chilopoda (centipedes), predators, 11,000 in Diplopoda (millipedes), 200 in Symphyla, 700 in Pauropoda, all saprophagous (Minelli and Golovatch, 2013). In Italy, nearly 100 species of Chilopoda in about 30 genera, 12 families and four orders are known. Diplopoda account for 473 species in 132 genera, 28 families and eight orders. 19 species in seven genera and two families and 43 species in 13 genera and three families are listed for Symphyla and Pauropoda, respectively.

6.7.24.5   Hexapoda

The subphylum Hexapoda – 1,063,532 species in 4 classes (Collembola, Protura, Diplura, Insecta) according to Zhang (2013) – is, notoriously, the most biodiverse group of both Arthropoda, representing about the 82% of the known species (1,242,040), and Metazoa, accounting for near 66% of the species (1,552,319). Hexapods also represent about 2/3 of all living organisms, that is, viruses, bacteria, protists, fungi, animals, and plants. The species diversity is obviously not uniform in the group and the most successful class are Insecta, with 1,053,578 known extant species; Collembola account for 8,163 species, Diplura 975, Protura 816 (Zangh, 2013) Coleoptera, represents about 40% of the insect species of the world, over one-third of all known species. Other major orders are Lepidoptera (accounting for 15% of the known insects species), Hymenoptera (13%), Diptera (12%), Rhynchota (10%); the remaining orders represent approximately 10% of the species (percentages from Sbordoni et al., 2004). According to the most recent compilation of the Italian fauna (Minelli et al., 1993–1995), at least 37.303 species of insects have been quoted in Italy but the present number is certainly larger, thanks to the increase in the faunistic and taxonomic research in this country in recent years. Moreover, Collembola are represented in Italy by near 400 species (about 8,100 species known worldwide), Diplura account for 76 species (975) and Protura for 30 (816). Of the insects orders/suborders represented in the Italian fauna, the largest ones are Coleoptera (12,005 species), Hymenoptera (7,509), Diptera (6,601), Lepidoptera (5,086), Homoptera (2,150) and Heteroptera (1,373), together accounting for the 93% of the total of the species, more or less reflecting the rates at global level. Other important orders are Trichoptera (367), Orthoptera (333), Mallophaga (243), Thysanoptera (214), Planipennia (153), Plecoptera (144), Psocoptera (102). A brief account of the diversity of the main insect orders of the Italian fauna here follows.

6.7.24.5.1   Coleoptera

Beetles represent the richer order of living organisms in terms of a number of known species. Globally, about 392,000 species in 400 genera and more than 160 families are currently known (Bouchard et al., 2011; Zhang, 2013) and new species are continuously discovered. Little less than 30,000 species are listed from Europe (Audisio et al., 2015).

The coleopterans fauna of Italy is represented by four suborders, Archostemata, Adephaga, Myxophaga, and Polyphaga, including near 140 families, 2,757 genera and near 12,000 species. Among Archostemata, only Crowsoniellidae are represented, with one species, the enigmatic *Crowsoniella relicta* Pace, 1975, a minute species (about 1.8 mm long), the only of the genus and of the family, only recorded from the type locality in Central Apennines (Lepini Mts).

Adephaga is represented by five families (365 genera, 1,885 species), with the most represented of which are Carabidae, the ground beetles, with 1,292 species belonging to 305 genera. Myxophaga is represented by two families, Microsporidae and Hydroscaphiade, with two genera and five species in all. Polyphaga are therefore the most represented suborder, including 132 families, 2,389 genera, and 10,479 species. Among this group, the utmost biodiverse family is Staphylinidae, the rove beetles, with 2,534 species in 365 genera (including Pselaphinae, originally regarded as a separate family, with 329 species in 51 genera). Another highly diverse group of this suborder is Curculionidae, the weevils, with 1,798 species in 412 genera (including Platypodinae, the ambrosia beetles, with one genus and two species, and Scolityinae, the bark beetles, with 41 genera and 129 species, both originally regarded as separate families).

Other three representative families of polyphagous beetles in Italy are Chrysomelidae, the leaf beetles, Cerambycidae, the long-horned beetles, and Tenebrionidae, the darkling beetles. Leaf beetles include 885 species in 163 genera (comprising Bruchinae, the bean weevils or seed beetle, with 72 species in 7 genera, originally regarded as a separate family), long-horned beetles account for 121 genera in 268 species, whereas darkling beetles fauna includes 108 genera and 258 species.

Families of Coleoptera Polyphaga with high levels of diversity in the Italian fauna are many other and among these one should remember Cholevidae (50 genera, 238 species), Elateridae (73 genera, 235 species), Melyridae (34 genera, 227 species), Cantharidae (17 genera, 210 species), Buprestidae (42 genera, 209 species), Apionidae (52 genera, 204 species), Scydmaenidae (31 genera, 185 species).

6.7.24.5.2   Hymenoptera

Sawflies, wasps, bees, ants and allied are represented in Italy by 72 families, 1,580 genera and 7,509 species. The most representative families or superfamilies are, in order, Ichneumonidae, a significant group of parasitoid wasp of other insects, including 1,880 species in 448 genera, Braconidae, an additional group of parasitoid wasp, with 867 species in 165 genera, Tenthredinidae (431 species in 82 genera), Chalcidoidea (at least 18 families with 1,130 species, but the real number should be doubled or more, in 351 genera; in terms of number of species, the most representative families are Eulophidae, 311 species in 66 genera, Pteromalidae, 253 species in 108 genera, Aphelinidae, 146 species in 70 genera). Among the most popular groups are also Vespidae, wasps and allied, including both eusocial and solitary representatives, with 21 species in 6 genera, Apoidea, bees and allies, also social or solitary, arranged in seven families, 62 genera and 943 species. Formicidae, ants, includes 60 genera and 226 species.

6.7.24.5.3   Diptera

Diptera are represented in Italy by 107 families, 1,706 genera and 6,601 species, divided in two suborders, Nematocera (30 families, 534 genera and 1,912 species) and Brachicera (70 families, 1,172 genera and 4,689 species). The richest families mostly belong to the suborder Brachicera. The larger family is Syrphidae, the hoverflies, a large group of colored dipterans, represented in the national fauna by 482 species in 94 genera. Other brachiceran representative families are Tachinidae, the tachinids, with 474 species in 222 genera, Bombyliidae, the bee-flies, 302 species, 47 genera. All belong to Brachicera, other large families such as Dolichopodidae (45 genera, 287 spp.), Muscidae (40 genera, 281 spp.), Empididae (42 genera, 269 spp.) and Agromyzidae (26 genera, 203 spp.). Among Nematocera, the largest families are Chironomidae, the nonbiting flies, with 427 species in 141 genera, Cecidomyiidae, the gall-midge, 324 species, 101 genera, and Limoniidae, 232 species in 70 genera.

6.7.24.5.4   Lepidoptera

This group includes moths, butterflies and skippers. Lepidoptera are represented in Italy by 5,086 species arranged in 1,435 genera and 89 families. The most representative family is Noctuidae, the owlet moths, with 816 species in 336 genera. Geometridae, the geometer moths, and Tortricidae, tortrix or leafrollers moths, are also well represented, the first with 619 species in 190 genera, the latter with 605 species in 117 genera. Other large families are Gelechidae, twirler moths (359 species, 80 genera), Crambidae, the crambid snout moths (293 species, 86 genera), Coleophoridae, the case-bearing moths (238 species, 4 genera), Pyralidae, the pyralid moths (212 species, 95 genera), Gracillariidae (142 species, 19 genera), Nepticulidae, the midget moths (138 species, 12 genera), Tineidae, the fungus moth (110 species, 33 genera), Depressariidae (100 species, 8 genera). The most popular group is, however, represented by Rhopalocera, butterflies and skippers, including nine families of diurnal species. In Italy, 275 species in 78 genera are included in the *Checklist of the species of the Italian Fauna* (Minelli et al., 1993–1995). The most representative families are Satyridae, the browns (79 species, 16 genera), Lycaenidae, the gossamer-winged butterflies (69 species, 21 genera), Nymphalidae (55 species, 16 genera) and Pieridae (26 species, 7 genera). Eight species in four genera of Papilionidae are also known.

6.7.24.5.5   Hemiptera

Hemiptera are represented in Italy by 3,523 species arranged, following the Fauna Europaea classification system, in four suborders, Cicadomorpha, Fulgoromorpha, Heteroptera and Sternorrhyncha. Heteroptera shows the higher diversity as are represented in the local fauna by 1,373 species in 549 genera and 39 families. Miridae, plant bugs (504 species, 186 genera), Lygeidae, milkweed bugs (233 species, 96 genera) and Pentatomidae, stink bugs (104 species, 97 genera) are the most biodiverse families. High values of diversity are also in Sternorrhyncha, that includes 1,283 species in 439 genera, 24 families and five superfamilies. Two of these superfamilies are the most representatives, Aphidoidea, the aphids, with 675 species, 209 genera and one family (Aphidae), and Coccoidea, the scale insects, with 343 species, 153 genera, 14 families, whose Diaspididae (114 species, 53 genera) and Pseudococcidae (114 species, 37 genera) are the richest. Superfamily Psylloidea account for 199 species, 46 genera in six families. Cicadomorpha includes 612 species in 198 genera, 4 families and three superfamilies. More than the 90% of the cicadomorphs in Italy is represented by Cicadellidae, the leafhoppers, with 562 species in 179 genera. Fulgoromorpha includes 255 species of plant-hoppers arranged in 100 genera and 10 families of which Delphacidae is the most representative group with 114 species in 54 genera.

6.7.25 Tardigrada

Tardigrada is a group of microscopic invertebrate including marine and freshwater species. Even if terrestrial species are known, these can be regarded as essentially aquatic animals, as they are associated with the water film on mosses, liverworts, and lichens. Tardigrada can colonize very extreme habitats and survive for long periods in dried state. 1,157 species are known all over the world. The Italian marine fauna all belong to the class Heterotardigrada and account for 75 taxa arranged in two orders, six families and 26 genera. Terrestrial and freshwater Italian Tardigrada fauna includes 192 species (44 in freshwaters) belonging to 36 genera and eight families.

6.7.26 Phoronidae

An exclusively marine group of vermiform, tubicolous, animals, infaunal, suspension feeders. Only 10 species are known. The Italian fauna includes five species all belonging to the genus *Phoronis*.

6.7.27 Bryozoa

A phylum of colonial aquatic microscopic animals. The majority are marine, but brackish-water and freshwater forms also exist. Near 5,500 extant species are known. The marine bryozoan fauna of the Italian seas presently includes 341 species/subspecies, belonging to 153 genera and 77 families arranged in two classes. The class Gymnolaemata with 295 species/subspecies, including

262 species of the order Cheilostomatida (115 genera, 54 families) and 33 species of the order Ctenostomatida (18 genera, 11 families) is dominant. The class Stenolaemata (consisting only of the recent Cyclostomatida order) includes 46 species (20 genera, 12 families) but it is likely that the number of the species of Stenolaemata is underestimated. Italian freshwater Bryozoa are represented by 11 species, belonging to two classes, two orders, 6 families, 7 genera.

6.7.28 Brachiopoda

Brachiopods (lamp shells) is a phylum of sessile, benthic, suspension-feeders, exclusively marine, animals. There are 443 known species. The present-day fauna of this phylum in Italy is poorly known. Twelve species arranged in four orders, 10 families/superfamilies and 12 genera, have been reported in Italian waters.

6.7.29 Chaetognatha

Chaetognatha are vermiform marine, mostly planktonic but also benthonic, organisms. 179 species are known. In Italy 20 species are known belonging to six genera and five families. See also Ghirardelli and Gamulin (2004) for more details on the Italian fauna.

6.7.30 Echinodermata

Echinoderms are one of the most important exclusively marine animal phyla. The extant species are grouped into five classes, Crinoidea, Holothuroidea, Asteroidea, Ophiuroidea, Echinoidea. Generally bentonic, some holoturians inhabit pelagic habitat. In Italian waters 120 species were recorded out of the 143 known in the Mediterranean Sea. Specifically, three Crinoidea (two families, two genera), 36 Holothuroidea (eight families, 19 genera), 27 Asteroidea (11 families, 18 genera), 28 Ophiuroidea (10 families, 16 genera) and 26 Echinoidea (13 families, 19 genera) were recorded from the Italian basins.

6.7.31 Hemichordata

Hemichordates comprise worm-like marine animals living in sand or mud. Two classes are recognized, Enteropneusta (= Balanoglossida), acorn-worms, and Pterobranchia; 120 species are worldwide known. Acorn worms are solitary organisms, generally living in burrows and are deposit feeders, pharyngeal filter feeders, or free-living detritivores. Pterobranchs are filter-feeders, mostly colonial, living in a collagenous tubular structure. Six species and four families belonging to both classes have been recorded in the Mediterranean. Pterobranchia is not yet found in the Italian seas. Enteropneusta comprise in Italy five species, arranged in one order, three families and four genera.

6.7.32 Chordata

Cephalochordata (amphioxi or lancelets) are exclusively marine animals. Only one species is represented in the Italian fauna, *Branchiostoma lanceolatum* (Pallas, 1774). Also Tunicata are exclusively marine animals. In the Italian seas 41 species of the class Appendicularia (appendicularians), in nine genera and three families, have so far been recorded. Class Thaliacea (salps, doliolids and pyrosomas) is represented by 20 species, arranged in 13 genera, 4 families and two orders. The Italian checklist of the class Ascidiacea (ascidians) includes 12 families, 40 genera and 129 species.

6.7.33 Agnatha, Chondrichthyes, Osteichthyes

According to Zerunian (2007) the agnathans diversity of the Italian fauna include four species of lampreys, arranged in two genera, one family and one order. Of these, two are euryhaline species, as Mediterranean populations do anadromous migrations, reproducing and developing in watercourses, whereas trophic adult stage takes place in the sea. The other two species live exclusively in fresh waters. In last decades Italian lampreys have experienced a drastic reduction in terms of numbers and geographic distribution (Zerunian, 2002, 2007).

The cartilaginous fishes of the Italian fauna (sharks, rays, skates, sawfish and chimeras) are all related to marine environments. This group is represented by 72 species under 43 genera, 26 families and 11 orders, that is almost all the species of Chondrichthyes in the Mediterranean; two species are considered alien in the Mediterranean (Vacchi and Serena, 2010).

Osteichthyes (bony fishes) of the Italian seas includes 468 species (about 78% of the species reported in the whole Mediterranean), arranged in near 280 genera, 132 families and 22 orders. Perciformes is the richer group, including 243 species (more than half of the Italian fauna of sea bony fishes), arranged in 43 families, with Gobiidae (47 species), Sparidae (21), Labridae (19), Blennidae (19) among the most diverse (Relini et al., 2017: Table 6.1). Most of the families of sea bony fishes represented in the Italian fauna include small numbers of species as near 70% contain up to three species and near 40% are monospecific; 417 species are considered native or possibly native, the others are occasional, marginal or nonnative in the Mediterranean Sea (Relini et al., 2017). Twelve species of euryhaline bony fishes should also be considered in the Italian seas (Rondinini et al., 2013; Relini et al., 2017). According to Zerunian (2002, 2007) the bony fish diversity of the Italian freshwaters include 44 indigenous species arranged in 12 orders, 17 families and 32 genera. The order Cipriniformes includes the main part of the species (18 species). Alien fish introduction, either intentional or unintentional, is among the main threat to the conservation of the Italian freshwater fish fauna (Zerunian 2002, 2007). Zerunian (2002) list 29 alien species in Italy, of these 12 are naturalized and widely distributed in the area, 8 are naturalized but localized, 9 are regarded as not naturalized as represented by small populations often supported by repeated introductions.

6.7.34 *Amphibia and Reptilia*

The most updated checklist of the Italian fauna (Sindaco et al., 2006) lists 38 species of amphibians and 58 species of reptiles. Three families, six genera and 17 species, are recorded for Amphibia suborder Urodela (or Caudata, salamanders and newts), whereas six families, seven genera and 23 species are registered under suborder Anura (frogs and toads). Reptiles include four families, six genera for Chelonii (or Testudnines, turtles, tortoises and terrapins), four families and 14 genera under Squamata Sauria (geckos and lizards) and two families and 10 genera for Squamata Serpentes (snakes and vipers).

However, after this account, many new taxonomic and faunistic studies have been published and these figures need to be updated. New species have been indeed recognized and some taxa have been canceled (see Speybroeck et al., 2010). For instance, the urodelan genera *Salamandrina* (spectacled salamanders) and *Speleomantes* (cave salamanders), have been revised (Mattoccia et al., 2005; Carranza et al., 2008). Two species are now recognized in *Salamandrina*, instead of one, and eight species, instead of six, in *Speleomantes*. The Javelin sand boa, *Eryx jaculus*, new species for the Italian fauna has been recorded in Sicily (Insacco et al., 2015). Lastly, a new species of viper has been described by Ghielmi et al. (2016), *Vipera walser*, a species restricted to Northwestern Alps, showing strong genetic and morphological differences from all other known European vipers.

According to Rondinini et al. (2013), Italy presently hosts 44 species of amphibians and 56 of reptiles. Amphibians account for 19 species of Caudata (including 11 endemics), 25 of Anura (3 endemics), whereas Italian reptiles include 11 species of Testudines (1 endemics) and 45 of Squamata (4 endemics).

6.7.35 Aves

A checklist of the birds recorded in Italy until the end of 2014 has been provided by Brichetti and Fracasso (2015). It includes 548 species, under 167 genera, 80 families and 25 orders. Most of the species are included in the large and diverse order Passeriformes, including in Italy 34 families, 87 genera and 216 species, near 39% of the whole Italian ornithofauna. The richest genera of this group are *Sylvia* (scrub warblers, Sylvidae, 18 species), *Emberiza* (buntings, Emberizidae, 13 species), *Phylloscopus* (leaf warblers, Phylloscopidae, 13 species), *Turdus* (thrushes, Turdidae, 11 species), mostly related to Mediterranean habitats. Other 332 species, under 80 genera and 46 families are listed in the remaining 24 orders of non-Passeriformes. Among these, worthy of note is the diverse order Charadriiformes, including in Italy 10 families, 44 genera and 108 species of small to medium-large sized birds. Scolopacidae (sandpipers, snipes, curlews), Laridae (gulls, terns) and Charadriidae (plowers) are the most richer families.

Although an evaluation of the Italian bird fauna according to the phaenology of the species is difficult mostly because of the different behavior between populations in the national territory, leading to a high number of categories of presence, 169 species (30%) are resident breeding (including 16 naturalized); 212 species (40%) are regular migratory (115, 22%, regular wintering, breeding in North Europe; 97 (18%), regular breeding in Italy, wintering in Africa); 28 (5%), are irregular migratory species, whereas 139 (25%) are listed as accidental.

6.7.36 Mammalia

According to the updated checklist by Gippoliti (2013), the total number of recognized species in Italy (excluding Mysticeti and Odontoceti, two sub-order of Cetacea, whales and dolphins respectively) is 130, 24 of which are endemics or near-endemics. Seven orders, 23 families and 55 genera are recognized in all. This number is far from being definitive, although it includes still unnamed bat species, while 'species' derived from feralization of domestic animals are excluded. Bats and rodents are the most biodiverse orders, with 42 and 36 species respectively to date recorded. According to Cagnolaro et al. (2015), eight species of cetaceans belonging to eight genera and four families regularly occur in the Italian seas of the 11 to date regularly recorded in the whole Mediterranean area. Details on biodiversity of Italian mammals are in Table 6.7.

Table 6.7   Biodiversity of the Italian Mammal Fauna

Orders	Families	Genera	Species
Erinaceomorpha	1	1	3
Soricomorpha	2	5	16
Chiroptera	4	11	42
Lagomorpha	1	2	5
Rodentia	6	18	36
Carnivora	7	13	19
Artiodactyla	3	6	9
Cetacea	4	8	8
Total	28	64	138

Sources: Gippoliti (2013); Cagnolaro et al. (2015).

6.8   Alien Species

Biological invasions, resulting from the intentional or accidental transfer of taxa outside their native range through man-made activities, have become one of the main drivers of global environmental changes and the second major cause of the loss of global biodiversity after habitat loss (Hurka, 2002). Introductions are not a recent phenomenon. It is quite hard nowadays to determine whether a certain species in the Mediterranean has expanded its distribution range naturally or with the help of human activities. Among the plant species *Lycium europaeum* L. and *Rhus coriaria* L. are examples of species whose native status in the Italian territory is still debated. The number of alien taxa registered in Italy is fairly large. As far as fauna is concerned some hundreds nonnative species occur in our country. Among these at least 60 species of vertebrates have been recorded. (Scalera and Celesti-Grapow 2007). In 2009 the Inventory of the nonnative flora of Italy was published by Celesti-Grapow et al. (2009) that included 1023 specific and subspecific taxa. Since then, several new records and the continuous update of systematic and nomenclatural knowledge brought the inventory to more than 1500 taxa (Galasso et al., 2016). An analysis of the data still being settled allow to anticipate that actually the nonnative flora of Italy numbers 1523 species and subspecies of which 134 (8.8%) are archaeophytes and 1389 are neophytes (91.2%). Only 39 taxa (2.56%) occur in all the administrative regions and are invasive in some part of the country. Among the most dangerous ones for their large dispersal capacity invading different environments are: *Ailanthus altissima* (Mill.) Swingle,

Robinia pseudoacacia L. and *Senecio inaequidens* DC. On the contrary 479 taxa (31.45%) have been recorded in a single region (191 casual and 288 naturalized) and occur only in heavily modified anthropogenic environments.

The "Italian Zoological Union" has been entrusted for the implementation and control of data concerning the terrestrial invertebrates. Starting point of the inventory has been the *Checklist of the Species of the Italian Fauna* (Minelli et al., 1993–1995), integrated with other European checklist (e.g., DAISIE, 2009; Roques et al., 2010). The database (updated to February 2017) contains information on more than 1,200 species, belonging to five phyla (Platyhelminthes, Nematoda, Annelida, Mollusca, Arthropoda). Arthropods (near 1,150 species in 7 classes, mostly Insecta Coleoptera) represent the largest group (94%). Among the most invasive insects species recently (since 1990s) introduced in Italy worth to note are *Harmonya axyridis* (Pallas) (Coleoptera Coccinellidae), *Rhynchophorus ferrugineus* Olivier (Coleoptera Dryophthoridae), *Aethina tumida* (Murray) (Coleoptera Nitidulidae), *Drosophila suzukii* (Matsamura) (Diptera Drosophilidae), *Tuta absoluta* (Meyrick) (Lepidoptera Gelechidae), *Dryocosmus kuriphilus* Yasumatsu (Hymenoptera Cynipidae), *Vespa velutina* Lepeletier (Hymenoptera Vespidae). The xenodiversity recorded by Gheradi et al. (2007) in inland Italian waters amount to 112 species (64 invertebrates and 48 vertebrates), which contribute for about 2% to the inland-water fauna in Italy. Northern and central regions are the most affected, and Asia, North America, and the rest of Europe are the main donor areas (Gherardi et al., 2007). Among invertebrates mostly are crustaceans and the Louisiana crayfish, *Procambarus clarkii* Girard, is the most problematic. Highly critical is the situation related to fishes, as 60% of the Italian freshwater ichtyofauna (near 67 species) have been introduced (Stoch, 2008). Among amphibians and reptiles some ten cases of introduction are known. The most problematic species is *Trachemys scripta* Schoepff, competing with the indigenous *Emys orbicularis* Linnaeus. Alien mammals are mostly rodents. Especially worrying is the nearctic *Sciurus carolinensis* currently widespread in NW Italy with a nucleus in central Italy (Umbria). Its presence is threatening the survival of the indigenous *Sciurus vulgaris* Linnaeus. Another invasive species is the neotropical *Myocastor coypus* (Molina), responsible for serious changes to the riparian ecosystems (Stoch, 2008). In sea habitats, at least 135 animal species have been recorded, mostly Annelida, Polychaeta, Mollusca, Gastropoda and Bivalvia, Crustacea Decapoda (GSA-SIBM, 2016). According to Occhipinti Ambrogi et al. (2011) alien species recorded in

Italian marine and brackish coastal habitats between 1945–2009 are mostly native from tropics and were mainly introduced in the 1980s and 1990s. The highest number of alien species has been observed in the Lagoon of Venice (North Adriatic Sea). Of the total number of alien marine species, 15% may be considered invasive.

6.9   Concluding Remarks

Italy is one of the richest countries of Europe in terms of biodiversity. This richness is supported by a long-standing tradition in natural researchers and a renewed enthusiasm for field investigations. The heavy exploitation of its territory in the last century for reasons of subsistence during the two world war events, and the wild building development, 30 years ago, have led to profound wounds in the territory that could hardly be remedied. In any case, it must be primary interest of the administrators at both national and local level to preserve high conservation areas and to create and maintain ecological corridors to keep them in contact. The protection of biodiversity has not to be seen only as a set of limitations to the use of the territory, but as an opportunity to improve, in the short-term, the so-called economic conditions of people leaving in agricultural and seminatural environments.

Acknowledgment

The authors wish to express their thanks to the colleagues that gave suggestions and materials for the preparation of this contribution. In particular, we thank Roberto Bianconi, Maurizio Broglio, Massimo Ceci, Carlo Cerrano, Massimiliano Di Giovanni, Beppe Di Gregorio, Enrico Lana, Caterina Longo, Maria Mercurio, Roberto Messori, Paolo Pantini, Valerio Sbordoni, and Robert Wagensommer for having provided a part of the photos accompanying the text.

Keywords

• alien species   • animal diversity

References

Abdelahad, N., & Bazzichelli, G., (2007). Freshwater algae. In: *Biodiversity in Italy,* Blasi, C., Boitani, L., La Posta, S., Manes, F., & Marchetti, M., (eds.). Roma, Italy, pp. 187–191.

Aleffi, M., (2007). Bryophytes. In: *Biodiversity in Italy,* Blasi, C., Boitani, L., La Posta, S., Manes, F., & Marchetti, M., (eds.) Roma, Italy, pp. 162–171.

Anderson, S., (2002). *Identifying Important Plant Areas.* Plantlife International, London.

Angelini, P., Bianco, P., Cardillo, A., Francescato, C., & Oriolo, G., (2009). *Gli Habitat in Carta Della Natura [Habitats in Map of Nature].* Ispra. Roma.

APAT., (2005). *La Realizzazione in Italia del Progetto Europeo Corine Land Cover 2000 –* Rapporti APAT 36/2005, Roma.

Audisio, P., Alonso Zarazaga, M. A., Slipinski, A., et al., (2015). Fauna Europaea: Coleoptera 2 (excl. series Elateriformia, Scarabaeiformia, Staphyliniformia and superfamily Curculionoidea). *Biodiversity Data Journal, 3,* e4750. doi:10.3897/BDJ.3.e4750.

Bartolucci, F., Peruzzi, L., Galasso, G., et al., (2018). An updated checklist of the vascular flora native to Italy. *Plant Biosystems 152*(2), 179–303. doi: 0.1080/11263504.2017.1419996.

Bianchi, C. N., (2004). Proposta di suddivisione dei mari italiani in settori biogeografici. *Notiziario SIBM, 46,* 57–59.

Bilz, M., Kell, S. P., Maxted, N., & Lansdown, R. V., (2011). *European Red List of Vascular Plants.* Luxembourg, Publications Office of the European Union.

Blasi, C., & Michetti, L., (2007). The climate of Italy. In: *Biodiversity in Italy*, Blasi, C., Boitani, L., La Posta, S., Manes, F., & Marchetti, M., (eds.). Rome, Palombi Editori, pp. 57–66.

Blasi, C., (2010). *La Vegetazione d'Italia Con Carta Delle Serie di Vegetazione in Scala 1:500.000.* Palombi ed, Roma.

Blasi, C., Boitani, L., La Posta, S., Manes, F., & Marchetti, M., (2007). *Biodiversity in Italy.* Palombi ed., Roma, Italy.

Blasi, C., Burrascano, S., Maturani, A., & Sabatini, F. M., (2010b). *A Thematic Contribution to the National Biodiversity Strategy. Old growth forest in Italy.* Rome.

Blasi, C., Capotorti, G., Copiz, R., Guida, D., Mollo, B., Smiraglia, D., & Zavattero, L., (2014). Classification and mapping of the ecoregions of Italy. *Plant Biosystems, 148,* 1255–1345.

Blasi, C., Marignani, M., Copiz, R., Fipaldini, M., & Del Vico, E., (2010a). *Le Aree Importanti per le Piante nelle Regioni d'Italia:* il presente e il futuro della conservazione del nostro patrimonio botanico. [The Importan Plant Areas in Italian Regions: present and future of conservation of our plant heritage] Roma, Italy.

Bouchard, P., Bousquet, Y., Davies, A. E., Alonso-Zarazaga, M. A., Lawrence, J. F., Lyal, C. H. C., et al., (2011). Family group names in Coleoptera (Insecta). *ZooKeys, 972*(88), 1–972.

Brichetti, P., & Fracasso, G., (2015). Check-list of Italian birds, updated to the end of 2014. *Rivista Italiana di Ornitologia [Italian Journal of Ornithology], 85*(1), 31–50.

Cabrini, M., (2010). Phytoplankton & microbenthos. *Biol. Mar. Medit., 17*(1), 685.

Cagnolaro, L., Notarbartolo di Sciara, G., & Podestà, M., (2015). Mammalia IV. Cetacea. *Fauna d'Italia [Fauna of Italy], vol. 49.* Calderini, Bologna.

Capotorti, G., Alós Ortí, M. M., Anzellotti, I., Azzella, M. M., et al., (2015). The MAES process in Italy: Contribution of vegetation science to implementation of European biodiversity strategy to 2020. *Plant Biosystems, 149*(6), 949–953.

Carranza, S., Romano, A., Arnold, E. N., & Sotgiu, G., (2008). Biogeography and evolution of European cave salamanders, *Hydromantes* (Urodela: Plethodontidae), Inferred from mtDNA sequences. *J. Biogeography, 35*(4), 724–738.

Celesti-Grapow, L., Alessandrini, A., Arrigoni, P. V., Banfi, E., Bernardo, L., et al., (2009). Inventory of the nonnative flora of Italy. *Plant Biosystems*, *143*, 386–430.

Cibic, T., & Facca, C., (2010). Microphytobenthos. *Biol. Mar. Medit.*, *17*(1), 754–800.

Conti, A., Abbate, G., Alessandrini, A., & Blasi, C., (2005). *An Annotated Checklist of the Italian Vascular Flora*. Palombi, Rome.

Cortini Pedrotti, C., (1992). Checklist of Mosses of Italy. *Fl. Medit.*, *2*, 119–221.

DAISIE, (2009). *Handbook of Alien Species in Europe*. Springer, Dordrecht.

Daly, M., Brugler, M. R., Cartwright, P., Collins, A. G., Dawson, M. N., France, S. C., et al., (2007). The phylum Cnidaria: A review of phylogenetic patterns and diversity 300 years after Linnaeus. In: *Linnaeus Tercentenary: Progress in Invertebrate Taxonomy*, Zhang, Z. Q., & Shear, W. A., (eds.). *Zootaxa*, *1668*, 127–182.

De Jong, Y., Verbeek, M., Michelsen, V., Bjørn, P., Los, W., Steeman, F., et al., (2014). Fauna Europaea – All European animal species on the web. *Biodiversity Data Journal*, *2*, e4034. doi:10.3897/BDJ.2.e4034.

De Smet, W. H., Melone, G., Fontaneto, D., & Leasi, F., (2015). Marine Rotifera. *Fauna d'Italia*. Calderini, Bologna, vol. 50

Furnari, G., Giaccone, G., Cormaci, M., Alongi, G., Catra, M., Nisi, A., & Serio, D., (2010). Macrophytobenthos. *Biol. Mar. Medit.*, *17*(1), 801–828.

Galasso, G., Conti, F., Peruzzi, L., et al., (2018). An updated checklist of the vascular flora alien to Italy. *Plant Biosystems, 152*(3): 556–592. doi: 10.1080/11263504.2018.1441197

Genovesi. P., Angelini, P., Bianchi, E., Dupré, E., Ercole, S., Giacanelli, V., Ronchi, F., & Stoch, F., (2014). Specie e habitat di interesse comunitario in Italia: Distribuzione, stato di conservazione e trend. ISPRA [Species and habitats of Communitary interest in Italy: Distribution, State of Conservation and Trends], *Serie Rapporti*, *194*.

Gherardi, F., Bertolino, S., Bodon, M., Casellato, S., Cianfanelli, S., Ferraguti, M., et al., (2008). Animal xenodiversity in Italian inland waters: Distribution, modes of arrival, and pathways. *Biological Invasions*, *10*(4), 435–454.

Ghielmi, S., Menegon, M., Marsden, S. J., Laddaga, L., & Ursenbacher, S., (2016). A new vertebrate for Europe: The discovery of a range restricted relict viper in the western Italian Alps. *J. Zool. Systematics and Evolutionary Res.*, *54*(3), 161–173.

Ghirardelli, E., & Gamulin, T., (2004). *Chaetognatha. Fauna d'Italia,* Calderini, Bologna, *39.*

Gippoliti, S., (2013). Checklist delle specie dei mammiferi italiani (esclusi Mysticeti e Odontoceti): Contributo per la conservazione della biodiversità Checklist of the species of Italian mammals (excluded mysticetes and odontocetes): Contribution to biodiversity conservation. *Bolletino del Museo Civico di Storia Naturale di Verona (Botanica e Zoologia)*, *37*, 7–28.

GSA-SIBM, (2016). *Specie Aliene Presenti nei Mari Italiani [Alien Species occurring in the Italian Seas]*. http://www.sibm.it.

Hurka, H., (2002). Evolutionary consequences of biological invasion. *Neobiota*, *1*, 203–204.

INFC, (2007). Le stime di superficie 2005 – Prima parte. *Inventario Nazionale Delle Foreste e Dei Serbatoi Forestali di Carbonio [Area Estimations 2005 – First Part. National Inventory of forests and Carbon Forest Accumulations]*, Tabacchi, G., De Natale, F., Di Cosmo, L., Floris, A., Gagliano, C., Gasparini, P., Genchi, L., Scrinzi, G., & Tosi, V., (eds.). MiPAF – Corpo Forestale dello Stato – Ispettorato Generale, CRA – ISAFA, Trento. http://www.inventarioforestale.org [Last accessed 01/06/2017].

Insacco, G., Spadola, F., Russotto, S., & Scaravelli, D., (2015). *Eryx jaculus* (Linnaeus, 1758): A new species for the Italian herpetofauna (Squamata: Erycidae). *Acta Herpetologica, 10*(2), 149–153.

IUCN, (2012). *IUCN Red List Categories and Criteria: Version 3. 1.* 2nd ed., Gland, Switzerland and Cambridge: IUCN Species Survival Commission, IUCN.

Landucci, F., Panella, L., Lucarini, D., Gigante, D., Domizia, D., Kell, S., Maxted, S., Venanzoni, R., & Negri, V., (2014). A prioritized inventory of crop wild relatives and wild harvested plants of Italy. *Crop Science*, doi: 10.2135/cropsci2013.05.0355.

Lucarini, D., Gigante, D., Landucci, F., Panfili, E., & Venanzoni, R., (2015). The Archive taxonomic Checklist for Italian botanical data banking and vegetation analysis: Theoretical basis and advantages. *Plant Biosystems, 149*, 958–965.

Magrini, S., Atzeri, P., Bacchetta, G., Bedini, G., Carasso, V., Carta, A., et al., (2016). The conservation of the endemic crop wild relatives in the RIBES seed-banks: Towards a national priority list for the Italian CWRs. In: *Atti del Convegno RIBES Una rete per la Biodiversità: 10 Anni di Conservazione*, Mariotti, M., & Magrini, S., (eds.). Cagliari, pp. 73–74.

Martellos, S., Michele, A., Roberta, T., Rodolfo, R., & Nimis, P. L., (2013). An information system on Italian liverworts, hornworts and mosses. *Plant Biosystems, 147*, 529–535.

MATTM (Ministero Ambiente Tutela Territorio e del Mare)., (2010). DECRETO 27 aprile 2010. Gazzetta Ufficiale Repubblica Italiana, *155*(1), Suppl. Ord. 115.

Mattoccia, M., Romano, A., & Sbordoni, V., (2005). Mitochondrial DNA sequence analysis of the spectacled salamander, *Salamandrina Terdigitata* (Urodela: Salamandridae), supports the existence of two distinct species. *Zootaxa, 995*(1), 1–19.

Médail, F., & Quézel, P., (1999). Biodiversity hotspots in the Mediterranean basin: Setting global conservation priorities. *Conservation Biology, 13*, 1510–1513.

Minelli, A., & Golovatch, S. I., (2013). Myriapods. In: *Encyclopedia of Biodiversity*, Levin, S. A., (ed.), Second edition, Waltham, MA, Academic Press, *vol. 5*, pp. 421–432.

Minelli, A., Ruffo, S., & La Posta, S., (1993–1995). *Checklist delle Specie della Fauna Italiana*. Fascicoli 1–110 Checklist of the species of Italian Fauna. Issues 1-110. Calderini, Bologna.

Minelli, A., Ruffo, S., & Stoch, F., (2006a). Endemism in Italy. In: *Checklist and Distribution of the Italian Fauna*, Ruffo, S., & Stoch, F., (eds.). Memorie del Museo civico di Storia naturale di Verona 2. *Serie, Sezione Scienze della Vita, 17*, 29–31.

Minelli, A., Ruffo, S., & Vigna Taglianti A., (2006b). The Italian fauna Provinces. In: *Checklist and Distribution of the Italian Fauna*, Ruffo, S., & Stoch, F., (eds.). Memorie del Museo civico di Storia naturale di Verona, 2. *Serie, Sezione Scienze della Vita, 17*, 37–39.

Nimis, P. L., & Tretiach, M., (1995). The lichens of Italy. A phytoclimatic outline. *Crypt. Bot., 5*, 199–208.

Nimis, P. L., (2016). *The Lichens of Italy. A Second Annotated Catalogue.* EUT, Trieste.

Occhipinti, A. A., Marchini, A., Cantone, G., Castelli, A., Chimenz, C., Cormaci, M., et al., (2011). Alien species along the Italian coasts: An overview. *Biological Invasions, 13*, 215–237.

Onofri, S., Bernicchia, A., Filipello, M. V., Perini, C., Venturella, G., Zucconi, L., & Ripa, C., (2007). Fungi. In: *Biodiversity in Italy,* Blasi, C., Boitani, L., La Posta, S., Manes, F., & Marchetti, M., (eds.). Palombi ed., Roma, Italy, pp. 172–181.

Pantini, P., & Isaia, M., (2016). *Checklist of Italian Spiders.* http://www.MuseoScienzeBergamo.it/web/index.php?option=com_content&view=category&layout=blog&id=96&Itemid=94.

Peruzzi, L., Conti, F., & Bartolucci, F., (2014). An inventory of vascular plants endemic to Italy. *Phytotaxa, 168,* 1–75.

Pieri, V., Martens, K., Meisch, C., & Rossetti, G., (2015). An annotated checklist of the recent nonmarine ostracods (Ostracoda: Crustacea) from Italy. *Zootaxa, 3919,* 271–305.

Pignatti, S., Guarino, R., & La Rosa, M., (2017-2018). Flora d'Italia & Flora digitale, 1–3. Edagricole, Milano.

Relini, G., (2008). Checklist della Flora e della Fauna dei mari italiani – Checklist of the flora and fauna of Italian seas. *Biologia Marina Mediterranea, 15*(1), 1–385.

Relini, G., (2010). Checklist della flora e della fauna dei mari italiani (Parte II) [Checklist of the flora and fauna of the Italian seas (Part II)]. *Biologia Marina Mediterranea, 17*(5–10), 387–828.

Relini, G., Tunesi, L., Vacchi, M., Andaloro, F., D'Onghia, G., Fiorentino, F., et al., (2017). *Lista Rossa dei Pesci Ossei Marini Italiani [Red list of Italian marine bony fishes].* Comitato Italiano IUCN e Ministero dell'Ambiente e della Tutela del Territorio e del Mare, Roma.

Rondinini, C., Battistoni, A., Peronace, V., & Teofili, C., (2013). *Lista Rossa IUCN dei Vertebrati Italiani [IUCN red list of the Italian vertebrates].* Comitato Italiano IUCN e Ministero dell'Ambiente e della Tutela del Territorio e del Mare, Roma.

Roques, A., Kenis, M., Lees, D., Lopez, V. C., Rabitsch, W., Rasplus, J. Y., & Roy, D. B., (2010). Alien terrestrial arthropods of Europe. *BioRisk, 4*(1), 1–570.

Rossi, G., Montagnani, C., Abeli, T., Gargano, A., et al., (2014). Are red lists really useful for plant conservation? The new red list of the Italian flora in the perspective of national conservation policies. *Plant Biosystems, 148,* 187–190.

Ruffo, S., & Stoch, F., (2006a). *Checklist and Distribution of the Italian Fauna. Memorie del Museo civico di Storia naturale di Verona,* 2. Serie, Sezione Scienze della Vita, *17,* 1–303 + CD Rom.

Ruffo, S., & Stoch, F., (2006b). Crustacea Malacostraca Amphipoda. In: *Checklist and Distribution of the Italian Fauna,* Ruffo, S., & Stoch, F., (eds.). Memorie del Museo civico di Storia natural di Verona, 2. Serie, *Sezione Scienze della Vita, 17,* 109–111 + CD Rom.

Ruffo, S., & Vigna Taglianti., (2004a). A brief history of research. In: *Wildlife in Italy,* Minelli, A., Chemini, C., Argano, R., & Ruffo, S., (eds.). Touring Editore, Milan and Italian Ministry for Environment and Territory. Rome, 18–23.

Ruffo, S., & Vigna Taglianti., (2004b). An overall view of Italian wildlife. In: *Wildlife in Italy,* Minelli, A., Chemini, C., Argano, R., & Ruffo, S., (eds.). Touring Editore, Milan and Italian Ministry for Environment and Territory. Rome, pp. 24–28.

Sbordoni, V., Bologna, M. A., & Vigna, T. A., (2004). *Gli Insetti e la Biodiversità. Atti XIX Congresso Nazionale Italiano Entomologia* (Catania 10–15 giugno 2002) [Insects and biodiversity. XIX Acta of the National Italian Congress of entomology (Catania 10-15 June 2002)], pp. 137–148.

Scalera, R., & Celesti, G. L., (2007). Alien species. In: *Biodiversity in Italy,* Blasi, C., Boitani, L., La Posta, S., Manes, F., & Marchetti, M., (eds.). Palombi ed., Roma, Italy, 128–130.

Segers, H., (2007). Annotated checklist of the rotifers (Phylum Rotifera), with notes on nomenclature, taxonomy and distribution. *Zootaxa, 1564,* 1–104.

Sindaco, R., Doria, G., Razzetti, E., & Bernini, F., (2006). *Atlante degli Anfibi e dei Rettili d'Italia/Atlas of Italian Amphibians and Reptiles*. Societas Herpetologica Italica. Edizioni Polistampa, Firenze, Italy.

Smiraglia, D., Capotorti, G., Guida, D., Mollo, B., Siervo, V., & Blasi, C., (2013). Land units map of Italy. *J. Maps, 9*, 239–244.

Speybroeck, J., Beukema, W., & Crochet, P. A., (2010). A tentative species list of the European herpetofauna (Amphibia and Reptilia) – An update. *Zootaxa, 2492*, 1–27.

Stoch, F., & Minelli, A., (2004). Il progetto 'Checklist delle specie della fauna italiana.' Atti Convegno "La conoscenza botanica e zoologica in Italia: Dagli inventari al monitoraggio," Università di Roma "La Sapienza," 14 dicembre 2001 [The project "Checklist of the species of Italian fauna". Acta Meeting "The Botanical and Zoological knowledge in Italy: from interventions to monitoring" University of Rome "La Sapienza" 14 December 2001]. *Quaderni di Conservazione Della Natura, 18*, 11–20.

Stoch, F., (2008). *La Fauna Italiana, dalla Conoscenza alla Conservazione. The Italian Fauna. From Knowledge to Conservation*. Ministero dell'Ambiente e della Tutela del Territorio e del Mare DPN, Direzione per la Protezione della Natura.

Vacchi, M., & Serena, F., (2010). Chondrichthyes. *Biol. Mar. Mediterr., 17*(1), 642–648.

Zerunian, S., (2002). *Condannati all'estinzione? Biodiversità, Biologia, Minacce e Strategie di Conservazione Dei Pesci D'acqua Dolce Indigeni in Italia Condemned to extinction? [Biodiversity, Biology, Treats and Conservation strategies of freshwater fish native to Italy]*. Edagricole, Bologna.

Zerunian, S., (2007). Problematiche di conservazione dei Pesci d'acqua dolce italiani [Problems of conservation of Italian freshwater fish]. *Biologia Ambientale, 21*(2), 49–55.

Zhang, Z. Q., (2011). Animal biodiversity: An introduction to higher – Level classification and taxonomic richness. *Zootaxa, 3148*, 7–12.

Zhang, Z. Q., (2013). Phylum Arthropoda. In: Zhang, Z. Q., (ed.). *Animal Biodiversity: An Outline of Higher-level Classification and Survey of Taxonomic Richness (Addenda 2013), Zootaxa, 3703*(1), 17–026.

CHAPTER 7

Biodiversity in Norway

KJETIL BEVANGER

Scientific Advisor and Senior Research Scientist, The Norwegian Institute for Nature Research (NINA), Pb. 5685 Torgarden, NO–7485 Trondheim, Norway, E-mail: Kjetil.bevanger@nina.no

7.1 Introduction

7.1.1 Geography, Geology, and Land Characteristics

Located in Northern Europe, Norway is one of the world's northmost countries. Together with Sweden, it constitutes the Scandinavian Peninsula. To the west, it borders the Atlantic Ocean (Norwegian Sea), the Barents Sea to the northeast, the North Sea to the southwest, and the Skagerrak inlet to the south. Norway shares a long border with Sweden to the east, and a short one with Finland and Russia to the northeast. The country has an extremely long coastline and looking at a map, the elongated shape is striking. The latitude of the southernmost part of Norway's mainland is at 58°N and the northernmost part at 71°N. The northern part of the Spitsbergen islands (Svalbard) is 81°N. The longitude covers 5°E (Solund) to 31°E (Vardø). The Jan Mayen and Kvitøya islands reach to 9°W and 33°E, respectively. The total land area is 324,220 km², of which 307,860 km² is the land area and 16,360 km² is water. When including the Jan Mayen and Svalbard islands, the total land area is 385,199 km². Norway is definitely one of the most mountainous countries in Europe. The average elevation of the Scandinavian mountains is 460 m and about one third (32%) of the mainland is located above the timberline (https://en.wikipedia.org/wiki/Geography_of_Norway) (Figure 7.1).

The Scandinavian geology is characterized by 400-million-years-old rocks, created when the Caledonian Mountain Range thrust up. At that time, very high mountains were created, much higher than anything we see today. However, over some 100 million years it was eroded down to a more or less flat plateau that remained from about 300 to about 50 million years ago. Increased geological movements heaved up the western side of the Scandinavian plate close to the present Norwegian coastline. The eastern side, i.e., Sweden, remained lower. Thus, the plate remained at an angle where the western side was significantly higher than the eastern side, allowing the rivers to flow eastwards. This mountain chain still remains and characterizes the entire Scandinavian Peninsula. Southern and Western Norway are part

Figure 7.1 Norway with Scandinavia.

of it, as is the border between Norway and Sweden in Northern Norway. The northern part is frequently referred to as the "Keel" (Kjølen), because it appears as the keel of an upturned boat. Every mountain peak in Scandinavia higher than 2000 meters is located along the Keel, with the Rondane mountain range at the northern edge. In the central and highest part of the Keel, we find Jotunheimen, which is Scandinavia's most impressive and spectacular mountain range (Holtedahl, 1960).

The biodiversity in Scandinavia is reflecting the fact that it has experienced several ice ages over the last 50 million years. In the west, the ice-eroding scars are particularly spectacular. During the last 2.5 million years there may have been as many as 40 separate ice ages. The glaciers, coming and disappearing, shaped the valleys and corries and scraped away most of the soil and gravel, leaving the rocks nude and smooth. The most recent ice age ended only some 10,000 years ago, having lasted for about 40,000 years. During this period, an ice sheet, in some areas being several thousand meters thick, covered Norway. When the climate improved and the ice melted, the ice movements formed the present topographic characteristics: rivers, lakes, mountains and islands. Deep valleys were carved out and filled with seawater, creating the famous Norwegian fjords. The Norwegian coastline consists of more than 50,000 islands, and the fjords are protruding far into the mainland, of which the Sognefjord, Hardangefjord, Geirangerfjord, Trondheimsfjord, and Vestfjord are among the most spectacular (https://en.wikipedia.org/wiki/Geography_of_Norway).

Although the last ice age was followed by a significant change in climate, with increased average temperatures allowing vegetation and trees to grow much higher up in the mountains than today, there have also been some setbacks. The last setback took place during the period 1300–1850, bottoming out around 1740 (Lamb, 1995). During this period, glaciers flowed forward and destroyed both agricultural areas and houses. The moraines are still very visible as the vegetation has yet not been able to recolonize them. Still, Norway has some of Europe's largest remaining glaciers, such as Folgefonna, Hardangerjøkulen, and Jostedalsbreen. These do not remain from the last ice age, however but are of more recent origin.

The Palearctic Taiga, also known as the boreal forest or snow forest, is the western part of the largest biome in the world. A large part of Norway belongs to the Taiga, characterized by coniferous forest consisting mostly of pine (*Pinus* spp.), spruce (*Picea* spp.) and larch (*Larix* spp.). The Scandinavian Montane Birch Forest and Grasslands ecoregion, or the Scandinavian and Russian Taiga, is part of the Alpine Tundra and Boreal Forest biomes, located in Norway, Sweden, and Finland. The Scandinavian Montane Birch Forest and Grasslands is one of the Global 200 ecoregions, and regarded as a high priority for conservation [as defined by World Wildlife Fund (WWF) classification (Ecoregion PA0608)] (http://www.worldwildlife.org/ecoregions/pa0608). It is situated between tundra in the north and temperate mixed forest in the south and covers about 2,156,900 km², being the largest ecoregion in the world.

This ecoregion follows the Scandinavian mountains, and spans 11 degrees of latitude from the south to the north. About two-thirds are located in Norway, about one-third in Sweden and a small area touches the northwesternmost part of Finland. Consequently, it is a varied ecoregion, 1,600 km long, with a total area of approximately 243,000 km² (slightly larger than

Great Britain). The growth period in the Taiga is usually slightly longer than the climatic definition of summer as the plants of the boreal biome have a lower threshold to trigger growth. In Canada, Scandinavia and Finland, the growing season is frequently estimated by using the period of the year when the 24-hour average temperature is +5°C or more.

The largest European mainland glaciers are located in this ecoregion (Jostedalsbreen, Svartisen), as are the highest mountains in northern Europe (Jotunheimen) and the largest mountain plateau (Hardangervidda). These mountains are part of the Keel and create a rain shadow, making the eastern part of Norway considerably drier than the western part. The southern part of this ecoregion, i.e., Norway's central mountains, stretches down to the western fjords, while the northern part follows the Keel, and stretches down to the western fjords as well. The western part of the ecoregion borders on the Scandinavian Coastal Conifer Forest ecoregion or directly on the fjords (the Norwegian Sea and the North Sea), both in the south and north. The eastern part borders on the Scandinavian and Russian Taiga ecoregion both in the south and the north. In the far south, it meets the Sarmatic Mixed Forest ecoregion, while the extreme northeasternmost part close to the Barents Sea borders on the Kola Peninsula Tundra ecoregion (http://www.worldwildlife. org/ecoregions/pa0608).

Parts of the Scandinavian Montane Birch Forest and Grasslands ecoregion are located in smaller mountain areas surrounded by lower elevation biomes, as is the case with coastal mountains in Norway. In some valley areas, this ecoregion meets the inland Taiga, belonging to the Scandinavian and Russian Taiga ecoregion, without mountain barriers. These include the Rauma valley connecting Åndalsnes to Lesja and Dombås, and the Namdalen valley connecting the Nord-Trøndelag county coast to the cold interior with a connection into Sweden. The pine forests in the north have some of the oldest trees in Scandinavia, some more than 700 years old in Forfjord valley at Hinnøya, the fourth largest island in Norway, located in the county of Nordland.

At high altitudes, high alpine tundra with very modest vegetation and bare rock, scree, snowfields, and glaciers occur. At lower altitudes, low alpine tundra is located, with continuous plant cover; dwarf birch (*Betula nana*) and willows (*Salix* spp.) up to 1 m tall and grasslands, as well as lakes and bogs. Further down is the adjacent montane birch zone with small (2–5 m) mountain downy birch (*Betula pubescens*) above the conifer tree line, some stunted spruce and pine, and several lakes and bogs. This part is regarded as part of the High Boreal (Sparse Taiga) Vegetation Zone. *Birch* forming the treeline is very rare outside Scandinavia (but is found in Iceland and at the Kamtchatka Peninsula).

At low elevations, the forests become closed-canopy (mid-boreal, south-boreal), denser and taller (8–20 m, exceptionally 30 m), with more species, including mature Scots pine (*Pinus sylvestris*) and aspen (*Populus tremula*). Much of the lowland forest near the fjords of western Norway also belong to this vegetation zone, as Norway spruce (*Picea abies*) is mostly absent here (both Norway spruce and Sitka spruce (*Picea sitchensis)* are commonly planted for economic reasons, of which Sitka spruce also is an alien species). In these temperate forests, species such as oak (*Quercus robur*), European ash (*Fraxinus excelsior*), European yew (*Taxus baccata*), holly (*Ilex aquifolium*) and small-leaved linden (*Tilia cordata*) in addition to Scotch pine, aspen and birch (*Betula verrucosa*) are located. Thus, there might be a very large span in environmental conditions within this ecoregion, from the temperate forest near the fjords with a six-month growing season to the highest mountains with glaciers and snowfields not melting until mid-summer (Gjærevoll, 1973).

7.1.2 Light and Climate Conditions

Because the country covers 13 latitude degrees, light conditions and climate differ considerably from the south to the north, as well as from the east to the west. The areas located north of the Arctic Circle, of which the position is not fixed (running at 66°33′46.4″ north of the Equator as of 15.10.2016), have midnight sun as well as winter darkness. In the city of Tromsø the sun disc appears above the horizon 24 hours a day from May 17 to July 25, and hides below between November 26 and January 15. The southern part of the country also experiences significant seasonal variations when it comes to daylight. In the capital, Oslo, the sun rises at 03.54 and sets at 22.54 at the summer solstice, but is above the horizon only between 09.18 and 15.12 at the winter solstice.

Because of the North Atlantic Current, a powerful ocean current continuing the Gulf Stream to the northeast, and its extension, the Norwegian Current, Norway has a more temperate climate compared to other areas at the same high latitudes. The warm water contributes to raising the air temperature, and prevailing winds from the southwest take the mild air inland. The average January temperature for the coastal city of Brønnøysund is 14.6°C higher than the January average in Nome, Alaska, sharing the same coastal environment and situated at the same latitude, i.e., 65°N.

The highest levels of precipitation in Europe are found in certain coastal areas in Western Norway, as a result of orographic lift. In the county of Sogn and Fjordane, the village of Brekke has the highest annual precipitation with 3,575 mm, although some mountainous areas in the west may exceed 5,000 mm per annum. The autumn and early winter periods experience the highest precipitation along the coastline, while April to June are the driest months. In the county of Oppland, we find the place on the mainland with the lowest average, Skjåk (Øygarden), having only 278 mm.

During winter, the coastal areas have a much milder climate than inland areas at the same latitude. The temperature difference between the coldest and the warmest month is frequently only 11–15°C. Inland areas are quite different, and certain parts of northernmost county, Finnmark, have locations with much higher differences. Finnmarksvidda (the biggest mountain plateau in Norway with its 22,000 km²), has the coldest winters on the Norwegian mainland; however, areas far to the south can also experience temperatures below minus 50°C. The coldest temperature recorded is minus 51.4°C at Finnmarksvidda (Karasjok). On the opposite side, the warmest temperature ever recorded in Norway was in Nesbyen in Buskerud County, southern Norway, with 35.6°C. The inland areas have a quite stable snow cover during the winter months, while some of the coastal areas in the west only experience shorter periods of snow (https://en.wikipedia.org/wiki/Geography_of_Norway).

7.2 Main Ecosystem Characteristics

Because of the latitudinal range, as well as the altitudinal, topographic and climatic variation, Norway has numerous ecosystem types. The country is commonly divided into a southeastern, southwestern, central and northern region. A landscape with forests, valleys and croplands characterizes the southeastern region. The southwestern region has a mountainous landscape with some peaks higher than 2000 m. The coastal areas consist of numerous islands and skerries as well as deep fjords protruding more than 200 km into the mainland. In southwest, however, there are also flat coastal areas, being important croplands. The central part has fjords and mountains as well and large lowland areas with arable land and fertile soil. The topography of the northern region is quite dramatic, dominated by saw-toothed mountains rising straight up from the sea. Due to the harsh weather conditions, the timberline is much lower compared to the southern and central parts of the country. In the far north (Finnmark), flat tundra dominates the landscape, with some areas completely treeless.

While organisms have been classified into species, families and kingdoms for hundreds of years, there is no such classification system for ecosystems and habitats. However, the Norwegian Biodiversity Information Centre has for more than ten years worked with a system called "Nature Types in Norway" (NiN). The system is supposed to be a dynamic system that can be adjusted according to increased knowledge and user needs (Halvorsen, 2015).

Ecosystems and habitat types are not well defined and recognizable entities like an animal or a plant species. Nature differs from area to area and there is nothing like two identical habitats with the same environmental conditions or species composition. A majority of the variation in nature derives from gradual differences in species composition, as a response to gradual deviations along ecological gradients. This recognition was the basic for the NiN design, covering the entire Norwegian mainland, the Arctic islands and the marine areas. The system can be described using three dimensions, i.e., scale, generalization and description systems. When it comes to scale, five nature type levels are used: region, landscapes and seascapes, landscape element, ecological system and microhabitat. Three levels of generalization are used: major type group, major type and descriptive system (based on sources of variation). In the description system, there are six sources of variation: local ecolines, condition or impact ecolines, regional ecolines, dominance (species or species groups), objects related to site condition and history and landform variation (Halvorsen, 2015).

7.2.1 Pelagic Communities in the Norwegian Sea

The ecosystems in the Norwegian Sea are characterized by huge differences regarding ocean depth, from the coastal plateau to the Deep Ocean and abyssal plain at 3000–4000 meters. There are also huge ocean currents and long natural oscillations in climate conditions. In spite of a temperature drop the past four-five years, the ocean has been warmer than the mean temperature since the year 2000. Cold and salt polar water is flowing south in the deep sea, while higher up the North Atlantic Current, as a prolongation of the Gulf Stream, brings saltier and more temperate water north (Skjoldal, 2004). Along the coastline from Møre to Finnmark some of the world's most important fish stocks are spawning.

The Norwegian part of the Norwegian Sea is home to more than 1 million breeding seabirds, of which approximately 50% are Atlantic puffins, while the others are pelagic feeding species like the northern fulmar (*Fulmarus glacialis*), northern gannet (*Morus bassanus*) and razorbill (*Alca torda*) with 7,500, 3,600 and fewer than 10,000 pairs, respectively. The magnitude of the populations during the winter period is less known; however, both local breeders as well as birds arriving from adjacent sea areas are present. Most seabird species can get old, become sexually mature late and produce few offspring, thus they are adapted to changing food conditions. While they can follow the prey migrations during the winter, they are dependent on finding food within a reasonable distance from the breeding colonies during the breeding period. Poor food conditions locally may result in bad breeding success (Anker-Nilssen et al., 2015). Sea-living mammals have the same life-history strategies as long-lived fishes and seabirds. They have few off-spring and live longer, and a depleted population will need a long time to recover (van der Meeren et al., 2015).

7.2.2 *Coastal Ecosystems*

The Norwegian coastline, including islands, islets and skerries, is more than 100,000 km long, equal to the distance around the Equator times two, making it to the second longest coastline in the world, after Canada. The coastal area reaches one nautical mile outside the sea boundary, and both the seabed and the water column above (the pelagic zone), harbor rich eco-systems. The contribution of nutrition from terrestrial areas with rivers, shallow waters getting light from the sun, and an alternation between stirring and stratification of water masses during differing seasons, are characteristics making the coastal areas very productive compared to the open sea waters further out.

When the plankton algae flourish during the spring, this brings food surplus for small single- and multicellular organisms like ciliates and copepods, eaten by carnivores like fish and jellyfish. However, not all of the food is eaten in the water column, and what sifts down on the seabed is food for bristleworms (Polychaeta), bivalves, and others. On rock substrate, and where it is possible to get grip, seaweeds and kelps are growing, creating the ocean forests dominating the lighted part of the seabed. These forests provide food and growth conditions as well as habitats to numerous other algae species, small and larger crustaceans, gastropods and fishes. Both molluscs, crustaceans and bristleworms hide in the vegetation and become a food source to fishes both inside and outside of the kelp forests. Several of the fish species live their whole lives along the coast, while others come to the coastal areas to feed or spawn. This huge abundance of food forms the basis for the rich populations of seabirds, seals and whales along the coast of Norway (Gundersen et al., 2015).

7.2.3 Freshwater

Approximately 5% (17,000 km²) of the Norwegian land area is freshwater. The waterways have typically several small lakes (<0.1 km²) and small fast flowing rivers (watersheds <10 km²). The ecosystem services connected to freshwater is subject to several conflicting interests, particularly because of their potential value as hydropower sources. A total of 2,800 animal species have their main distribution connected to Norwegian freshwater, during their whole lives or part of their lives (Aagaard and Dolmen, 1996). Invertebrates are the dominating animal groups, with most species belonging to chironomids, rotifers and water bugs. Norway has 42 freshwater fish species; however, the number depends on how freshwater/saltwater species are defined. Freshwater bodies are also important feeding and/or breeding areas for approximately 80 bird species and 6 mammals. The number of plant species connected to freshwater is unknown. Phytoplankton (i.e., microscopic algae drifting in the water column) is the most diverse species group with about 1050 taxa (i.e., species, genera or higher taxonomic levels). Stationary algae is a diverse group in flowing water with about 900 taxa. There are 125 species of water plants, including Charales (an order of freshwater green algae in the division Charophyta). Thus, in total about 5,000 animal and plant taxa are known (Schartau et al., 2015).

7.2.4 Wetlands

A large part of the land area in Norway is covered with wetlands. Wetlands are characterized by soils being saturated with water permanently or periodically, requiring special adaptations among the organisms living in these habitats. Wetlands constitute a long range of habitats, from sea level to high up in the mountains. Bogs cover the largest area; however, springs, waterfalls, meadows and floodplains (swamps, water borders, dried up riverbeds, etc.) are other examples of wetlands. It is however, difficult to make a sharp distinction between wetlands and other ecosystem and habitat types. Typical animal groups connected to wetlands are amphibians (Bjerke et al., 2015).

An interesting wetland type is the palsas, i.e., low, oval frost heaves of peat with a permanently frozen ice lens. Palsas typically occur in areas with discontinuous permafrost and are created during cold periods when the summer is too cold to offer complete thawing conditions. Several palsas were probably created during the Little Ice Age, reaching its maximum around 1740. Palsa peat bog is regarded as a threatened habitat type, due to significant thawing during the last 100 years (Hofgaard and Myklebost, 2016). The reason for the thawing could be increased summer precipitation reaching down to the ice lens and melting it. It could also relate to increased winter snow depths, creating an isolating cover not allowing cold air to go down, resulting in increased summer and winter temperatures. Erosion is, however, also a natural development process of older palsas.

7.2.5 Forests

Forest and areas with trees characterize about 44% of the Norwegian land area. However, there are significant geographical differences when it comes to how forest dominates the landscape. Many of the forested areas are mountainous forests suitable for, e.g., firewood, and forested bogs or other types of low productive forest (Storaunet and Framstad, 2015). Thus, the productive forest areas only occupy 86,000 km^2, or about 27% of the forested land areas. Since the 1920s, the productive forest area has increased by about 10%. At the same time, the standing timber volume has increased from about 300 million m^3 to about 900 million m^3 (Tomter and Dalen, 2014). This significant increase is particularly connected to modern forestry practices, and reflects an intention of establishing dense forest stands that utilize the growth potential of the areas more effectively than previous practices of limit- and selection-felling did.

The forest production of living and dead organic matter, together with the huge variation in local environmental conditions and development after disturbances like fires and storm felling, has created growth conditions to a huge number of species. It is estimated that about 60% of the number of known species in Norway (being about 40,000) is connected to forest habitats. In the latest Red Data List for species (Henriksen and Hilmo, 2015), approximately 50% of the threatened or near threatened species are forest species. For many of these the industrialized forestry practice is assessed as the most important impact factor.

Today, Norwegian forestry is mainly based on the stand forestry model, with open logging, i.e., clear felling combined with leaving crop trees. The reference basis for forests used by the Nature Index, i.e., the former state that the present condition is compared with, has its own definition. It is defined as forest close to nature with minor human activities and encroachments, where the natural disturbance processes (e.g., forest fires, storm-felled trees, insect outbreaks) with following succession stadiums are present on all forested area (Pedersen and Nybø, 2015). Such a situation has not been seen in Norway for many hundred years, as humans have exploited the forest and the outlying fields for a very long time. There are, however, some small areas with forest being close to nature in protected areas, and some areas, which are difficult to access. Data from the Norwegian Central Forest Monitoring Database (Landskogstakseringen) is used as part of the basis to evaluate the reference state (Nilsen et al., 2010). Forestry is regarded as the main human-induced disturbance in Norwegian forests together with land use change and road and power-line construction, as well as the increased input of nitrogen and climate change.

7.2.6 *Mountain Ecosystems*

About 28% or 119,000 km^2 of the Norwegian land area is mountainous habitats. Stones, blocks or nude rocks with a sparse vegetation cover, dominates the majority of these areas. Although there are some huge continuous mountain areas, many of them are naturally fragmented in smaller or bigger areas containing fjords and forested valleys. Plants and animals living in the mountains are partly exposed to extreme environmental conditions with stress factors like low temperatures, dehydration, wind, a short growing season, as well as ice and snow cover during the winter. Mountainous habitats have poorly developed soils with restricted nutrient access. Changes in these factors may have significant consequences for abundance and survival for mountain species. Over the last 100 years, mountain habitats are mainly affected by human activities like reduced summer farming, increased tourism, road and railway construction, hydropower development, pollution, hunting and climate change (Pedersen and Aarrestad, 2015).

As a result of road and railway construction, previously unbroken mountainous areas have become fragmented, which in turn have contributed negatively to one of the species for which Norway has a special responsibility, the reindeer. Roads and railways, as well as power lines, can act as barriers to reindeer and restricted habitat access. Hydropower development and huge water reservoirs have seriously affected important reindeer habitats (NFR, 2002; Bevanger and Jordhøy, 2004; Strand et al., 2006). Grouse and gallinaceous birds have proved to be vulnerable to collisions with both power lines and fences, which may affect the population development locally (Bevanger, 1995, 1998; Bevanger and Brøseth, 2000).

Over hundreds of years, farmers have used the mountain areas all over Norway to collect the outlying resources. Thus, many areas have become cultural impacted habitats. However, over the last 100 years, the use of the mountains for summer farming and as grazing pastures for goats, sheep and cattle have significantly decreased, and many habitats depending on this farming system have been reduced or disappeared (Austrheim et al., 2009,

Fremstad and Moen, 2001). To what extent this affects the mountain biodiversity is difficult to assess (Danell et al., 2006), however, grazing reduction and summer farm activities (e.g., logging for firewood) will inevitably result into clogging and a higher timberline (Linkowski and Lennartsson, 2005). In the end, this can affect typical mountain species connected to open areas in connection to the mountainous forest. Hunting and trapping have particularly long traditions in Norway; however, today it is mainly aiming at edible species like reindeer, willow and rock ptarmigan. Hunting and trapping of these species is strictly regulated within a defined hunting season and the harvest is based on sustainable management criteria. It may nevertheless affect the populations, as it is not easy to assess what a sustainable outtake is (Pedersen et al., 2015).

7.2.7 Open Lowland

Open lowland is open areas below the timberline with natural and seminatural vegetation. The habitat types originate from long-term, extensive and traditional land use, such as grazing, cutting, and in some areas burning activities. The areas may have been cleared of stones and forest, and only have insignificant footprints after plowing, use of herbicides or cultivation for grain or grass production (Halvorsen et al., 2015). These seminatural habitat types are characterized by several light dependant species subject to some type of care, and with a small contribution of introduced species. Such habitats are found all over the country, and the species composition in different geographical areas will vary along natural gradients like lime content, moisture and management practices and change of use.

Semi-natural habitats may be both home fields and outlying fields, and includes areas with important grazing resources for domestic animals. They are important to several biodiversity services other than grazing and hay production. For instance, such old cut and grazing areas are important living habitats for several groups of pollinating insects. Open lowlands also have important soil resources that may easily contribute to future food production and which may play an important role in recreation and nature experiences. The agricultural landscape contains a major part of the species diversity, and is part of the cultural history. The major threats against this ecosystem types are clogging because of changes in the cultural traditions, as well as fertilization, cultivation, forestation and infrastructure development and dismantling (Johansen et al., 2015; Norderhaug and Johansen, 2011).

7.3 Organisms and Taxonomic Groups

According to the Norwegian Biodiversity Information Centre about 44,000 species living in Norway are registered as wild; however, the real number is likely to be close to 55,000. The present knowledge of these species varies considerably (http://www.miljostatus.no/tema/naturmangfold/arter/) (Table 7.1). For some groups, rather good and reliable estimates are available, while for others, like invertebrates, we know very little. Norway has few endemic species, however some 16 plants and about 30 terrestrial and marine invertebrates are known. In the following subchapters comments on important plant and animal groups are made.

7.3.1 Vegetation

As previously stressed, the topographic and climatic diversity in Norway, as well as the longitudinal, latitudinal and altitudinal range, creates a high number of ecosystems and niches, being reflected in the vegetation diversity. Moreover, the edaphic conditions vary, from calciferous rocks to hard acid rocks of granitic composition (Holtedal, 1960, 1968). The number of species is not high compared to temperate and tropical areas, however quite a few species for a cold climatic zone, of which the vascular species is best

Table 7.1 Number of Known and Expected Species Within Main Taxonomic Groups in Norway

Taxon	Expected number of species
Insects	16,000
Algae	20,000
Lichen	1,800
Mosses	1,050
Vascular plants	2,800
Fungi	7,000
Birds	250 (nesting)
Mammals	90
Fish (freshwater)	45
Fish (saltwater)	150
Invertebrates (freshwater)	1,000
Invertebrates (saltwater)	3,500

Based on https://en.wikipedia.org/wiki/Geography_of_Norway.

recorded. Botanists (e.g., Grytnes et al., 2004) have revealed that the total species richness of ground-dwelling vascular plants, bryophytes, and lichens in western Norway is peaking at intermediate altitudes.

In general, there are few tree species in Norway compared to similar climatic regions in, e.g., North America. The reason is connected to the fact that the north-south dispersal routes in Europe after the last ice age are more complex, with water bodies (e.g., the Baltic Sea and the North Sea) as well as mountains creating barriers. America is a continuous continent with mountains arranged in a north-south direction. It is, however, interesting that recent DNA-studies of core lake sediments have revealed that conifers survived the ice age in refuges free of ice far to the north (Andøya in the Troms county).

Approximately, one third of the Norwegian mainland is covered by forests, and so far the majority of the plant species recorded (i.e., 60%) are associated with forests. Other species are depending on special conditions connected to, e.g., cultural landscapes and wetlands. During the short history of man in Norway, i.e., after the ice age (about 9,000 years) people have used most of mainland Norway for farming, hunting and other activities, which still characterizes vegetation and species diversity patterns.

7.3.1.1 Vascular Plant Species

The number of vascular plant species recorded in Norway is 2880 (Elven, 2004). However, only 1313, i.e., 46%, of these are supposed to have a natural Norwegian origin. The number of established introduced species is 1567 species, totaling 54%, of which 693 species are considered to occur permanently.

7.3.1.2 Mosses

The west coast of Norway, the eastern border of the Taiga, displays a unique number of bryophyte elements, although more eastern areas have a rich moss flora as well (Störmer, 1969). Several of the species occurring here are at their northern limit in Europe. Mosses in a long range of boreal forest types constitute an abundant component of the forest floor vegetation (Bach et al., 2009). The net primary production (NPP) of bryophytes may equal or even exceed the overstory NPP (excluding trees). Moss species with a featherlike form ("feathermosses"), e.g., *Hylocomium splendens, Pleurozium schreberi* and *Ptilium crista-castrensis*, contribute to the biomass in particular (Bach et al., 2009). The high number of moss species at the western coastal areas is of course connected to the year-round moist and mild climate. There are, however, several moss species living in mountainous areas in southern Norway that can be found in the lowland areas in northern Norway, so many species are highly adaptable.

A checklist for Norwegian moss species was made in 1995 (Frisvoll et al., 1995). This list describes 2 hornworts, 270 liverworts and 794 mosses in 260 genera, and a total of 1066 species on the mainland. In Svalbard, 85 liverworts and 288 mosses in 137 genera, 373 species in total, have been recorded. The checklist includes 2 hornworts in 2 genera, 277 liverworts in 73 genera and 823 mosses in 191 genera; a total of 1102 species in 266 genera.

7.3.1.3 Lichen

Lichen is found in most terrestrial habitats, but they mainly grow in places where the competition from mosses and higher plants is low, such as trunks and rocks.1985 lichen species are currently known in Norway, although new species are frequently discovered and described (Timdal et al., 2010). The foliose and fruticose lichen make up 477 species of the Norwegian lichen flora, and the remainder is crustose lichen. The present knowledge varies, and the taxonomy of fruticose lichen is particularly poorly understood.

7.3.1.4 Fungi

The fungi kingdom of organisms is a complex group and the classification changes frequently. The number and distribution of fungi species in Norway are unknown, however the situation is steadily improving. In the 2010 version of the Norwegian Red List (Kålås et al., 2010), a total of 900 species are included, of which 418 are considered threatened (CR, EN, VU), 302 near threatened, (NT), 177 have deficient data (DD) and 3 regionally extinct (RE). A total of 298 lignicolous, 118 grassland, 259 mycorrhizal and 226 saprotrophic fungi species are included in the 2010 issue of the Norwegian Red List, respectively (Kålås et al., 2010).

7.3.2 Animals

7.3.2.1 Terrestrial Mammals

The moose (*Alces alces*), the red deer (*Cervus elaphus*), the roe deer (*Cervus capreolus*), and the reindeer (*Rangifer tarandus tarandus*) are the four deer species to be found in Norway. In many ways they represent cornerstone species, as they are important both to how the vegetation develop, as hunting objects for man as well as food to carnivores. Wildlife hunting generates significant income to several local communities. To keep these populations at a level suitable for the grazing and browsing areas available, it is necessary to regulate the populations by annual outtake through hunting and culling as the Norwegians more or less eradicated the four big carnivore species more than 100 years ago. This applies to the brown bear (*Ursus arctos*), the wolf (*Lupus lupus*), the lynx (*Lynx lynx*) and the wolverine (*Gulo gulo*). Consequently, it became possible to leave the sheep, goats and cows unattended in the fields during the summer to harvest the outlying field resources. Now the situation is changed, and international agreements on biodiversity have committed Norway to restore viable populations of these carnivore species. As in several other countries, this has created significant conflicts between farmers, politicians and conservationists.

The moose is found all over the country with the exception of some areas at the west coast and the outermost coastal areas from Lofoten and northwards. Although this species probably entered Norway quite soon after the great inland ice melted, it was probably never as numerous as it is today. This is a wanted development as the moose is very important and popular as a hunting object, and each year close to 40,000 specimens are shot. The meat is mainly used for local consumption.

The red deer has expanded its distribution range considerably over the last 50 years. Earlier it was mainly found in coastal areas in the west and southwest, but can now be found in more continental areas and north to Nordland County. Like the moose, it is an important hunting object, and in 2017 approximately 40,000 specimen were shot. The roe deer is regarded as a popular sports-hunting object as well, and categorized among the small wildlife species. It is distributed all over the country and the annual outtake is about 30,000 specimens.

The reindeer has a special status in Norway, as it is the only European country still having viable populations of the species, although there is a small population in Russia, on the Kola Peninsula. During the last ice age, the reindeer was forced towards the southern parts of Europe. However, as soon as the ice started to thaw, the reindeer followed the ice border to the north and was probably among the first animals to colonize Norway some 10,000 years ago. This also made it possible for humans to follow and colonize, as the reindeer are both easy to trap and easy to domesticate.

An ethnic minority in Norway, the Sami people, started to domesticate reindeer quite early, and have had reindeer herds in several areas both in southern and northern Norway. The DNA of the domesticated reindeer is slightly different from the mountain or tundra reindeer as it is mixed with a subspecies mainly connected to forested areas in the east (*Rangifer tarandus fennicus*). Today wild tundra reindeer is only found in southern Norway, while the Sami people and their domesticated reindeer herds mainly use the areas north of the Trondheimsfjord. However, even in the southern part of Norway, the only area with original tundra reindeer DNA is restricted to the areas in the Dovrefjell region. The present population size of the wild reindeer is about 30,000.

The wild reindeer populations are also managed by hunting, and each year 5,000–10,000 animals are taken out during the hunting season in August–September. Even in the National Parks it is allowed, and necessary, to hunt the species to keep the population at a sustainable level. The high mountain pastures are vulnerable, and the lichen, being the most important food for the reindeer, can easily be overgrazed. As it takes up to 20 years for these lichens to regenerate, overgrazing could have disastrous consequences.

The Sami people have their stronghold in Northern Norway, particularly in Finnmark. Reindeer husbandry and herding is important to several families, and an important part of the Sami culture. It is difficult to know the exact number of domesticated reindeer in Norway, but it is likely around 200,000. Apart from the reindeer on the Norwegian mainland, there is also a reindeer subspecies at Svalbard (*Rangifer tarandus plathyrhyncus*). It is smaller, have a denser pelt and rounder head compared to the mainland reindeer (Bevanger and Jordhøy, 2004).

Another ruminant living in the central Norwegian mountains is the muskox (*Ovibos moschatus*). This Pleistocene survivor lived in Scandinavia during the last interglacial stage. Around 1900 whalers and Norwegians, hunting seals in the Greenland coastal areas, started to bring home muskox calves they caught on the Greenland mainland. Over a period of nearly 50 years, several attempts were made to establish a muskox population in the central Norwegian mountains, i.e., Dovre. However, it was not until after World War II it was successful. The population increased very slowly for many years. Today, however, there is a viable population close to 300 animals. The muskox has, of course, become a big tourist attraction (Bevanger, 2005).

There are 15 carnivore species in Norway; of which four are canids (i.e., species belonging to the Canidae family), i.e., wolf, Artic fox (*Vulpes lagopus*), red fox (*Vulpes vulpes*), and raccoon dog (*Nyctereutes procyonoides*). The raccoon dog is an alien species and now and then stray animals are observed. There are two bear species (the Ursinae family): the brown bear and the Polar bear (*Ursus maritimus*). The Polar bear has actually been observed on the mainland now and then, however, that is very rare. Thus, it is necessary to go to the Arctic islands like Svalbard to see it. Only one species in the Felidae family exist, which is the lynx. It lives in small numbers in most parts of the country.

The Mustelidae family is represented with eight species, of which one, the American mink, is introduced. The smallest mustelid species is the least weasel (*Mustela nivalis*), which actually is the smallest carnivore species in the world. Its main prey is rodents, particularly lemmings, and the weasel is small enough to follow the lemmings in their underground burrows. The stoat (*Mustela erminea*) is slightly bigger; however, it is also an eminent rodent hunter. The European polecat (*Mustela putorius*) is of the same size as the stoat, and has a restricted distribution in the southeastern part of the country. The European pine marten (*Martes martes*) is also a small mustelid that can be found more or less all over the country, although it prefers forested areas. The otter (*Lutra lutra*), being widely distributed, occurs particularly along the coastline, but also in inland wetland habitats. The two largest species are the European badger (*Meles meles*) and the wolverine. The badger is omnivorous and a generalist predator, having adapted even to urban environments, while the wolverine is mainly connected to mountainous areas where it scavenges on reindeer remains during the winter, but also remains from sheep killed during the summer period (Bevanger, 2012).

Rodents are cornerstone species in several arctic ecosystems like the ones found in Norway. There are several species (Table 7.2), that can appear in huge numbers and migrate over vast distances, however, the most important being the Norway lemming (*Lemmus lemmus*). The species is able to reproduce very fast and the population growth rate is tremendous. After reaching a peak, the population collapses. Like several other rodent species, it is common that the population reaches a high point every four years. The "huge" lemming years are not that frequent, however. Lemmings are conspicuously colored and behave aggressively towards predators and humans. Many theories have been put forward to explain the lemming population cyclisity; from antipredator toxin accumulating in the vegetation to predators. However, it is still not well understood why lemming populations fluctuate the way they do. When the lemming population builds up, it means a tremendous source of food to several mammalian and avian predators. Species like the Arctic fox and the snowy owl (*Bubo scandiacus*) manage to raise huge litters during the top lemming population cycle, after which the populations can survive some years without reproducing at all.

Table 7.2 Norwegian Small Mammals

Insectivorous Vertebrates	
Eurasian pygmy shrew	*Sorex minutus*
Eurasian least shrew	*Sorex minutissimus*
Common shrew	*Sorex araneus*
Laxmann's shrew	*Sorex caecutiens*
Taiga shrew	*Sorex isdodon*
Eurasian water shrew	*Neomys fodiens*
Western European hedgehog	*Erinaceus europaeus*
Bats	
Daubenton's bat	*Myotis daubentonii*
Brandt's bat	*Myotis brandtii*
Whiskered bat	*Myotis mystacinus*
Natterer's bat	*Myotis nattereri*
Common noctule	*Nyctalus noctula*
Northern bat	*Eptesicus nilssonii*
Serotine bat	*Eptesicus serotinus*
Parti-colored bat	*Vespertilio murinus*
Soprano pipistrelle	*Pipistrellus pygmaeus*
Common pipistrelle	*Pipistrellus pipistrellus*
Nathusius's pipistrelle	*Pipistrellus nathusii*
Brown long-eared bat	*Plecotus auritus*
Barbastelle	*Barbastella barbastellus*
Muroids	
Harvest mouse	*Micromys minutus*
Yellow-necked mouse	*Apodemus flavicollis*
Wood mouse	*Apodemus sylvaticu*
House mouse	*Mus musculus*
Brown rat	*Rattus norvegicus*
Black rat	*Rattus rattus*
Norway lemming	*Lemmus lemmus*
Wood lemming	*Myopus schisticolor*
Northern red-backed vole	*Myodes rutilus*
Bank vole	*Myodes glareolus*
Grey Red-Backed Vole	*Myodes rufocanus*
Northwestern water vole	*Arvicola amphibius*

Table 7.2   *(Continued)*

Muskrat (introduced)	*Ondatra zibethicus*
Field vole	*Microtus agrestis*
Sibling vole (Svalbard only)	*Microtus rossiaemeridionalis*
Root vole	*Microtus oeconomus*
Northern birch mouse	*Sicista betulina*

7.3.2.2   Marine Mammals

Marine mammals are generally grouped into whales (Cetacea), seals (Pinnipedia) and sea cows (Sirenia). In Norway, there are mainly whales and seals (Table 7.3). Like terrestrial mammals, marine mammals are warm-blooded and do not differ particularly from other mammals, although they have several physiological adaptations enabling them to live and survive in the water. Terrestrial mammals need a skeleton and legs, while a submerged animal get support and buoyancy, and the skeleton serves mainly as a basis for the body muscles. Marine mammals also have to breathe; however, they have to go to the surface to do so, and must store the oxygen for some time. There are also a long range of other adaptations connected to how to survive in a continuously cold environment, such as hearing and vision.

In Norwegian waters, there are approximately 20 whale species. Some of them can be seen all year, whereas others only occasionally when they are using peaks in the coastal water nutritional production. Some are irregular guests. Norway has long traditions in whale hunting, a tradition having been a part of the Norwegian coastal culture for centuries. Today, however, the hunt only involves the Minke whale. The whale meat is used for human consumption, generally in Northern Norway. The Minke whale hunting has occurred since the early twentieth century (Bjørge et al., 2010).

There are mainly two seal species in Norwegian coastal waters, the Harbor seal (*Phoca vitulina*) and the gray seal (*Halichoerus grypus*). The harbor seal is stationary living in breeding colonies along the entire coast and in some fjords. The gray seal breeds in several colonies from the central Norwegian coastline to the border of Russia. The animals disperse over wider areas outside the breeding season to find food. Occasionally ringed seals (*Phoca hispida*) and harp seals (*Pagophilus groenlandicus)* migrate southwards along the Norwegian coast.

The seals are regarded important elements of the Norwegian marine biodiversity and several Marine Protected Areas have been established to

Table 7.3 Whale Species Observed in Norwegian Waters (Based on Øien, 2009)

Balaenidae (Baleen whales)	Barents Sea	Norwegian Sea	North Sea	Occurrence
Baleen whales (Mysticeti)				
Bowhead whale *Balaena mysticetus*	X	X	X	F
North Atlantic right whale *Eubalaena glacialis*		X		N
Balaenopteridae				
Common minke whale *Balaenoptera acutorostrata*	X	X	X	N
Sei whale *Balaenoptera borealis*		X	X	N
Fin whale *Balaenoptera physalus*	X	X		N
Blue whale *Balaenoptera musculus*	X	X		N
Humpback whale *Megaptera novaeangliae*	X	X		N
Toothed whales (Odontoceti)				
Monodontidae (Narwhvales)				F
Belugawhale *Delphinapterus leucas*	X			
Narwhvale *Monodon monoceros*	X	X		F
Phocoenidae (Porpoise)				
Harbor porpoise *Phocoena phocoena*	X	X	X	F
Delphinidae (Dolphins)				
White-beaked dolphin *Lagenorhynchus albirostris*	X	X	X	F
Atlantic white-sided dolphin *Lagenorhynchus acutus*		X		F
Risso's dolphin *Grampus griseus*			X	G
Common bottlenose dolphin *Tursiops truncatus*			X	F
Striped dolphin *Stenella coeruleoalba*			X	G
Short-beaked common dolphin *Delphinus delphis*		X	X	G
Killer whale *Orcinus orca*	X	X	X	F
Long-finned pilot whale *Globicephala melas*		X		N
Ziphiidae (Beak whales)				
Sowerby's beaked whale *Mesoplodon bidens*		X	X	F

Table 7.3 (Continued)

Balaenidae (Baleen whales)	Barents Sea	Norwegian Sea	North Sea	Occurrence
Northern bottlenose whale *Hyperoodon ampullatus*		X		N
Physeteridae (Sperm whales)				
Sperm whale *Physeter macrocephalus*		X		N

X = Regular occurring;

N = Makes regular nutritional migrations to the area (summer months);

F = Spend the whole life cycle in the area;

G = Guest, observed now and then.

ensure that the coastal seal populations will survive, particularly to prevent disturbance during the breeding season, but also to protect the species habitats. However, the seal populations are also subject to a sustainable use and harvest of marine resources. This is in accordance with Article 10 of the Convention on Biological Diversity. From time to time, coastal seal populations are conflicting with fishing interests. Apart from the traditional coastal fisheries, fish farming has become increasingly important to the Norwegian economy. Coastal seals are also the host of cod worm, a parasitic nematode infecting coastal cod and other fish species. Particularly, when the harp seal migrates in huge numbers into coastal waters, it may result in significant economic losses to the fishermen. To deal with this conflict the Norwegian management authorities both use conservation and harvesting tools. Norway has defined annual quotas for seal hunting, worked out, and implemented management plans for the coastal seal populations (Bjørge et al., 2010).

7.3.2.3 Birds

Up to 2014, approximately 500 bird species have been observed in Norway, including Svalbard. Species from 19 orders are observed (Table 7.4). This includes several species that occurs occasionally, set free intentionally, or escaped from farming. Altogether 259 species have been observed breeding on the Norwegian mainland, and if Svalbard is included, the number is 269. It is, however, a smaller number that may be regarded as regular breeding species, probably around 230.

The Norwegian avifauna is complex in the sense that it is a mixture of resident species, migratory species, and species that make regional move-

Table 7.4   Number of Norwegian Bird Species Within the 74 Families Recorded, a
Total of 473 Species, Including One Extinct, Seven Globally Threatened,
and Two Introduced

S. No.	Bird Family	Number of Species	S. No.	Bird Family	Number of Species
1.	Gaviidae	4	32.	Pteroclidae	1
2.	Podicipedidae	6	33.	Columbidae	6
3.	Diomedeidae	2	34.	Cuculidae	3
4.	Procellariidae	7	35.	Tytonidae	1
5.	Hydrobatidae	4	36.	Strigidae	12
6.	Pelecanidae	2	37.	Caprimulgidae	1
7.	Sulidae	2	38.	Apodidae	5
8.	Phalacrocoracidae	2	39.	Alcedinidae	1
9.	Ardeidae	10	40.	Meropidae	1
10.	Ciconiidae	2	41.	Coraciidae	1
11.	Threskiornithidae	2	42.	Upupidae	1
12.	Phoenicopteridae	1	43.	Picidae	8
13.	Anatidae	50	44.	Alaudidae	9
14.	Pandionidae	1	45.	Hirundinidae	4
15.	Accipitridae	20	46.	Motacillidae	14
16.	Falconidae	6	47.	Regulidae	2
17.	Tetraonidae	5	48.	Bombycillidae	1
18.	Phasianidae	4	49.	Cinclidae	1
19.	Gruidae	2	50.	Troglodytidae	1
20.	Rallidae	7	51.	Prunellidae	3
21.	Otididae	2	52.	Turdidae	14
22.	Haematopodidae	1	53.	Locustellidae	5
23.	Recurvirostridae	2	54.	Acrocephalidae	11
24.	Burhinidae	1	55.	Phylloscopidae	11
25.	Glareolidae	3	56.	Sylviidae	7
26.	Charadriidae	14	57.	Muscicapidae	23
27.	Scolopacidae	45	58.	Panuridae	1
28.	Stercorariidae	4	59.	Aegithalidae	1
29.	Laridae	17	60.	Paridae	7
30.	Sternidae	13	61.	Sittidae	1
31.	Alcidae	7	62.	Certhiidae	1

Table 7.4   (Continued)

S. No.	Bird Family	Number of Species	S. No.	Bird Family	Number of Species
63.	Remizidae	1	70.	Calcariidae	2
64.	Oriolidae	1	71.	Cardinalidae	3
65.	Laniidae	6	72.	Icteridae	5
66.	Corvidae	9	73.	Fringillidae	18
67.	Sturnidae	3	74.	Passeridae	3
68.	Parulidae	1			
69.	Emberizidae	14		**Total**	**473**

Based on https://en.wikipedia.org/wiki/List_of_birds_of_Norway.

ments. Some species could also be called invasive species as they can occur in huge numbers some years, coming from Siberia, for example. In addition, some migratory species could happen to overwinter along the coastline in mild winters. A majority of the breeding species, however, leaves the country in autumn when the frost and snow come, as they are not adapted to find food in these conditions.

The spring and bird migration is an interesting phenomenon, as millions of birds move to the south to escape the winter and return to favorable food conditions during spring and summer. Several species actually go to Africa, although many spend the winter in central and southern Europe. Why this migration has evolved is an interesting story, and it is no doubt that it is connected to the fact that Scandinavia was covered with ice until approximately 10,000 years ago. As the ice withdrew northwards, plants and animals started to follow. The huge organic production in several Norwegian habitats during summer makes it worthwhile to take the journey. For most species, there is no need to fight due to the food surplus.

Although there are many passerine species living in the forest and other habitats, from the lowland to the mountains, what is most striking about the Norwegian avifauna are the seabirds. The long Norwegian coastline harbors millions of seabirds like auks and kittiwakes, becoming particularly abundant from central Norway northwards. The southernmost bird cliffs with huge colonies are found at the Runde Island close to the central Norwegian city Ålesund. Along the coastline from central Norway up to the northernmost county, Finnmark, there are numerous seabird-breeding colonies with hundreds of thousands of birds, particularly the Atlantic puffin (*Fratercula arctica*), guillemot (*Uria aalge*), razorbill (*Alca torda*) and several gull species.

7.3.2.4 Freshwater Fishes

To classify fishes as either belonging to freshwater or saltwater is not easy, as some species live partly in both fresh- and salt waters. That is why the number of freshwater fishes, as well as saltwater fishes belonging to the Norwegian fauna differ among authors. In general, a fish is regarded as a freshwater species when it is spawning in freshwater, and mainly stay in freshwater during its life cycle. Species depending on salt water to spawn and to develop offspring are regarded as saltwater species. Some freshwater species, like the salmon, may migrate into salt water during some periods of its life cycle, and some are able to reproduce and live both in fresh- and salt-water, like the three-spined stickleback (*Gasterosteus aculeatus*). Species found in freshwater during some periods, but spawn in salt water, should not be regarded as freshwater species, such as the eel and European flounder (*Platichthys flesus*). Thus, in general Norway is regarded to have 44 or 45 freshwater species, however, in Table 7.5, 54 species are listed (http://fish. mongabay.com/data/Norway.htm#q0cBhuXs0If1OxEX.99).

The most common freshwater fish species in Norway is the trout, being found all the way from sea level to high up in the mountains (sometimes higher than 1400 m). The freshwater fish species are commonly categorized based on how and when they arrived in the country after the last glaciation. During the glaciation, the freshwater species survived in some refuges in the south or southeast, before spreading northwards along the waterways created when the ice melted. Species being tolerant against cold waters arrived first and could migrate furthest into the mainland. On the west coast, few and mainly cold tolerant species arrived through salt water from the west, including salmon, trout and Arctic char (*Salvelinus alpinus*). Species coming in from the east were stopped by the high mountain barriers, thus the main numbers of freshwater fish species are confined to the eastern and northern parts of the country, among other several cyprinid species like the carp. The distribution of the different freshwater fish species is, however, highly influenced by the fact that humans have brought different species with them.

7.3.2.5 Marine Fishes

It is not a surprise that there are about five times the number of saltwater fish species compared to freshwater in Norway, i.e., approximately 250 species.

Table 7.5 Freshwater Fish Species in Norway

Order	Family	Species	Status	Fishbase Name
Cypriniformes	Cyprinidae	*Abramis brama*	Native	Common bream
Acipenseriformes	Acipenseridae	*Acipenser sturio*	Native	Sturgeon
Cypriniformes	Cyprinidae	*Alburnus alburnus*	Native	Bleak
Clupeiformes	Clupeidae	*Alosa alosa*	Native	Allis shad
Siluriformes	Ictaluridae	*Ameiurus melas*	Introduced	Black bullhead
Siluriformes	Ictaluridae	*Ameiurus nebulosus*	Introduced	Brown bullhead
Anguilliformes	Anguillidae	*Anguilla anguilla*	Native	European eel
Cypriniformes	Cyprinidae	*Aspius aspius*	Native	Asp
Cypriniformes	Cyprinidae	*Blicca bjoerkna*	Native	White bream
Cypriniformes	Cyprinidae	*Carassius auratus auratus*	Introduced	Goldfish
Cypriniformes	Cyprinidae	*Carassius carassius*	Native	Crucian carp
Mugiliformes	Mugilidae	*Chelon labrosus*	Native	Thicklip gray mullet
Salmoniformes	Salmonidae	*Coregonus albula*	Native	Vendace
Salmoniformes	Salmonidae	*Coregonus lavaretus*	Questionable	Common whitefish
Scorpaeniformes	Cottidae	*Cottus gobio*	Native	Bullhead
Scorpaeniformes	Cottidae	*Cottus poecilopus*	Native	Alpine bullhead
Cypriniformes	Cyprinidae	*Cyprinus carpio carpio*	Introduced	Common carp
Esociformes	Esocidae	*Esox lucius*	Native	Northern pike
Gasterosteiformes	Gasterosteidae	*Gasterosteus aculeatus aculeatus*	Native	Three-spined stickleback
Cypriniformes	Cyprinidae	*Gobio gobio gobio*	Introduced	Gudgeon

Table 7.5 (Continued)

Order	Family	Species	Status	Fishbase Name
Perciformes	Percidae	Gymnocephalus cernuus	Native	Ruffe
Petromyzontiformes	Petromyzontidae	Lampetra camtschatica	Questionable	Arctic lamprey
Petromyzontiformes	Petromyzontidae	Lampetra fluviatilis	Native	European river lamprey
Petromyzontiformes	Petromyzontidae	Lampetra planeri	Native	European brook lamprey
Cypriniformes	Cyprinidae	Leucaspius delineatus	Introduced	Belica
Cypriniformes	Cyprinidae	Leuciscus idus	Native	Ide
Cypriniformes	Cyprinidae	Leuciscus leuciscus	Questionable	Common dace
Pleuronectiformes	Pleuronectidae	Liopsetta glacialis	Native	Arctic flounder
Mugiliformes	Mugilidae	Liza aurata	Native	Golden gray mullet
Mugiliformes	Mugilidae	Liza ramado	Native	Thinlip mullet
Gadiformes	Lotidae	Lota lota	Native	Burbot
Perciformes	Centrarchidae	Micropterus dolomieu	Not established	Smallmouth bass
Perciformes	Centrarchidae	Micropterus salmoides	Questionable	Largemouth bass
Salmoniformes	Salmonidae	Oncorhynchus gorbuscha	Introduced	Pink salmon
Salmoniformes	Salmonidae	Oncorhynchus keta	Introduced	Chum salmon
Salmoniformes	Salmonidae	Oncorhynchus mykiss	Introduced	Rainbow trout
Osmeriformes	Osmeridae	Osmerus eperlanus	Native	European smelt
Perciformes	Percidae	Perca fluviatilis	Native	European perch
Petromyzontiformes	Petromyzontidae	Petromyzon marinus	Native	Sea lamprey

Table 7.5 *(Continued)*

Order	Family	Species	Status	Fishbase Name
Cypriniformes	Cyprinidae	*Phoxinus phoxinus*	Native	Eurasian minnow
Pleuronectiformes	Pleuronectidae	*Platichthys flesus*	Native	Flounder
Gasterosteiformes	Gasterosteidae	*Pungitius pungitius*	Native	Ninespine stickleback
Cypriniformes	Cyprinidae	*Rutilus rutilus*	Native	Roach
Salmoniformes	Salmonidae	*Salmo salar*	Native	Atlantic salmon
Salmoniformes	Salmonidae	*Salmo trutta trutta*	Native	Sea trout
Salmoniformes	Salmonidae	*Salvelinus alpinus alpinus*	Native	Charr
Salmoniformes	Salmonidae	*Salvelinus fontinalis*	Introduced	Brook trout
Salmoniformes	Salmonidae	*Salvelinus namaycush*	Introduced	Lake trout
Perciformes	Percidae	*Sander lucioperca*	Native	Pike-perch
Cypriniformes	Cyprinidae	*Scardinius erythrophthalmus*	Questionable	Rudd
Cypriniformes	Cyprinidae	*Squalius cephalus*	Native	European Chub
Salmoniformes	Salmonidae	*Thymallus thymallus*	Native	Grayling
Cypriniformes	Cyprinidae	*Tinca tinca*	Introduced	Tench
Scorpaeniformes	Cottidae	*Triglopsis quadricornis*	Native	Fourhorn sculpin

Based on http://fish.mongabay.com/data/Norway.htm#q0cBhuXs0If1OxEX.99.

However, it is difficult to find the exact number and it is not a rare event to discover new species, normally living in warmer waters in the south. This could partly be a response to higher water temperatures during the last years. The Norwegian coastline, as well as both the shallow and deep-sea areas represents a high number of different ecosystems and living conditions, reflecting the diversity of fish species.

Along the coastline from Møre to Finnmark some of the world's most important fish stocks are spawning, like the skrei cod (*Gadus morhua*), northeast Arctic pollock (*Pollachius virens*), haddock *(Melanogrammus aeglefinus*) and Atlantic herring (*Clupea harengus*). These sea areas are also the main feeding place for some of the world's largest pelagic school-ers, particularly pelagic feeders (plankton feeders) like the Norwegian spring-spawning herring and mackerel, but also blue whiting (*Microme-sistius poutassou*) and other species. The dramatic oscillation in these fish stocks has been known for more than 1,000 years. The Norwegian Sea is also an important feeding area for young Atlantic salmon (van der Meeren et al., 2015).

7.3.2.6 Amphibians and Reptiles

Norway has only six amphibian species: the northern crested newt (*Triturus cristatus*), the smooth newt (*Lissotriton vulgaris*)**,** common frog (*Rana temporaria*), moor frog (*Rana arvalis*), common toad (*Bufo bufo*) and pool frog (*Rana lessonae*). The northern crested newt is distributed in three separate areas, Central Norway, Southwest Norway and Southeast Norway. The smooth newt nearly occurs in all counties along the coast-line, north to Nordland (Vefsn local authority). The common frog occurs all over the country, except from some islands, and has been found up to elevations of 1350 meters. Amphibians lay their eggs in water where the larvae lives until they are able to walk ashore in the autumn. As adults, they live most of their lives on land, where the majority also overwin-ter. Thus, amphibians depend on both wetlands and terrestrial habitats, including the migrating routes between the breeding locality and the sum-mer habitats.

Of the five Norwegian reptile species, only one is venomous, the viper (*Vipera berus*). This species has a wide distribution and can be found from the sea level to at least 1200 m a.s.l. The other two species, the grass snake (*Natrix natrix*) and the smooth snake (*Coronella austriaca*), are harmless and has a quite restricted distribution in the southern parts of the country. There are also two other species, the common lizard (*Zootoca vivipara*) and the slow worm (*Anguis fragilis*), both being quite widespread.

7.3.2.7 Insects

As in other countries, the number of insects in Norway is overwhelming, but largely unknown. Due to efforts made by, e.g., the Norwegian Biodiversity Information Centre, the knowledge is, however, improving. So far, about 15,000 species have been recorded, but the expected number is likely to be

around 23,000 species. The University of Oslo has compiled a list with the
number of species recorded within different orders, and the expected number (cf. Table 7.6) (http://www.nhm.uio.no/english/research/collections/
zoological/insect/facts/).

Table 7.6 Number of Insect Species Recorded Within Different Insect Orders up to
2006 and Expected Existing Number

Order	Published 1993	Published/ unpublished 1993	Registered 2006	Expected
Thysanura	5	5		5
Diplura	2	2		3
Protura	2	2		2
Collembola	280	283		300
Ephemeroptera	45	45	48	48
Odonata	44	44	48	48
Plecoptera	35	35	35	35
Orthoptera	28	28	29	31
Dermaptera	3	3	3	3
Blattodea	6	6	10	10
Psocoptera	17	51		57
Phthiraptera	41	246		525
Hemiptera (Homoptera + Heteroptera)	1059	1143	1221	1320
Thysanoptera	90	96		120
Megaloptera	5	5	5	5
Raphidioptera	3	3	3	4
Neuroptera	55	55	57	57
Coleoptera	3293	3375	3495	3800
Strepsiptera	3	6		7
Mecoptera	5	5	5	5
Siphonaptera	51	55		60
Diptera	2654	3955	4639	6029
Lepidoptera	2092	2092	2194	2400
Trichoptera	189	195	199	200
Hymenoptera	1272	2959		8158
Total	**11279**	**14694**		**23222**

Based on http://www.nhm.uio.no/english/research/collections/zoological/insect/facts/.

7.4 Species For Which Norway Has a Special Responsibility

Norwegian authorities have defined a species for which the country has a special responsibility as a species with its main distribution and abundance in Norway, meaning where 25% or more of the European distribution is in Norway. There are particularly many special responsibility species (SRS) connected to open seawaters, both pelagic and seabed (21), and mountainous areas (14). Among these are important fish species like Greenland halibut (*Reinhardtius hippoglossoides*), Atlantic halibut (*Hippoglossus hippoglossus*), Atlantic salmon (*Salmo salar*), common ling (*Molva molva*), lodde (*Mallotus villosus*) and Atlantic mackerel (*Scomber scombrus*). It also applies to the kelp – a big brown alga also known as tangle (*Laminaria hyperborea*) and the brown seaweed (*Saccharina latissimi*). Mammals like the harbor seal (*Phoca vittulina*) and mink whale (*Balaenoptera acutorostrata*), as well as the Atlantic puffin (*Fratercula arctica*) are also SRS. In the mountain ecosystem the rock ptarmigan (*Lagopus mutus*), rough-legged hawk (*Buteo lagopus*) and gyrfalcon (*Falco rusticolus*) and mammals like reindeer (*Rangifer tarandus*) and wolverine (*Gulo gulo*) are SRS. Among plant species Arctic poppy (*Papaver radicatum*), glacier buttercup (*Ranunculus glacialis*) and the turgid aulacomnium moss (*Aulacomnium turgidum*) are SRS (http://www.environment.no/topics/biodiversity/species-in-norway/).

7.5 National Laws and International Obligations

In 2009, Norway got its Nature Diversity Act, which is now the core environmental legislation and management tool, covering the conservation of biological, landscape and geological diversity. Moreover, it applies to all societal sectors responsible for managing biodiversity and the environment, and those making decisions that may influence biodiversity. The act includes requirements on species management, protected areas, alien species, habitats and priority species. The Wildlife Act and the Salmonids and Freshwater Act are also designed to maintain biodiversity.

Together with several other nations, Norway has ratified a number of environmental agreements, setting requirements to the conservation and sustainable use of biodiversity and ecosystem services. The Convention on Biological Diversity is the agreement with the most wide-ranging scope and objectives. Like other parties to the convention, Norway is required to establish protected areas, take steps to eradicate, contain and control invasive alien species, safeguard threatened species and promote cooperation with other countries on biodiversity conservation.

7.6   Biodiversity Threats

Globally it is agreed that land use change, climate change, invasive alien species, pollution, and overexploitation are the major threats to biodiversity. In Norway land use change, resulting in habitat conversion is the most important factor and driver to both species and habitat biodiversity loss. Land use and land-use change may affect habitats in several ways, e.g., by deforestation and open areas being forested, drained, converted to farmland and through infrastructure development. Land use changes are considered a threat to 87% of the red-listed, threatened species in Norway. Still unclear effects of climate change will obviously also affect Norwegian ecosystems, making species depending on ecosystems at high altitudes, as well as in Arctic areas like Svalbard, particularly vulnerable (Kålås et al., 2010).

The last Norwegian Red List for Species (Henriksen and Hillmo, 2015) lists 2355 species as threatened. It is, however, a huge knowledge gap with respect to the true number of species in Norway and the status of many of these species.

Invasive alien species is a growing threat to the indigenous fauna and flora in Norway. In 2006 the first Norwegian Black List was published (Gederaas et al., 2007), and a revised version was published in 2012 (Gederaas et al., 2012). Non-native species are discovered regularly, which in part can be explained by the globalization of the economy and accompanying increase in travel and trade. One of the most serious introductions in Norway took place as early as in 1927 with the arrival of the American mink (*Neovison vison*). At that time fur-farming started and became big business in several regions. However, animals frequently escaped from the farms and within 60 years the species has colonized the whole country (Bevanger and Henriksen, 1995). To have a new predator like the mink is something of the biggest disaster an ecosystem can experience. The American mink is particularly adapted to wetlands, both freshwater and marine. In Norway, this has resulted in a serious threat to many ground-nesting seabird species, and ground-nesting birds in general.

The Black List contains an overview of alien species recorded in Norway, as well as an assessment of their impact to indigenous species and habitats. The aliens are categorized according to how serious a threat they pose. The Biodiversity Information Centre has so far registered 2320 alien species in Norway and risk assessment has been carried out for 1180 species already reproducing in the wild. 217 species have been categorized as having "severe impact" and "high impact," including the American mink, Spanish slug (*Arion vulgaris*) and garden lupin (*Lupinus polyphyllus*).

7.7 The Nature Index

A Nature Index (Framstad et al., 2015) developed to show the state and development of biodiversity in Norway has recently come into use. The index is based on summarized information of 301 indicators from nine main ecosystems; oceanic seabeds, oceanic pelagic zone, seabed in coastal waters, pelagic zone in coastal waters, freshwater, wetlands, forest, mountains and open lowland. Experts from several Norwegian research institutions have handled and evaluated the indicators. The indicators consist mainly of species populations, but may also be combined indexes of several species as well as indicators for important resources for many species. It is defined as a state of reference for the ecosystems with intact local biodiversity. The indicator values are scaled against their respective values in their state of reference, from one in an intact ecosystem to zero in a complete destroyed ecosystem.

The Nature Index has revealed substantial variation in the biological diversity state between the ecosystems. In 2014 the index values were highest in freshwater (0.75) and marine ecosystems (0.62–0.72) and lowest for the forest (0.37) and open lowland (0.47). The developments for open waters (pelagic zone) in the ocean and the coastal areas have been positive until 2010. The index values for wetland and open lowland are decreasing significantly, while a weak increase is registered for the forest. For the rest of the ecosystems, there is no clear trend (Framstad et al., 2015).

7.8 Protected and Conservation Areas

The Norwegian protected areas are categorized into national parks, protected landscapes, nature reserves, habitat management areas or marine protected areas. Norway administers around 2,700 conservation areas on the mainland and Svalbard, and in total, these areas constitute about 30% of the national land mass. About 17% of mainland Norway is protected under the Nature Diversity Act. A major proportion, however, consists of low productive mountainous areas. Other habitat types, particularly in coastal and marine areas, are not yet adequately represented.

The *National parks* are quite large areas with natural habitats with ecosystems and/or landscapes without major infrastructure and development.

There are national parks on both the mainland and Svalbard. *Protected land-scapes* are landscapes of cultural or natural interests that are important in ecological or cultural terms. Buildings and other elements of the cultural heritage contributing to its distinctive character are considered part of the landscape. The most strictly protected areas are *Nature reserves*. They contain endangered, rare or vulnerable species, communities, habitats or landscape types. Nature reserves are also established to take care of specific habitat types and habitats with high importance for biodiversity, or if it has a distinctive geological feature or are of a specific scientific interest. *Habitat management areas* are protecting the habitats of a particular animal or plant species and that fulfill specific ecological functions for one or more species. Such areas could be spawning or nursery areas, migration routes, feeding areas, molting grounds, display grounds or mating or breeding areas.

Marine protected areas are supposed to safeguard areas with high marine conservation value, but can also apply to areas necessary to terrestrial species. The same criteria apply for these as for nature reserves. Marine protected areas are also established to maintain distinctive or representative ecosystems lacking major infrastructure or areas important to specific ecological functions for one or more species. These protected areas are mainly implemented to protect the seabed, the water column, the surface water, or a combination of these. There are also other types of protected areas, established based on the 1970 Nature Conservation Act. These were mainly aimed at "natural monuments," i.e., geological formations or botanical or zoological features, bird reserves and other protected areas for particular species. Additionally, the Wildlife Act provided for a form of habitat protection. All these sites are still protected, however, if the protection regulations need to be revised, new regulations will apply under the 2009 Nature Diversity Act and adjusted to bring them in line with the new protection categories (http://www.environment.no/topics/biodiversity/protected-areas/).

Keywords

- biodiversity threats
- conservation areas
- national laws
- Nature Index

References

Aagaard, K., & Dolmen, D., (1996). *Limnofauna Norvegica: A Catalog of the Norwegian Freshwater Fauna*. Tapir Academic Press, Trondheim, Norway. (In Norwegian with english introduction and summary).

Anker-Nilssen, T., Barrett, R. T., Lorentsen, S. H., Strøm, H., Bustnes, J. O., Christensen, D., S., et al., (2015). *SEAPOP*. The ten first years. Key document 2005–2014. Report made by the Norwegian Institute for Nature Research, The Norwegian Polar Institute, the Arctic University of Norway, The Norwegian University of Science and Technology. (In Norwegian).

Austrheim, G., Bråthen, K. A., Ims, R. A., Mysterud, A., & Ødegaard, F., (2009). Mountain. – Environmental conditions and impact on red listed species. *Facts. The Norwegian Biodiversity Information Centre*, 107–117.

Bach, L. H., Frostegård, Å., & Ohlson, M., (2009). Site identity and moss species as determinants of soil microbial community structure in Norway spruce forests across three vegetation zones. *Plant Soil*, *318*, 81–91.

Bevanger, K., & Brøseth, H., (2000). Reindeer fences as a mortality factor for ptarmigan. *Wildlife Biology*, *6*, 121–127.

Bevanger, K., & Henriksen, G., (1995). The distributional history and present status of American mink (*Mustela vison*) in Norway. *Annales Zoologici Fennici*, *32*, 11–14.

Bevanger, K., & Jordhøy, P., (2004). Reindeer – The mountain nomad. *Naturforlaget, Oslo, Norway*.

Bevanger, K., (1995). Estimates and population consequences of tetraonidae mortality caused by collisions with high tension power lines in Norway. *J. Appl. Ecol.*, *32*, 745–753.

Bevanger, K., (1998). Biological and conservation aspects of bird mortality caused by electricity power lines: A review. *Biol. Conserv.*, *86*, 67–76.

Bevanger, K., (2005). Interactions between muskox and reindeer. An overview. *NINA Report 7*. (In Norwegian with an English abstract).

Bevanger, K., (2012). Norwegian carnivores. *Cappelen Damm, Oslo, Norway*.

Bjerke, J. W., Skarpaas, O., & Dervo, B. K., (2015). Wetland. Nature index for Norway 2015. State and Development for Biodiversity. *Norwegian Environment Agency Report M-441*, 68–76. (In Norwegian with English abstract).

Bjørge, A., Lydersen, C., Skern-Mauritzen, M., & Wiig, Ø., (2010). *Marine Mammals*. Institute of Marine Research, Bergen, Norway.

Danell, K., Bergström, R., Duncan, P., & Pastor, J., (2006). *Large Herbivore Ecology, Ecosystem Dynamics and Conservation*. Cambridge University Press, Cambridge.

Elven, R., (2004). *Norwegian Flora, edition 7*. Det norske samlaget, Oslo, Norway.

Framstad, E., (2015). Nature index for Norway 2015. State and Development for Biodiversity. *Norwegian Environment Agency Report M-441*, 68–76. (In Norwegian with English abstract).

Framstad, E., Pedersen, B., & Nybø, S., (2015). The design and data platform for the Nature Index. Nature index for Norway 2015. State and Development for Biodiversity. *Norwegian Environment Agency Report M-441*, 68–76. (In Norwegian with English abstract).

Fremstad, E., & Moen, A., (2001). Threatened vegetation types in Norway. *The Norwegian University of Science and Technology, The Natural History Museum. Report Bot. Ser. 4.* (In Norwegian with English abstract).

Frisvoll, A. A., Elvebakk, A., Flatberg, K. L., & Økland, R. H., (1995). Checklist of Norwegian bryophytes. Latin and Norwegian nomenclature. *NINA Temahefte 4*. (In Norwegian with English abstract).

Gederaas, L., Moen, T. L., Skjelseth, S., & Larsen, L. K., (2012). *Alien Species in Norway – With Norwegian Blacklist 2012*. Norwegian Biodiversity Information Centre, Trondheim, Norway.

Gederaas, L., Salvesen, I., & Viken, Å., (2007). *Norwegian Black List – Ecological Risk Analysis of Alien Species*. Norwegian Biodiversity Information Centre, Trondheim, Norway.

Gjærevoll, O., (1973). *Plant Geography*. Scandinavian University Books.

Grytnes, J. A., Kapfer, J., Jurasinski, G., Birks, H. H., Henriksen, H., Klanderud, K., et al., (2014). Identifying the driving factors behind observed elevational range shifts on European mountains. *Global Ecology and Biogeography, 23*, 876–884.

Gundersen, H., Norderhaug, K. M., Rinde, E., Johnsen, T. M., Van der Meeren, G., Nilssen, K. T., & Lorentsen, S. H., (2015). *Coast*. Nature index for Norway 2015. State and development for biodiversity. *Norwegian Environment Agency Report M-441*.

Halvorsen, R., (2015). *NiN Nature System Level – Translation From NiN Version 1., and Norwegian Red List for Nature Types 2011 to NiN version 2. 0. – Nature in Norway, Document 4 (Version 2.0.4)*. Norwegian Biodiversity Information Centre, Trondheim, Norway, (http://www.artsdatabanken.no).

Halvorsen, R., Bryn, A., Erikstad, L., & Lindgaard, A., (2015). *Nature in Norway – NiN Version 2.0.0*. Norwegian Biodiversity Information Centre, Trondheim, Norway. (http:// www. artsdatabanken. no/nin).

Henriksen, S., & Hilmo, O., (2015). *The 2015 Norwegian Red List for Species*. Norwegian Biodiversity Information Centre, Trondheim, Norway.

Hofgaard. A., & Myklebost, H., (2014). *Monitoring of Palsa Peatlands*. Second reanalysis at Haukskardmyrin and Haugtjørnin, Dovre, Norway, changes from 2005 to 2015. *NINA Report 1035*. (In Norwegian with English abstract).

Holtedahl, O., (1960). *Geology of Norway*. The Geological Survey of Norway, Trondheim Norway, Publication No. *208*.

Holtedal, O., (1968). How our land was created. An overview of the geology of Norway. *J. W. Cappelen, Oslo, Norway*.

Johansen, L., Hovstad, K. A., & Åström, J., (2015). *Open Lowland*. Nature Index for Norway 2015. State and Development for Biodiversity. Norwegian Environment Agency Report M-441. (In Norwegian with English abstract).

Kålås, J. A., Viken, Å., Henriksen, S., & Skjelseth, S., (2010). T*he 2010 Norwegian Red List for Species*. Norwegian Biodiversity Information Centre, Trondheim, Norway.

Lamb, H. H., (1995). *The Little Ice Age, Climate, History and the Modern World*. Routledge, London.

Linkowski, W., & Lennartsson, T., (2005). *Biological Diversity in Mountain Birch Forest – A Compilation of Information*. Swedish Biodiversity Centre, Swedish University of Agricultural Sciences. (In Swedish).

NFR, (2002). (The Research Council of Norway), Report from the REIN-project. *NFR Report, Oslo, Norway*. (In Norwegian).

Norderhaug, A., & Johansen, L., (2011). *Cultivated Land and Boreal Heath. Red List for Ecosystems and Habitat Types*. Norwegian Biodiversity Information Centre, Trondheim, Norway. (http://www.biodiversity.no/Pages/135568).

Øien, N., (2009). Distribution and abundance of large whales in Norwegian and adjacent waters based on ship surveys 1995–2001. *NAMMCO Sci. Publ., 7*, 31–47.

Pedersen, B., & Nybø, S., (2015). Nature Index for Norway 2015. Ecological frame-work, computational methods, database and information systems. *NINA Report 1130*.

Pedersen, H. C., & Aarrestad, P. A., (2015). *Mountain*. Nature Index for Norway 2015. State and Development for Biodiversity. Norwegian Environment Agency Report M-441.

Pedersen, H. C., Follestad, A., Gjershaug, J. O., & Nilsen, E., (2016). Status of harvestable small game species in Norway. *NINA Report 1178.*.

Schartau, A. K., Pedersen, B., Van Dijk, J., & Solheim, A. L., (2015). *Freshwater.* Nature Index for Norway 2015. State and Development for Biodiversity. Norwegian Environment Agency Report M-441.

Skjoldal, H. R., (2004). *The Norwegian Sea Ecosystem.* Tapir Academic Press, Trondheim, Norway.

Storaunet, K. O., & Framstad, E., (2015). *Forest.* Nature Index for Norway 2015. State and Development for Biodiversity. Norwegian Environment Agency Report M-441.

Störmer, P., (1969). *Mosses With a Western and Southern Distribution in Norway.* The University Press, Oslo.

Strand, O., Bevanger, K., & Falldorf, T., (2006). *Wild Reindeer Habitat Use at Hardangervidda.* Final report from the Hw7 Project. NINA Report 131.

Timdal, E., Bratli, H., Haugan, R., Holien, H., & Tønsberg, T., (2010). *Lichens.* The 2010 Norwegian Red List for Species. Norwegian Biodiversity Information Centre, Trondheim, Norway.

Tomter, S. M., & Dalen, L. S., (2014). *Sustainable Forestry in Norway.* Norwegian institute of Bioeconomy Research (NIBIO) Report, Ås, Norway.

Van der Meeren, G. I., Lorentsen, S. H., Ottersen, G., & Jelmert, A., (2015). *The Ocean.* Index for Norway 2015. State and Development for Biodiversity. Norwegian Environment Agency Report M-441. (In Norwegian with English abstract).

Photo 7.1   Fly amanita (*Amanita muscaria*).

Photo 7.2   Snow measuring lichen in boreal birch forest (*Melanohalea olivacea*).

Photo 7.3   Spring pasqueflower (*Pulsatilla vernalis*).

Photo 7.4   Scheuchzer's Cottongrass (*Eriophorum scheuchzeri*).

Photo 7.5   Lapland rosebay (*Rhododendron lapponicum*).

Photo 7.6   Great alpine rockfoil (*Saxifraga cotyledon*).

Photo 7.7 Yellow Alpine Milkvetch (*Astragalus frigidus*).

Photo 7.8 Norway lemming (*Lemmus lemmus*).

Photo 7.9 Muskox (*Ovibos moschatus*).

Photo 7.10 Tundra reindeer (*Rangifer tarandus*).

Photo 7.11 Mountain hare (*Lepus timidus*).

Photo 7.12 Common European viper (*Vipera berus*).

Photo 7.13 Purple sandpiper (*Calidris maritima*).

Photo 7.14   An Atlantic puffin (*Fratercula arctica*) colony at Røst Lofoten.

Photo 7.15   Red admiral (*Vanessa atalanta*).

Photo 7.16   Bee beetle (*Trichius fasciatus*).

Photo 7.17   Old spruce forest.

Photo 7.18   Palsa peat bog at Dovre Mountain.

Photo 7.19   Looking down to the the Geirangerfjord.

Photo 7.20 The Nordkapp Plateau in Finnmark.

Photo 7.21 Grey alder forest (*Alnus incana*) with *Matteuccia struthiopteris* and *Anemone nemorosa*.

Photo 7.22   Highland pine forest in Nordland.

Photo 7.23   Abandoned cod fish flake in Finnmark.

Photo 7.24 Blanket bogs at Runde Island.

Photo 7.25 Central Norwegian coastal landscape.

Photo 7.26 Cultural landscape in Central Norway.

CHAPTER 8

Biodiversity in Serbia

BILJANA PANJKOVIĆ and **SLOBODAN PUZOVIĆ**

Institute for Nature Conservation of Vojvodina Province, Serbia,
E-mail: biljana.panjkovic@pzzp.rs, slobodan.puzovic@pzzp.rs

8.1 Introduction

Serbia is located in Southeast Europe, in the northcentral part of the Balkan Peninsula (about 75% of the territory) and in Central Europe, in the southeastern part of the Pannonian Plain (25%). The country lies between 41°53′ and 46°11′ north latitude and 19°49′ and 23°00′ east longitude. The total area covers approximately 88,499 km² with moderate continental climate (Statistical Yearbook of Serbia, 2016).

Serbia consists of two distinct geographical and orographic parts. The northern lowland part is located in the Pannonian Plain which is predominately flat and southern mountain-valley part of the Balkan Peninsula. They are separated by the rivers Sava and Danube, but they are also interconnected by lowland montane Peripannonian Serbia. They give both the character of the biodiversity of Serbia and differences between the plant and animal world distributed in these two microregions (Stevanović et al., 1999).

More than 55% of Serbian land is arable, the large part of which is located in Vojvodina, where all types of natural habitat are threatened by reduction, fragmentation, and isolation (Panjković and Szabados, 2012). The mountains belong to the systems of Rhodope, Carpathian, Balkan, Dinaric and the Skardo-Pindus massifs. The mountain landscape is rich in canyons, gorges, and caves, as well as preserved forests. Serbia is a medium-forested country, out of the total territory, 29.1% is under forests. Other forest land, which according to the international definition includes thickets and shrubberies, stretches over 4.9% (Banković et al., 2009).

Serbia, with its rich natural heritage, represents one of the important centers of biological diversity in Europe, where almost all zonal types of climate, land, and biomes of Europe alternate (Stevanović, 1999).

Besides, regarding the floristic richness, Serbia as a part of Balkan Peninsula is one of the global centers of floristic diversity with relatively large percentage of endemic species 15% of which are the Balkan endemic species. Although it takes only 2.1% of European territory, biodiversity of various groups of living organisms is high. Out of all species diversity in Europe, in

Serbia there are: 39% vascular flora, 51% of fish fauna, 49% of reptile and amphibian fauna, 74% of bird fauna and 67% of European mammal fauna. Agrobiodiversity comprises 4,238 varieties of crop plant species and around 100 breeds of domestic animals, 14 of which are autochthonous (Radović and Kozomara, 2011).

8.2 Ecosystem Diversity

Biogeographically, Horvat et al. (1974) state that Serbia belongs to Moesian province of central European region and Pannonian province of Pontic-South-Siberian region. According to Stevanović et al. (1999) Serbia is divided into the Mediterranean, central European Pontic-South-Siberian, circumboreal and central-south-European mountain region, with subregions and provinces.

Stevanović V and Stevanović B (1995) state that there are two basic types of zonal climate in Serbia: a typical moderate continental climate and a dry continental climate. Serbia has a varied heritage of biodiversity encompassing a wide spectrum of habitats divided into major European biomes which are present in Serbia from steppe zonobiome on steppes in loess plateaus, through deciduous forests zonobiome in lowland, hilly and mountain zones, coniferous forests zonobiome in sub-Alpine mountain belts, to the alpine and oro-Mediterranean biome distributed in high mountain regions above the timber line of the Dinaric Alps, Scardo-Pindic and Balkan mountains. There are also sub-Mediterranean deciduous forests and scrubs which are influenced by Mediterranean climate. A range of cross and mutual impacts occur between these zonobiomes, due to the geographic, petrographic and orographic characteristics of Serbian territory. High level of biodiversity of Serbia is also due to versatile climate zonal vegetation, including a large number of extrazonal, intrazonal and azonal ecosystems such as wetlands, peats, salt, and sands vegetation. During the ice age Serbia represented refugia for numerous species. Therefore, Serbia has got a large number of relic and endemo-relic species. Flora of Serbia comprises 547 vascular plant species (14.94%) Balkan endemics and 59 local endemic species (1.5%) (Stevanović et al., 1999; Radović and Kozomara, 2011).

Serbia is characterized by a diversity of habitats, hence a diversity of biocenoses that makes this region as one of the most significant European centers of diversity of ecosystems (Lakušić, 2005). The ecosystem diversity of Serbia is characterized by great wealth (Amidžić et al., 2014).

According to Lakušić (2005) in Serbia, there have been registered 2,370 vegetation units which are classified into 1,399 associations and 971 subassociations, 242 alliances, 59 classes and 114 vegetation orders. Potential vegetation is typologically diverse and comprised of a relatively large number of communities or higher vegetation units. The real (recent) ecosystems of Serbia are comprised between 700 and 800 associations. There are 10 basic types of climate zonal ecosystems and three pedoclimate types of land ecosystems. The basic types of zonal ecosystems are: thermophilic sub-Mediterranean-Mediterranean deciduous forests, mesophytic deciduous forests, thermophilic deciduous oak (*Quercus* spp.) forests, thermophilic deciduous forests in the forest-steppe region, xerophytic steppes, hygrophilic lowland forests, coniferous forests of boreal type, coniferous forests of Balkan endemorelic pines, sub-Alpine scrub vegetation, and alpine grasslands, pastures, and rocky ground. (Stevanović, 1992; Radović and Kozomara, 2011)

Ecosystems in Serbia are characterized by the presence of both endemic and relic plant associations. Most phytocenoses with endemic features are found within rocky areas, mountain glades and rock screes which are the most important local and regional centers of ecosystem diversity. These include thermopile serpentine stone grounds, mountain peatland, high greenery and mountain mesophile meadows (Stenanović et al., 1995; Radović and Kozomara, 2011). Ecosystems of the Pannonian part in Serbia do not have the same rank as fragile high mountain ecosystems in Serbia. Aquatic ecosystems, pools and marshes are disappearing and today they represent a refuge for many species, remnants of the old tertiary flora. Saline lakes, waterlogged habitats and bottomlands are of special value because they represent marshes of international significance for conservation (Ramsar sites). Fragile saline and steppic ecosystems, which are the most beautiful and best conserved, and especially complex systems of continental saline lands that are interwoven in the zone of partially still well-conserved steppes. Especially significant are the species which are the symbol of rare and endangered flora and of specific zonal steppe and forest-steppe vegetation in Vojvodina like boreal xerothermic relics (Butorac and Panjković, 2007).

The basic types of habitats in Serbia are defined in 2005 (Lakušić et al., 2005) and adapted to the international community standards on the basis of the EUNIS classification system. Basic natural habitat types are the following: continental water habitats; marshland, peatland and bog habitats; grassland habitats and habitats dominated by tall greens, moss and lichens; heathland, shrubland habitats and high alpine tundra; forests and other forest habitats as well as habitats without vegetation or with poorly developed vegetation.

According to the data of CORINE Land Cover for 2012, forested areas in Serbia cover 30% of the territory. The most dominant forests are deciduous forests (91.27%), followed by mixed forests (5.12%) and coniferous forests (3.61%). Oak forests (32%) and beech forests (29.3%) are predominant. Among the coniferous forests, the most common are pine forests (5.6%) and spruce forests (3.8%) (Popović, 2016).

8.3 Species Diversity

Mountains of Balkan Peninsula, along with the Alps, the Carpathian Mountains, and mountains of Iberian and Apennine Peninsula, are described as one of several "centers" of floristic diversity in Europe. That is, as one of 156 such centers in the world (IUCN-WWF Plant Conservation Program and IUCN Threatened Plant Unit 1987). Balkan flora is estimated at over 8,000 species, among which there are 2,600 to 2,700 endemic plants (Stevanović, 2005). It is worth mentioning here that Balkan Peninsula is most abundant with fauna of land vertebrates (Džukić, 1995; Džukic and Kalezić, 2004; Vasić, 1995a; Krystoufek, 2004).

Species diversity in Serbia has not been researched or documented enough (Radović and Kozomara, 2011). In Serbia 44,220 taxa have been discovered and classified at the species and subspecies level. According to real estimations, that figure is bigger and is around 60,000 taxa. Among registered taxa, the most numerous are insects (35,000) followed by vascular plants (3,662) and freshwater algae (1,400). Serbia is characterized by a high diversity of fauna on a European scale, especially birds (Aves), mammals (Mammalia) and amphibians (Amphibia), Higher vertebrates are relatively well researched, while it is necessary to significantly improve the knowledge of lower vertebrates and especially a group of invertebrates (Mijović et al., 2012).

There are no precise data on the number of prokaryotic species belonging to the *Monera* kingdom, 220 species of cyanobacteria are listed for Serbia and Montenegro. Limited information is available concerning the diversity of freshwater algae (1,400 species) and *Rhizopoda* – amoebas with shells (236 species). Fungi of Serbia are relatively poorly known, especially when it comes to Ascomycota. According to the data available, in Serbia it has been registered presence of about 3,000 species so far, out of which the greater part, about 1,700 species, belong to Basidiomycota, although it is realistic to expect that the diversity of Ascomycetes is much higher because they account for 75% of all hitherto described species in the world (Savić, 2016). Recent research on lichen diversity (*Lichenes*) indicates that there are 586 species of lichens found. The *Plantae* kingdom is probably the most researched kingdom in Serbia. There are 400 species of moss widespread (*Bryophyta*) and a total of 3,730 taxa of vascular flora (Tomović, 2007).

The following individual animal groups are known as particularly rich in Serbia, as follows: (1) Invertebrates: about 12,000 species/subspecies, which accounts for about 10% of the described species in Europe; (2) Fish (Pisces): 98 types, which is 16% of the type described in Europe; (3) Amphibians (Amphibia): 21 species, accounting for 25% of the described species in Europe; (4) Reptiles (Reptilia): 24 species, accounting for 16% of the described species in Europe; (5) Birds (Aves): 352 species, representing 67% of the described species in Europe; (6) Mammals (Mammalia): 96 types, which represents 37% of the species described in Europe (Stevanović and Vasić, 1995; Radović and Kozomara 2011).

Flora of vascular plants in Serbia contains 3,662 taxa, i.e., 3,272 species and 390 subspecies, which makes Serbia the country with the greatest floristic diversity and density per unit area, among European countries. The most recent data indicate that the presence of many species of 3,730 (Tomović, 2007), but in reality, it is close to 4,000. Vascular plant species are classified into 141 families and 766 genera. The families with largest number of species are Asteraceae (366), Poaceae and Fabaceae (250), Caryophyllaceae (205), Cruciferae (194), Scrophulariaceae (161), Labiatae (116), Umbelliferae (142), Ranunculaceae (121), Liliaceae (116), Cyperaceae (115), Rosaceae (111) and Orchidaceae (66). The richest genera are *Hieracium* (89 species), *Carex* (79), *Trifolium* (58), *Silene* (57), *Centaurea* (55), *Ranunculus* (54), *Dianthus* (46), *Veronica* (46), *Campanula* (35), *Euphorbia* (35), *Allium* (33) and *Viola* (33). Flora of Serbia is affected by Central European, Alpine, Sub-Mediterranean and Pontic impacts, i.e., structural differences at family level are in correlation with geographical position, orographic characteristic and the history of the flora deriving from the Tertiary and Ice Age up to date (Stevanović et al., 1999).

According to Turil (1927), on Balkan Peninsula, there are 1,754 endemic species or 27% out of total flora of Balkan Peninsula (Stevanović et al., 1999). Latest research indicates that there are 2,600 to 2,700 endemic plants (Stevanović et al., 2005). In the flora of Serbia Gajić (1984) and Diklić (1987) recorded the presence of 197 endemic species and subspecies, whereas Stevanovic' et al. (1995) concluded that the number of the Balkan endemic in Serbia is higher (287 taxa). According to the data, there are 492 Balkan endemic species and subspecies in Central Serbia and Kosovo regions, representing approximately 18% of the total number of 2,660–2,700 endemic taxa in the Balkan Peninsula (Stevanović, 2005; Stevanović et al., 1995).

The Balkan endemics in the researched territory belong to 48 families and 167 genera. There are no endemic families, although there are two monotypic Balkan endemic genera: *Paramoltkia* Greuter and *Halacsya* Dörfler. Ferns (Pteridophyta) do not have any endemics; in gymnosperms (Gymnospermae), there are only three endemic species and endemic monocots (Monocotyledones) are only represented by 37 taxa (7.5%). The families richest in endemic taxa are Asteraceae, Caryophyllaceae, Fabaceae, Brassicaceae, and Lamiaceae (Tomović et al., 2014)

The number of local endemics is significantly lower (59 taxa) which makes 1.5% out of total flora. Local endemics are mostly tertiary relicts (*Achillea alexandri-regis, Euphorbia pancicii, Picea omorica, Wulfenia blecicii, Centaurea kosaninii,* etc.).

The basic type of endemics is high mountain one due to geographical isolation (Stevanović, 1995). In Serbia edaphic endemism is also distinct, especially on serpentines in western and central Serbia and Metohija, where there is ophiolitic endemic flora (Stevanović, 1996). Vojvodina, the Pannonian part of Serbia, is characterized by Pannonian endemics. There are also Balkan subendemics, which belong to various floral elements.

Centers of the diversity of endemic flora are high mountains where 31–90 endemic species live on 100 km², mountain and hill regions of Serbia, gorges and canyons. Special value is given to flora of Serbia by relicts, namely tertiary, glacial, boreal and xerothermic or steppe relicts. Endemic and relict plants indicate a specific florogenesis, which accounts for the richness and diversity of Serbian flora, which makes it standing out in comparison with other parts of Europe (Stevanović et al., 1999).

Serbian territory within the kingdom of animals (Animalia) is according to present knowledge inhabited with 139 species of roundworm (Nematodes), a total of 18 species from the group Anostraca, Notostraca, and Conchostraca, as well as 33 species of Amphipoda (Radovic and Kozomora, 2011). In Serbia, there are about 193 species of butterflies (Jakšić, 2008), which make 44% of known species in Europe. About 155 hoverflies species (Diptera, Syrphidae) were identified as important for conservation in Serbia (Radišić et al., 2016). In wetlands of Vojvodina 175 species have been evidenced during the long-term investigation (Šimić et al., 2009). The most numerous group is insects, among which, as pollinators, the most significant are bees (Hymenoptera: Apiformes) out of ca. 2,000 species is known in European fauna. It is estimated for Serbia that there are about 800–1,000 species. Out of almost 900 species of hoverflies (Syrphidae) in European fauna, over 400 have been described in Serbia. About 200 species of daily butterflies (Hesperoidea, Papilionoidea) have been described, while moths are many times more numerous but less researched; it is estimated that there are 1,000 species of Macrolepidoptera and more than 1,100 species of Microlepidoptera (Stanisavljević et al., 2016).

Serbia is inhabited by 21 amphibians: 8 caudates and 13 anurans (Kalezić et al., 2015). Serbia is inhabited by 24 species of reptiles (Reptiles), including three species of turtles, 11 species of lizards and 10 species of snakes (Tomović et al., 2015). In the wetlands of all three river basins of Serbia, lives a total of 98 species of fish (94) and jawless fishes (4) from 23 families (Mijović et al., 2012), which is about a fifth of the total number of freshwater fish species in Europe. Birds are probably the best researched faunistic groups, with up to 352 now established species (Šćiban et al., 2015), of which 240 belong to contemporary nesting species (Puzović et al., 2015). Fauna of mammals has 96 known species, among them, the presence of 30 species of bats (Chiroptera) has been found so far, which makes the high 58% of the known species in Europe (Paunović, 2016).

Status and trends of distribution and population of individual groups/ species of animals in Serbia are different. It is best known in the bird fauna, where it was established for the period 2000–2013 that 64% of the bird fauna has a stable population, 10.4% of the species has a declining trend and 15.4% of the population is with increasing trend. A very important fact is that in the period 2008–2013 even 54 bird species have only 99 pairs in Serbia, which makes them very sensitive and vulnerable (Puzović et al., 2015).

In the 2008–2013 period, there were significant differences in the richness of breeding bird species among some regions in Serbia (Puzović et al., 2015). In the nonpasserine group, the richest region was Vojvodina with 111 species, followed by Northeastern Serbia (96 species). The four other regions had a significantly lower diversity of nonpasserine birds, in a range of 72–76 species. The situation is quite different in the group of passerines: Vojvodina is the poorest region regarding the number of species (87) as well as Western Serbia (88), while the richest regions are Kosovo and Metohija (102), Eastern Serbia (100) and Southeastern Serbia (99). These are realistic relations of the number of species between regions, which apparently do not change significantly over time, since analysis from the 2008–2013 period almost completely coincide with the data from the 1990–2002 period regions allocated at that time (Puzović et al., 2003).

Concerning butterflies, around 10% of the species has a trend of increasing population while in a third species declining trend occurred (Mijović et al., 2012).

Some species of bats (Chiroptera) in Serbia are registered so far in 24.2% UTM square grid 10× 10 km. Population trends of 23 out of 30 known species are stable (Paunović, 2016). Centers of the diversity of amphibians in Serbia are situated in Pomoravlje, in northwestern Serbia and South Banat (Vukov et al., 2013), while the centers of diversity of reptiles are in southeast Serbia and in Kosovo and Metohija (Tomovic et al., 2014). The populations of most species of amphibians and reptiles is showing a downward trend.

8.4 Genetic and Crop Diversity

Genetic resources in Serbia include a large number of autochthonous culti-vated plant and domestic animal species. The genetic potential of plant and animal species has yet to be thoroughly investigated or estimated. Researches determined genetic differentiation of species of genera *Asyneuma*, *Cerastum*, *Edraianthus*, *Hypericum*, *Thymus*, *Ramonda*, *Vaccinium*, etc. as well as for some species of mosses. The level of genetic differentiation between popu-lations is known for some protected wild species (e.g., *Vipera ammodytes*, *Rana synklepton esculenta*, *Capreolus capreolus*, etc.). Genetic diversity has been researched for the fish species *Salmo trutta*, as well as for some species of the *Barbus* genus (Tošić et al., 2014; Simonović et al., 2013; Radović and Kozomara, 2011)

Agro-biodiversity in Serbia includes species and habitats of cultivated plants and animals, as well as species and ecosystems of importance to the pro-duction of food. The number of cultivated plant species exceeds 150. More than 1,200 sorts of agricultural plants have been developed in Serbia. The national collection of the Plant Gene Bank is initiated in the 1990s, as part of the Direc-torate for Nationally Referent Laboratories. The most important autochthonous breeds of domestic animals are included in the List of autochthonic breeds of domestic animals in the Republic of Serbia ("Official Gazette of the Republic of Serbia," Issue 38/10). Serbia also has an autochthonous breed of bee, *Apis melifera carnica,* with its varieties. It is one of the most valuable breeds of hon-eybee in the world (Radović and Kozomara, 2011).

The total number of medicinal and aromatic plant species in Serbia is estimated at around 700, of which 420 have been officially registered and 280 are traded as commodities (Amidžić et al., 1999). The greatest impor-tance among genetic resources of medicinal and aromatic herbs is given to the genetic diversity of commercially important species (chamomile, mint, sage, hypericum, yarrow, oregano, bearberry, valerian, plantain, primula, etc.) (Radović and Kozomara, 2011).

8.5 Threatened Biodiversity

In Serbia, four Red books were published – *RedData Book of Flora of Serbia 1* – *extinct and critically endangered taxa* (Stevanović,1999) and it accounts 171 plant taxa, which comprise approximately 5% of the total flora in the Republic of Serbia. Of that number, four taxa have become extinct (*Althaea kragujevacensis, Althaea vranjensis, Scabiosa achaeta* and *Trapa annosa*); 46 taxa have been extinct from Serbia, but can be found in neighboring areas, 121 species are highly endangered, with a high probability of disappearing from the region in the near future or becoming extinct if not given appropriate attention. According to Preliminary Red List of Flora of Serbia (Stevanović, 2002), vascular flora of Serbia includes more than 1,000 threatened vascular plant species. Some plant species were rediscovered for flora of Serbia after they had been considered extinct in this territory. Therefore, they were included in the publication of Red Book of Flora of Serbia in the category of Extinct taxa (EX) (Tomović et al., 2009).

*Red Book of Butterflies for Serbia (*Jakšić, 2003), evaluated the conservation status of 57 species of butterflies, accounting for 34% of butterfly fauna in Serbia. One species is listed as extinct (*Leptidea morsei*) and 11endangered species, including *Pyrgus andromedae, Papilio machaon, Euchloe ausonia, Erebia alberganus, E. manto, Parnassius apollo, Colias myrmidone, Apatura ilia* and *A.iris.*

Red book of Fauna of Serbia I – Amphibians (Kalezić et al., 2015) of 21 species of amphibians in Sebia (8 caudates and 13 anurans), four are at the risk of extinction, according to IUCN criteria, 8 species are considered threatened according to the DELH criteria. That situation is almost certainly more a consequence of the lack of information on populations and trends than a really low level of threats to Serbian amphibians survival. The diversity of amphibian species in Serbia is relatively high compared to other Balkan countries, Serbia is the second in diversity of amphibians in the Balkan (Vukov et al., 2013).

Red Book of Fauna of Serbia II – Reptiles (Tomović et al., 2015) is dealing with 24 species of Reptiles: 11 lizards, 10 snakes, and 3 chelonians. According to the IUCN criteria 8 species of reptiles are at high risk of extinction, i.e., according to the DELH criteria 11 reptiles taxa are highly threatened.

Currently, several institutions and experts are working on the preparation of the Red Book of Birds in Serbia that needs to process more than the 100 bird species whose populations are threatened. The publication of this book is expected by the end of 2017.

Preliminary List of Species of Vertebrate Red Book was published in 1991 (Vasić et al., 1991). This list provides a starting point for the development of a comprehensive vertebrate red list for Serbia. The list identifies one species of cyclostomes and 30 species of fish, 22 species of amphibians, 21 species of reptiles, 72 species of mammals and a large number of birds (353 species) as being threatened and in need of conservation attention (Figures 8.1).

Under the protection there are 2633 wild species of plants, animals and fungi in Serbia, out of which 1,760 are strictly protected and 868 protected (The Rulebook on proclamation and protection of strictly protected and protected wild species of plants, animals and fungi, The Official Gazette of the Republic of Serbia, No. 5/2010, 47/2011, and 98/ 2016). Out of the species listed above, 112 species of plants and animals are under use and trade control. Almost all mammals, birds, amphibians and reptiles are under a protection regime, as well as a large number of insects and plants. Over 50% of strictly protected species is on the list of international Conventions and EU Directives (Popović, 2016).

There are numerous factors threatening the biodiversity in Serbia, which influence individual species and their habitats. Among the most important should be noted the destruction and fragmentation of habitats, cutting ecological corridors, pollution of water, soil, and air, change of way of use of land containing valuable natural habitats, persecution and unsustainable collection of species, inadequate enforcement of regulations and funding for biodiversity protection with insufficient sanctioning offenders. A particular problem is the lack of harmonization of sectoral policies and a relatively small percentage of Serbian territory under protected areas and ecological network.

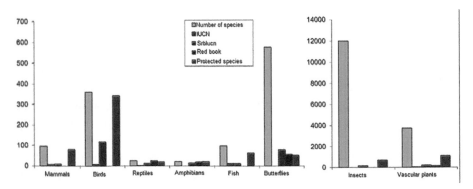

Figure 8.1 Threatened and protected plant and animal species (*Source:* SEPA, Popović, 2016).

Figure 8.2 Serbian Spruce (*Picea omorica* Panč.) on locus classicus Zaovine site in the National Park Tara (Photo by B. Panjković).

8.6 Protected Areas and Ecological Network in Serbia

The protection of biodiversity in Serbia is achieved through the system of protected natural goods: protected areas, protected species and protected natural documents, i.e., through the implementation of measures for protection and improvement of species, their populations, natural habitats and ecosystems by Law on Nature Conservation (The Official Gazette of the Republic of Serbia, No.36/2009, 88/2010 91/2010, and 14/2016; Sekulić, 2013).

According to the data of the Institute for Nature Conservation of Serbia (INCS, 2017), the number of protected areas in Serbia is 474, out of which 5 national parks, 16 natural parks, 20 landscapes of exceptional features, 70

Figure 8.3   Special Nature Reserve Slano Kopovo,-Pannonian saline site , important habitat type for protection in Serbia and EU (Photo by S. Pozović).

Figure 8.4   Black Stork (Ciconia nigra L.) important breeding bird species in Serbia (Photo by S. Puzović).

nature reserves, 4 protected habitats, 311 natural monuments and 38 natural areas around the cultural monuments or protected environments. The total protected surface is 579.199 ha, which represents 6.54% of the total area of Serbia.

Within the UNESCO "Man and Biosphere" program are protected registered Golija Studenica (2001) as part of Golia mountain in the west Serbia and Bačko Podunavlje (2017) part of flooded Danube Basin in the north part of Vojvodina.

Legal nature conservation of Serbia dates back in fourteenth century, with the first written law (Dušan's Code) which controls deforestation, determines pastures, forbids plowing of forest pastures as well as building settlements on them. The first protected area was established in 1874 when Obedska Bara was put under strict protection. After the World War II and after The Institute for Nature Conservation was founded (1948), the first object were put under protection (strict natural reserve "Mustafa," "Felješana" (1948), "Oštrozub," "Pod Crnim vrhom" and "Rošijana" (1950), etc. In 1960 the first national park "Fruška Gora" was declared in Serbia (Amidžić, 1998).

The Decree on the ecological network (The Official Gazette of the Republic of Serbia, No.102/10) defines the ecological network in Serbia which was based on the principles of Pan-European Ecological Network (PEEN). Certain protected areas have been established by the Law on Nature Conservation, with a primary goal of conservation of biodiversity as well as areas of international importance. Emerald network sites, IBA, IPA, PBA, Ramsar areas, border areas of ecological importance and certain areas of habitat types and wild species habitats are a part of the ecological network of Serbia. The national ecological network is currently being developed in Serbia and consists of 101 ecologically important areas or about 20.9% of the Serbian territory. In Vojvodina province, ecological network is elaborated in detail, built in spatial-planning documentation. On the territory of Autonomous Province of Vojvodina (APV) nature conservation has been established on 16.5% of the area, 665 habitats which are outside the protected areas have been described (INCV, 2017). The registered habitats, with the role of areas protected as priority habitats in Europe, corridors or buffer belts depending on the results of valorization, represent a part for establishing national ecological network NATURA 2000. When areas of protected regions, natural resources in the process of protection and regions prioritized for valorization and putting under protection are analyzed, it can be noticed that about 11% of AP Vojvodina territory can be put under protection. Within national ecological network (including corridors and protected zones of the network) and EU ecological networks Natura 2000, it can be expected that maximally up to 18% of APV area is included. With the upcoming process of valorization of defined natural habitats it should be determined which habitats offer the conditions for permanent conservation of natural values, for which there are other priorities of sustainable development (Szabados et al., 2009, Panjković and Szabados, 2012).

The Law on Nature Conservation states that the ecological network in Serbia should be determined and become part of the Natura 2000 Network. The Natura 2000 Network should be established in the Republic of Serbia by EU accession date.

About 61 areas have been selected for European Emerald Network in Serbia. These areas are considered as particularly important for the protection and conservation of wild species and their habitats. Their total surface is 1,019,269.31 ha, or 11.54% of the Serbian territory (Sekulić and Sekulić – Šinžar, 2010).

In Serbia, 42 areas of international importance for the conservation of bird diversity have been selected by applying the Important Bird Area (IBA) with the total coverage of 1,259,624 ha or 14.25% of the territory of Serbia (Puzović et al., 2009). 61 areas of Important Plant Areas (IPA) have been identified in Serbia (8.5%) (Stevanović, 2005b). On the territory of Serbia 40 Prime Butterfly Areas (PBA) have been selected. These areas occupy 903,643 ha, that is 10.23% of Serbian territory. About 40 PBA have been selected and occupy 903,643 hectares (10.23%) (Jakšić, 2008).

Based on the Ramsar Convention on Wetlands of International Importance, 10 areas are protected in Serbia which are also protected on national level. These include: "Obedska bara," "Carska bara," "Ludaško jezero," "Slano kopovo," "Labudovo okno" "GornjePodunavlje," "Zasavica" and "Koviljsko petrovaradinski rit" in Vojvodina Province, and "Vlasinsko jezero" and "Karajukića bunari" in the central part of Serbia.

Additionally, the remaining part of the territory of Serbia (93.7%), i.e., beyond protected areas (although the national ecological network has been identified and is in a significant part overlapping with the boundaries of the protected areas) has a more complex situation in relation to the pressure on the fauna and the needs of active protection of animals and their habitats (Radović and Kozomara, 2011; Mijović et al., 2012, Puzović et al., 2015).

So far, in Serbia active measures to protect animal and plant species and their habitats were applied selectively and with a small investment of funds. Examples of good practice can be found on the territory, especially in protected areas (Puzović and Panjković, 2015), and in other parts of Serbia as a partial protection of the Griffon Vulture (*Gyps fulvus*) and the establishment of its feeding stations (Grubač, 2016).

Scientific institutions and associations of citizens have an important role in the monitoring and protection of fauna, which is especially evident at the Society for Bird Protection and Study Serbia NGO related birds (Šćiban et al., 2015), Mustela NGO related Mammals (Paunović et al., 2011), The

HabiProt NGO related butterflies (Popović and Đurić, 2011), and Protego NGO related Amphibians and reptiles (Mesaroš, 2011), which has developed a substantial number of activities in this field.

Based on the analysis of size, shape, and structure of the protected area, it has been noticed that the size and structure of the protected area is directly dependent on the degree of fragmentation of natural habitat types. Most protected areas do not have enough surface for conservation of internal, unchanged parts of habitats and specialized species in relation with them. An unfavorable form of protected areas with expressed border effect enables deeper penetration of negative influences from the direct surroundings, for which the protected area in Vojvodina has an important role in preventing negative influences from the surroundings. Fragmentation of natural habitats is a process, which takes place continuously due to mismatching of the existing law legislative in the area of agriculture, forestry and water management with nature protection, as well as because of economic interests (Panjković and Szabados, 2012)

The most threatened species are in most cases evidenced in half natural or secondary (anthropogenic) habitats which represent a suitable "replacement" habitats because of the destruction of the vast natural regions. Conservation of these habitats cannot be realized by the declaration of protected natural resources, but instead, they should become parts of ecological network (Panjković and Szabados, 2012).

The most important task is to protect the biodiversity in Serbia in the future, especially referring to the preservation and improvement of areas designed as centers of biodiversity. There is a need to apply the necessary measures of active protection, process management and coordination of sectoral policies through spatial planning. It should significantly increase the surface area in Serbia under the protected areas and establish an international ecological network Natura 2000.

Keywords

- crop diversity
- ecological network
- protected areas
- threatened biodiversity

References

Amidžić, L., (1998). Flora and vegetation protection of Serbia. *Protection of Nature, 50*, 115–124.

Amidžić, L., Bartula, M., & Cvetković, D., (2014). The State of Biodiversity in Serbia. *Natural Areas Journal*, pp. 222–226.

Amidžić, L., Dražić, S., Kostić, M., Maksimović, S., Panjković, B., Mandić, R., Menković, N., Panjković, B., Popov, V., Radanović, D., Roki, D., Sekulović, D., Stepanović, B., & Tasić, S., (1999). Strategy of Medicinal Herbs Protection in Serbia. Ministry of Environment Protection of the Republic of Serbia. Special edition.

Banković, S., Medarević, M., Pantić, D., & Petrović, N., (2009). National Forest Inventory of the Republic of Serbia: Forest Fund of the Republic of Serbia. Ministry of Agriculture, Forestry and Water Management of the Republic of Serbia, Forestry Directorate.

Butorac, B., & Panjković, B., (2007). Florist and vegetation diversity of Serbia and its protection. The Congress of Biologists of Serbia with international participation. KOBIS. Summary of the summary. Plenary papers Synopsis. Palić. 18–23.

Diklić, N., (1987). Endemic species in the plant world of Serbia. Non-threats of vulnerability and protection. Academy of Sciences and Arts of Bosnia and Herzegovina. Special Editions LXXXIII. Department of Primary and Mathematical Sciences, *14*, 13–18.

Džukić, G., (1995). Diversity of amphibians and reptiles of Yugoslavia with an overview of the species of an international trade–In: Biodiversity of Yugoslavia Review of the Type of International Character, Stevanović, V., & Vasić, V., (eds.). Faculty of Biology, University of Belgrade Ecolibri, Belgrade.

Džukic, G., & Kalezić, M., (2004). Biodiversity of amphibians and reptiles in the Balkan Peninsula. In: *Balkan Biodiversity,* Griffiths, I. H., Krzštofek, B., & Reed, J. M. (eds.). *Pattern and Process in the European Hotspot.* Kluwer Academic Publisher. pp. 167–192.

Gajić, M., (1984). Floral Elements of Serbia. In: Vegetation of SR Serbia, Saric, M., (ed.). Belgrade: Serbian Academy of Arts and Sciences, *vol. 1*, pp. 317–397.

Grubač, B., (2016). *Griffon Vulture (Gyps fulvus).* Institute for Nature Conservation of Serbia, Belgrade.

Horvat, I., Glavač, V., & Ellenberg, H., (1974). *Vegetation Südosteuropas.* VEB Gustav Fischer Verlag, Stuttgart.

IUCN-WWF, (1987). *Plant Conservation Program and IUCN Threatend Plant Unit: Centers of Plant Diversity – A Guide and Strategy for their Conservation.-IUCN TPU,* The Green, Kew, Richmod, Surrey, UK.

Jakšić, P., (2003). *Red Data Book of Serbian Butterflies.* Institute for Nature Conservation of Serbia, Belgrade.

Jakšić, P., (2008). *Prime Butterfly Areas: A Tool for Nature Conservation in Serbia.* HabiProt, Belgrade.

Kalezić, M., Tomović, L., & Džukić, G., (2015). Serbian Academy of Arts and Sciences, Scientific meetings of books. CXI, Department of Chemical and Biological Sciences, Book 2, 75-104, Belgrade.

Kryštufek, B., (2004). A quantitative assessment of Balkan mammal diversity. In: *Balkan Biodiversity,* Griffiths, I. H., Krzštofek, B., & Reed, J. M., (eds.). *Pattern and Process in the European Hotspot,* Kluwer Academic Publisher. pp. 79–133.

Lakušić, D., (2005). Odnos specijskog i ekosistemskog diverziteta. In: *Biodiverzitet na Početku Novog Milenijuma,* Anđelković, M., (ed.). Zbornik radova sa naučnog skupa, Srpska akademijanauka i umetnosti, Naučni skupovi knj. CXI, Odeljenje hemijskih i bioloških nauka, knjiga *2*, 75–104, Beograd.

Lakušić, D., Blaženčić, J., Ranđelović, V., Butorac, B., Vukojičić, S., Zlatković, B., Jovanović, S., Šinžar-Sekulić, J., Žukovec, D., Ćalić, I., & Pavićević, D., (2005). Habitats of Serbia–Manual with descriptions and basic data. In: *Habitats of Serbia, Results of the project "Harmonization of National Nomenclature in Habitat Classification with International Standards,"* Lakušić, D., (ed.) Institute of Botany and Botanical Garden "Jevremovac", Faculty of Biology, University of Belgrade, Ministry of Science and Technology protection of the environment of the Republic of Serbia.

Mesaroš, G., (2011). *Wetland Habitats in Vojvodina.* Protego, Subotica.

Mijović, A., Sekulić, N., Popović, S., Stavretović, N., & Radović, I., (2012). *Biodiversity of Serbia: Status and Perspectives.* Institute for Nature Conservation of Serbia, Belgrade.

Panjković, B., & Szabados, K., (2012). Nature and perspectives of nature protection in AP Vojvodina in the process of European integration (SRB). Proceedings, Scientific Conference, Zasavica. 6-16. Sremska Mitrovica.

Paunović, M., (2016). *Bats of Serbia. PhD theses.* Faculty of Biology, Belgrade.

Paunović, M., Karapandža, B., & Ivanović, S., (2011). Blind mice and environmental impact assessment Methode guidelines for environmental impact assessment and strategic assessment of impacts on the environment. *Bats Environmental Impact Assessment.* Wildlife Conservation Society Mustela, Belgrade.

Popović, M., & Đurić, M., (2011). *Diurnal Butterflies of Serbia.* HabiProt, Belgrade.

Popović, S., (2016). Indicator of biodiversity in the Republic of Serbia year. Ministry of Agriculture and Environmental Protection. Agency for Life Insurance,–Electronic Journal.: – Beograd: Reports on the state of environment in the Republic of Serbia for 2015. http://www.sepa.gov.rs/download/Izvestaj_2015.Pdf.

Puzović, S., & Panjković, B., (2015). *Natural Heritage Management in Vojvodina.* Provincial secretariat for urban planning, construction and environmental protection and Institute for Nature Conservation of Vojvodina Province, Novi Sad.

Puzović, S., Radišić, D., Ružić, M., Rajković, D., Radaković, M., Pantović, U., Janković, M., Stojnić, N., Šćiban, M., Tucakov, M., Gergelj, J., Sekulić, G., Agošton, A., & Raković, M., (2003). *Birds of Serbia and Montenegro: Breeding Population Estimates and Trends: 1990–2003.* Ciconia, *12*, 35–120.

Puzović, S., Radišić, D., Ružić, M., Rajković, D., Radaković, M., Pantović, U., Janković, M., Stojnić, N., Šćiban, M., Tucakov, M., Gergelj, J., Sekulić, G., Agošton, A., & Raković, M., (2015). *Birds of Serbia: Breeding Population Estimates and Trends for the Period 2008–2013.* Bird Protection and Study Society of Serbia, Department of Biology and Ecology, PMF, Novi Sad.

Puzović, S., Sekulić, G., Stojnić, N., Grubač, B., & Tucakov, M., (2009). *Important Bird Areas in Serbia – IBA.* Ministry of environment and spatial planning, Provincial secretariat for urban planning, construction and environmental protection and Institute for Nature Conservation of Serbia, Belgrade.

Radišić, D., Vujić, A., Radenković, S., Nikolić, T., Trifunov, S., & Andrić, A., (2016). Prime hoverfly (Diptera, Syrphidae) areas (PHA): Review and proposal for protection. *II International Symposium "Nature Conservation – Experiences and Perspectives."* Institute for Nature Conservation of Vojvodina Province, Book of abstracts, Novi. Sad., pp. 55–56.

Radović, I., & Kozomara, M., (2011). *Biodiversity Strategy of the Republic of Serbia for the Period 2011–2018.* Ministry of Environment and Spatial Planning of Republic of Serbia, Belgrade.

Savić, D., (2016). The diversity of mushrooms divides Ascomycota in the Fruška gora area with a special emphasis on Helotiales. Doctoral dissertation, Faculty of Natural Sciences and Mathematics, University of Novi Sad.

Šćiban, M., Rajković, D., Radišić, D., Vasić, V., & Pantović, U., (2015). *Birds of Serbia – Critical list of Species.* Institute for nature conservation of vojvodina province and bird protection and study society of Serbia, Novi Sad.

Sekulić, G., (2013). Recommendations for implementing the IUCN protected area management categories in Serbia. In: *Protected Areas in Focus: Analysis and Evaluation,* Getzner, M., & Jungmeier, M., (eds.). Verlag Johannes Heyn. Klagenfurt.

Sekulić, N., & Sekulić-Šinžar, J., (2010). *Emerald Ecological Network in Serbia.* Ministry of environment and spatial planning, Institute for Nature Conservation of Serbia, Dexin, Belgrade.

Šimić, S., Vujić, A., Radenković, S., Radišić, P., & Nedeljković, Z., (2009). *Hoverflies (Diptera, Syrphidae) of Wetlands in Vojvodina.* Matica Srpska, Novi Sad.

Simonović, P., Tošić, A., Vassilev, M., Apostolou, A., Mrdak, D., Ristovska, M., Kostov, V., Nikolić, V., Škraba, D., Vilizzi, L., & Copp, G. H., (2013). Risk identification of nonnative freshwater fishes in four countries of the Balkans region using FISK. *Mediterranean Marine Science, 14*(2), 369–376, DOI10.12681/mms.337.

Stanisavljević, L., Vujić, A., Jakšić, P., Markov, Z., & Ćetković, A., (2016). Functional-ecological status, vulnerability and economic evaluation of insect pollinators in Serbia. SANU, Academic Committee for the Study of Fauna of Serbia, Scientific Meeting "The ecological and economic significance of fauna of Serbia. Book of Abstracts. 28–30.

Statistical Yearbook National of the Republic of Serbia (2016), RZC Statistical office of the Republic of Serbia, Belgrade.

Stevanović, V., & Stevanović, B., (1995). Basic climatic, geological and pedological factors of biodiversity of the contaminated ecosystems of Yugoslavia. In: *The Biodiversity of Yugoslavia A Summary of the International Meaning,* Stevanović, V., Vasić, V., (eds.). Ecolibri and Faculty of Biology, University of Belgrade,.

Stevanović, V., & Vasić, V., (1995). *Biodiversity of Yugoslavia with overview of Species of International Importance.* University of Belgrade, Faculty of biology, Ekolibri, Belgrade.

Stevanović, V., (1992). Floristic division of the territory of Serbia with an overview of higher chorions and appropriate floristic elements. *Flora Srbije 1,* Sarić, M. R., (ed.), Beograd, Srpska akademija nauka, pp. 47–56.

Stevanović, V., (1995). Biogeographical division of the territory of Yugoslavia. In: *Biodiversity of Yugoslavia with Overview of the Type of International Importance,–Ecolibri,* Stevanović, V., & Vasić, V., (eds.). Belgrade, Biological Faculty in Belgrade, pp. 117–127..

Stevanović, V., (1996). Analysis of the Central European and Mediterranean orophyte element on the mountain of the Central Balkan Peninsula, with special reference to endemics. *Bocconea, 5,* 77–97.

Stevanović, V., (1999). *The Red Data Book of Flora of Serbia 1 – Extinct and Critically Endangered Taxa.* Ministry of environment of the republic of Serbia, Faculty of biology – University of Belgrade, Institution for Protection of Nature of the Republic of Serbia, Belgrade.

Stevanović, V., (2002). Preliminary Red List of vascular flora of Serbia and Montenegro-according to IUCN criteria Belgrade.

Stevanović, V., (2005a). Evaluation of biodiversity - from interpretation to conservation. An example of the endemic vascular flora of the Balkan peninsula. In: *"Biodiversity at the Beginning of the New Millennium,"* Proceedings of the Scientific Scientific Society, Anđelković, M., (ed.). Belgrade, Serbian Academy of Sciences and Arts, Scientific Conferences CXI. Department of Chemical and Biological Sciences 2. pp. 1–30.

Stevanović, V., (2005b). IPAs in Serbia. In: *Important Plant Areas in Central and Eastern Europe – Priority Sites for Plant Conservation,* Anderson, S., Kušik, T., & Radford, E., (eds.). Plant life International, UK. pp. 74–75.

Stevanović, V., Jovanović, S., Lakušić, D., & Niketić, M., (1995). Biodiversity of the vascular flora of Yugoslavia with an overview of species of international significance. In: *Biodiversity of Yugoslavia With Overview of the Type of International Significance,* Stevanović, V. & Vasić, V., ed. Faculty of Biology, University of Belgrade, Ecolibri, Belgrade. pp. 183–217.

Stevanović, V., Jovanović, S., Lakušić, D., & Niketić, M., (1999). Phytogeographic affiliation of flora and vegetation of Serbia. In: *The Red Book of the Flora of Serbia 1 - Searching for Extremely Endangered Taksoni,* Stevanović, (ed.). Ministry of Environment of the Republic of Serbia, Faculty of Biology Univ. in Belgrade and the Institute for the Protection of Nature of the Republic of Serbia.

Stevanović, V., Jovanović, S., Lakušić, D., & Niketić, M., (1999). Characteristics and peculiarities of the flora of Serbia and its phytogeographical position on the Balkan peninsula and in Europe: 9–18. In: *Red Book of Flora of Serbia 1 - Thoughts IKrajnje Threatened Taksoni,* Stevanović, V., ed. Ministry of Environmental Protection of the Republic of Serbia, Faculty of Biology, University of Belgrade, Institute of Nature Protection of Serbia, Belgrade.

Stevanović, V., Kit, T., & Petrova, A., (2005). Size, distribution and phytogeographical position of the Balkan endemic flora.- *XVII International Botanical Congress,* Vienna, Austria, July, pp. 66.

Szabados, K., Panjković, B., Kiš, A., Stojčić, V., Dobretić, V., Timotić, D., Delić, J., Stanišić, J., Galamboš, L., Tucakov, M., Kicošev, V., Stojnić, N., Pil, N., Kovačev, N., & Perić, R., (2009). Establishment of an ecological network in AP Vojvodina - a review of the situation, analyzes and possibilities. Institute for the Protection of Nature of Serbia, Novi Sad.

Szabados, K., Panjković, B., Marinković, L., Kiš, A., Kicošev, V., Cvijić, D., Tucakov, M., & Čalakić, D., (2011). Establishment of a national ecological network in Vojvodina (Serbia). *Scientific Meeting with International Participation / International Conference "Nature Protection in the 21st Century" Proceedings of Plenary Sessions* (Book No.1). 185–192.

Tomović, G., (2007). Fitogeografska pripadnost, rasprostranjenje i centri diverziteta Balkanske endemične flore. Doktorska disertacija, Univerzitet u Beogradu, Biološki fakultet.

Tomović, G., Niketić, 'M., Lakušić, D., Ranđelović, V., & Stevanović, V., (2014). Balkan endemic plants in Central Serbia and Kosovo regions: Distribution patterns, ecological characteristics, and centers of diversity. *Bot. J. Linn. Soc., 176,* 173–202.

Tomović, G., Zlatković, B., Niketić, M., Perić, R., Lazarević, P., Duraki, Š., Stanković, M., Lakušić, D., Anačkov, G., Knežević, J., Szabados, K., Krivošej, Z., Prodanović, D., Vukojičić, S., Stojanović, V., Lazarević, M., & Stevanović, V., (2009). Threat status revision of some taxa from The RedData Book of Flora of Serbia 1. *Botanica Serbica., 33*(1), 33–43.

Tomović, L., Kalezić, M., & Džukić, G., (2015). *Red Book of Fauna II – Reptiles.* University of Belgrade—Faculty of Biology, Institute for nature conservation of Serbia. Belgrade.

Tošić, A., Škraba, D., Nikolić, V., Mrdak, D., & Simonović, P., (2014). New mitochondrial DNA haplotype of brown trout *Salmo trutta* L. from Crni Timok drainage area in Serbia. *Turkish J. Fisheries and Aquatic Sciences*, *14*, 37–42. DOI: 10.4194/1303–2712-v14_1_05.

United Nations Disaster Assessment and Coordination (UNDAC), (2014). Report on flooding handed over to the Government of Serbia.

Vasić, V., (1995a). Diversity of Birds of Yugoslavia with an overview of species of international significance. In: *Biodiversity of Yugoslavia With a Review of the Species of International Importance*, Stevanović, V., & Vasić, V. (eds.). Faculty of Biology, University of Belgrade and Ecolibri, Belgrade.

Vasić, V., (1995b). Biodiversity in sensitive ecosystems and areas of international importance. In: *Biodiversity of Yugoslavia With a Review of the Type of International Significance*, Stevanović, V., & Vasić, V., (eds.). Faculty of Biology, University of Belgrade and Ecolibri, Belgrade.

Vasić, V., Džukić, G., Janković, D., Simonov, N., Petrov, B., & Savić, I., (1991). *Preliminary List of Species of Serbian Vertebrate Red Book*, Protection of nature, Belgrade.

Vukov, T., Kalezić, M. L., Tomović, L., Krizmanić, I., Jović, D., Labus, N., & Džukić, G., (2013). *Amphibians in Serbia: Distribution and Diversity Patterns*. Bulletin of the Natural History Museum, Belgrade, *6*, 90–112.

Additional References

The Fifth Report on the State of the Environment in the Republic of Serbia for 2014, 2011. Ministry of Environment, Mining and Spatial Planning of Republic of Serbia, Agency for Environmental Protection, Belgrade.

The Rule book on proclamation and protection of strictly protected and protected wild species of plants, animals and fungi, "The Official Gazette of the Republic of Serbia," No. 5/2010, 47/2011, and 98/ 2016).

The List of autochthonic breeds of domestic animals in the Republic of Serbia ("Official Gazette of the Republic of Serbia," Issue 38/10).

Law on Nature Conservation.("The Official Gazette of the Republic of Serbia," No. 36/2009, 88/2010 91/2010, and 14/2016, Sekulić, 2013).

The Decree on ecological network ("The Official Gazette of the Republic of Serbia," No. 102/10).

CHAPTER 9

Biodiversity in Slovakia

**LIBOR ULRYCH,[1] ANNA GUTTOVÁ,[2] PETER URBAN,[3]
ALICA HINDÁKOVÁ,[2] JAROMÍR KUČERA,[2] VIKTOR KUČERA,[2]
JOZEF ŠIBÍK,[2] and MARTA MÚTŇANOVÁ[1]**

[1]State Nature Conservancy of Slovak Republic, Tajovského 28B, 974 01 Banská Bystrica,
Slovakia, E-mail: libor.ulrych@sopsr.sk, marta.mutnanov@sopsr.sk
[2]Plant Science and Biodiversity Centre, Slovak Academy of Sciences, Dúbravská cesta 9,
845 23 Bratislava, Slovakia, E-mail: anna.beresova@savba.sk, alica.hindakova@savba.sk,
jaromir.kucera@savba.sk, viktor.kucera@savba.sk, jozef.sibik@savba.sk
[3]Matej Bel University, Faculty of Natural Sciences, Department of Biology and Ecology,
Tajovského 40, 974 01 Banská Bystrica, Slovakia, E-mail: peter.urban@umb.sk

9.1   Introduction

Slovakia is situated in the Central Europe between 47° 44' and 49° 37' northern latitude and between 16° 50' and 22° 34' eastern longitude. The area of Slovakia covers 49,036 km^2. Territory of Slovakia belongs to the Palearctic realm, Euro-Siberian region, Pannonian Province and the Middle European Forest province (Udvardy, 1975). Two main geomorphological phenomena create the background for biodiversity – mountains of the Carpathian massif and lowlands outreaching from Pannonian region. The altitude ranges from 94 m to 2,655 m a.s.l. This highest point (Gerlach peak) is also the highest peak of the whole Carpathian massif. The highest average temperature is in Pannonian region, in its southwest part. In January, which the coldest month, is average temperature ranges from –1 to 2°C, in July, which is the warmest month, is average temperature from 18 to 21°C. Annual average temperature in Pannonian region is from 9 to 11°C. According to altitude the annual average temperature increases in 1000 m a.s.l. to 4–5°C, in 2000 m a.s.l. to –1°C. The lowest temperature recorded ever is –41°C from locality Vígľaš-Pstruša (11.2. 1929), the highest is 40.3°C from Hurbanovo meteorological station (20 July 2007). (http://www.shmu.sk).

The general geological structure is complex. It has undergone several stages of development, including tectonic and morphological processes, which results in a complicated structure and lithological content. With regard to age, the structure consists of a complete range of rocks from the oldest of Precambrian era to recent rocks (ca. 50,000 years old vulcanites). Dominant rocks include sedimentary formations (limestones, slates, flysh, sedimentary quartz), but the cores of the mountains are formed by granites and metamorphites (Straka, 1998).

9.2   Ecosystem Diversity

Despite the fact that Slovakia is relatively a small country, diversity of plant and animal communities in this territory is high. Various climatic conditions, long period of land use activities, miscellaneous geological bedrock together with rugged relief and the position on the border zone of the Carpathians and the Pannonia region and the position on the crossroad in Central Europe affect the total variability of vegetation types including those that are bound to the refugial areas. Relict arctic-alpine communities with a high presence of arctic-alpine elements can be found in the mountains, while many thermophilous communities are confined to foothills and lowlands in the southern part of Slovakia in the Pannonian region.

The central European region, where Slovakia is situated, similarly to other parts of Northern Hemisphere were affected by the Quaternary climatic oscillations over a period of approximately 2.4 million years (Comes and Kadereit, 2003). While the Alps were almost entirely covered by a large ice sheet, only localized glaciers were present on the highest summits of the Carpathians (Puşcaş et al., 2008; Šibíková et al., 2010). Oldalpine flora of the high mountains was pushed down from the peaks to broader and lower areas, which allowed the interchange of alpine floras (or their parts) between isolated mountain ranges, such as between the Carpathians and the Alps. In addition, many new species came to Central Europe with the tundra or subarctic vegetation; hence the individuals of the arctic flora were mixed together with the alpine flora of European mountains. Diverse, ecologically homogenous arctic-alpine flora was formed, although with various genesis. This was the basis for the subsequent formation of broad arctic-alpine disjunctions (Hendrych, 1984; Šibíková et al., 2010). The recolonization of large, unoccupied territories after deglaciation in the Alps most likely arose from nearby refugia, although this is unlikely for the Carpathians, in which large areas in all the massifs remained ice-free, even during the last glacial maximum (Puşcaş et al., 2008; Šibíková et al., 2010).

When looking at overall species richness of individual plant communities, the highest numbers of species were recorded in seminatural grasslands, in which the new world records for fine-scale species richness were reported in Slovakia – 52 species of vascular plants in 0.25 m^2 and 63 species 0.5 m^2, in the Kopanecké lúky meadows (Figure 9.1) in the Slovenský raj Mts (Chytrý et al., 2015). The most species-rich grasslands and forests in Slovakia are concentrated in regions with base-rich soils in the Western Carpathians, especially in the flysch zone of the Czech-Slovak borderland and in limestone and volcanic areas in central Slovakia. The richest types of

Figure 9.1 Kopanecké lúky meadows, locality of world records for fine-scale species richness – 52 species of vascular plants in 0.25 m² and 63 species 0.5 m² (Photo by P. Olekšák).

nonforest vegetation include semidry base-rich meadows (*Bromion erecti* and *Cirsio-Brachypodion pinnati*), base-rich pastures and mesic meadows (*Cynosurion cristati* and *Arrhenatherion elatioris*), *Nardus stricta* grasslands (*Violion caninae* and *Nardo strictae-Agrostion tenuis*) and some wet meadows and natural sub-Alpine grasslands (Chytrý et al., 2015). Forest record was sampled in relict oak forest with *Fagus sylvatica* and *Pinus sylvestris* (the alliance *Quercion pubescenti-petraeae*) in Veľká Fatra Mts near Blatnica village in Tlstá Nature Reserve and in *Acer pseudoplatanus-Fagus sylvatica-Picea abies* forest of the *Cortuso-Fagetum* (the *Fagion sylvaticae*), in the alluvium of the stream Gaderský potok. 109 species of vascular plants were recorded on the plot size of 400 m² and 118 species on the plot size of 500 m², respectively (Chytrý, et al., 2015).

Based on broad habitat-type designations we can sort the communities of Slovakia into subsequent groups: (i) aquatic, shoreline and swamp vegetation; (ii) springs, fens and bogs; (iii) rock fissures and screes; (iv) high-altitude vegetation; (v) grasslands and meadows; (vi) scrub vegetation; (vii) forests; and (viii) anthropogenic vegetation (cf. Jarolímek and Šibík, 2008).

The first group of communities – aquatic, shoreline and swamp vegetation, represented by those ones that can be found in the rivers, ponds, lakes and wetland areas consist of various vegetation types strongly dependent on water level. They are mainly species-poor communities usually dominated by a single species such as reeds and sedges. Species-typical for these habitats play a vital role in sedimentation of mesotrophic and eutrophic wetlands and reflect the water regime dependent on duration of floods, depth of flooding waters, quality of soil conditions resulting from the flooding disturbance and the surrounding riparian forests (Oťaheľová et al., 2001). Five vegetation classes are recognized within this group of azonal communities: *Phragmito-Magnocaricetea* (reed swamp, sedge bed and herbland vegetation of freshwater or brackish water bodies and streams of Eurasia), *Lemnetea* (free-floating duckweed vegetation of still and relatively nutrient-rich freshwater bodies), *Charetea intermediae* (submerged macroalgal stonewort swards), *Potamogetonetea* (vegetation of rooted floating or submerged macrophytes of stagnant mesotrophic, eutrophic and brackish freshwater bodies and slowly flowing shallow streams); and *Isoëto-Nanojuncetea* (pioneer ephemeral dwarf-cyperaceous vegetation in periodically freshwater flooded habitats).

Vegetation of springs, fens and bogs consists of three phytosociological units (classes), namely *Montio-Cardaminetea* (vegetation of water springs), *Scheuchzerio palustris-Caricetea fuscae* (sedge-moss vegetation of fens, transitional mires and bog hollows) and *Oxycocco-Sphagnetea* (dwarf-shrub, sedge and peat-moss vegetation of the Holarctic ombrotrophic bogs and wet heath on extremely acidic soils). While spring communities comprise plant communities of flowing water habitats, so-called crenal habitats (Valachovič et al., 2001), the second group represents natural heliophilous communities dominated by hygrophytic sedges and mosses growing on permanently flooded habitats with a high concentration of organic matter undergoing processes of peat formation. Floristic composition is driven by nutrient status of soil water dependent on geological bedrock, content of organic sediments and the water level determined by climatic, hydrochemical and geomorphological conditions (Háberová and Hájek, 2001). Bog vegetation occupies sites saturated with water in areas where precipitation exceeds evaporation. Bog water has a very low pH because of humic acids releasing from organic matter and active acidification (Hájková et al., 2011). It is typical by the dominant role of mosses, mainly from the genus *Sphagnum,* which affects basic ecological conditions of the stands. Raised bogs are typical by lack of endemic species due to extreme oligotrophic conditions that homogenize sites from both physiognomic as well as floristic point of view (Šoltés et al., 2001).

Plant communities of rock fissures and screes represent the pioneer communities on the sites with poorly developed soils where the ecological succession usually starts or the stands are typical by a high level of natural disturbances due to very extreme conditions. Chasmophytic vegetation of crevices, rocky ledges, and faces of rocky cliffs and walls classified within the *Asplenietea trichomanis* are azonal communities with low total cover bound to rock cliffs, walls and boulders. In such habitats, vascular plants are confined to crevices and small ledges. Bryophytes and lichens are common, the former colonizing both ledges with soil accumulation and bare rock surfaces, the latter extending also to rock surfaces. In this habitat, plants experience various forms of stress, including nutrient and water shortage and large temperature fluctuations (Chytrý, 2009). Vegetation of scree habitats and pebble alluvia of the *Thlaspietea rotundifolii* is influenced by unstable rocky substrata formed of small to mid-sized rock fragments (Chytrý, 2009), which affects species composition as well as functional types occupying these habitats. In high altitudes, the scree communities are characteristic by the high occurrence of West Carpathian endemic taxa (Šibíková et al., 2010) mainly due to the island character of the sites that facilitate isolation of target taxa.

9.2.1 High-Altitude Vegetation

High-altitude vegetation presents a specific group of communities occurring in higher elevations; mostly above a timberline (Figure 9.2). Plant species composition of alpine habitats is influenced by the availability of soil water, depth and duration of snow cover, animal activities, wind, temperature, precipitation and soil development and nutrient content dependent on geological bedrock (Chapin III. and Körner, 1996; Körner, 2003; Kliment et al., 2010). The specific conditions in high mountains have given rise to a diverse mosaic of vegetation types, with an abundance of rare, relic and endemic taxa (Kliment et al., 2010). Alpine habitats usually serve as the biodiversity hotspots. One of the most important condition for speciation of new taxa (including endemics), as well as for preservation of relic taxa (depending on the plasticity of individual taxa), is diversity of geological bedrock and rugged relief, which directly influence habitat diversity and in turn affect the diversity of flora and vegetation (Kliment et al., 2011). Based on geological bedrock we can identify climax communities on basiphillous substrata – the class *Elyno-Seslerieta* and acidophilous alpine grasslands of the *Juncetea*

Figure 9.2 Mosaic of forest, krummholz and alpine meadows create typical landscape in many Slovakian mountains. (Krivánska Malá Fatra Mts) (Photo by J. Šibík).

trifidi on siliceous bedrock (Figure 9.3). The tall-herb vegetation in nutrient-rich habitats moistened and fertilized by percolating water at high altitudes is classified within the *Mulgedio-Aconitetea*. Snow bed communities of the S*alicetea herbaceae* represent arctic and alpine-subnival sheltered snow-bed vegetation at high altitudes of the mountain ranges. On the other side of the gradient there are the *Carici rupestris-Kobresietea bellardii* communities of wind-exposed sites with low snow cover. These fellfields and dwarf-shrub graminoid tundra communities are very valuable survivors from glacial period recently bounded only to Vysoké and Belianske Tatry Mts where they are creating mosaics within other vegetation types. Arctic-boreal dwarf-shrub mountain tundra of the *Loiseleurio-Vaccinietea* can be found on the soils with a thick layer of raw humus mainly on siliceous bedrock.

In Slovakia, we can identify grasslands and meadows as vegetation that is predetermined mainly by environmental conditions and wild herbivores (cf. Hejcman et al., 2013). Another factor shaping the seminatural grasslands is associated with long-term human activity from the beginning of agriculture during the Mesolithic–Neolithic transition; and improving (intensification) of the grasslands, a product of modern agriculture based on sown and highly

Figure 9.3 Siliceous alpine vegetation (Vysoké Tatry Mts), typical relief formed by glaciers. (Photo by L. Ulrych).

productive forage grasses and legumes (Hejcman et al., 2013). Based on these reasons we recognize following vegetation classes: Dry grasslands of the *Festuco-Brometea* (Figure 9.4), fringe communities of the *Trifolio-Geranietea*, lowland and submontane hay meadows and pastures of the *Molinio-Arrhenatheretea* and the *Nardetea strictae*, acidophilous sub-Atlantic heathlands of the *Calluno-Ulicetea* and saline vegetation of the *Crypsietea aculeatae, Festuco-Puccinellietea*; and *Scorzonero-Juncetea gerardii*. The origin and existence of grasslands and meadows depend on human activities, such as mowing and grazing of the various intensities. Combination of ecological factors and habitat management impacts are strongly reflected in species composition of particular vegetation types (Hegedüšová-Vantarová and Škodová, 2014).

9.2.2 Scrub Vegetation

Scrub vegetation is characterized by plant communities dominated by low woody plants, which usually form very dense bush of conspicuous

Figure 9.4   Xerotherm limestone vegetation—Zádiel Gorge (Photo by L. Ulrych).

physiognomy. The class *Crataego-Prunetea* consists of scrub and mantle vegetation seral or marginal to broad-leaved forests. Plant communities of *Franguletea* represents willow carrs of the sub-Atlantic regions of Central Europe. Recognition of this class is based on principles of zonality/azonality and separation of forest/wood and scrub communities into separate classes (Mucina et al., 2016). Sub-Alpine and subarctic herb-rich alder and willow scrub and krummholz of the *Betulo carpaticae-Alnetea viridis* occur on sites where the woody plants are able to rapidly and effectively colonize disturbed habitats. They occupy extreme habitats above the (edaphically limited) tree line (Kliment and Valachovič, 2007). Pine krummholz (*Pinus mugo*) in the sub-Alpine belts of nemoral mountain ranges (*Roso penduli-nae-Pinetea mugo*) create distinctive zone between alpine meadows and montane spruce forests both on acidic and basiphilous substrata (Figures 9.2 and 9.5). Four general ecological types can be distinguished in whole area of the occurrence: (i) a dry, rocky type on basiphilous bedrock; (ii) a moist type on nutrient-rich soils on basiphilous, as well as silicate bedrock; (iii) an acidophilous, oligotropic, species-poor type; (iv) an oligotropic, windswept

Figure 9.5 Rugged relief of the Vysoké Tatry Mts was formed by glaciers during the ice age. Montane spruce forests replaced by *Pinus mugo krummholz* represent typical zonal vegetation in the lower parts of mountain valleys and slopes (Photo by J. Šibík).

type at the transition between sub-Alpine and alpine belt on silicate bedrock (Šibík et al., 2010).

Forests are the most common vegetation type in Slovakia covering approximately 40% of the surface (Gubka et al., 2013). Ecological amplitude of most woody plants is by far wider than forbs and reflects mainly meso- and macroclimatic factors that can be seen as different forest types in individual altitudinal ranges (cf. Šibík, 2007) reflecting different altitudinal zonation that exists due to varying climatic conditions. In the low altitudes the zonal oak, mixed deciduous oak and conifer woods of warm regions of the *Quercetea pubescentis* (Figure 9.6) occur in the cool-temperate Nemoral zone. Acidophilous oak and oak-birch forests on nutrient-poor soils of the *Quercetea robori-petraeae* are replaced by mesic deciduous and mixed forests of the *Carpino-Fagetea sylvaticae* in higher elevations (Figure 9.7). These are typical prevailing forest dominated by *Fagus sylvatica* which grow both on acidic and limestone substrata (Figure 9.8). Recently, a multidimensional classification of basiphytic beech forests was proposed, including both ecological and geographical criteria as equally valid concepts

Figure 9.6 Thermo-xerophilous Oak forest, Plášťovce (Photo by L. Ulrych).

Figure 9.7 Landscape with broadleaf forests and scattered meadows (Považský Inovec Mts) (Photo by L. Ulrych).

Figure 9.8   Beech forest, Rokoš Mt (Photo by L. Ulrych).

which may be used alternatively depending on the purpose and context of the classification (Willner et al., 2017). Ravine forests (the *Tilio platyphylli-Acerion*) occur in ravines and on steep slopes on disturbed areas by soil erosion and falling rocks (Chytrý, 2012). Holarctic coniferous forests of *Vaccinio-Piceetea* (Figure 9.5) represent extrazonal occurrence of the taiga biome in our geographical latitudes that is typical by cold winters and short growing seasons (Chytrý, 2013). Over the last few decades, windthrow disturbances have increased in frequency and severity in European coniferous forests (Havašová et al., 2017) usually followed by bark beetle (*Ips typographus*) outbreak (Figure 9.9). Forests in the Western Carpathians were predisposed to recent severe disturbance events as a result of synchronized past disturbance activity, which partly homogenized size and age structure and made recent stands more vulnerable to bark beetle outbreak (Janda et al., 2017). These findings show the ecosystems currently have high ecological resilience to disturbance (Janda et al., 2017). These processes should be

Figure 9.9   The wind disturbance and following spruce beetle outbreak are natural components positively affecting diversity (Photo by L. Ulrych).

recognized as a natural part of ecosystem dynamics in the mountain forests of Central Europe (Thom and Seidl, 2015; Janda et al., 2017). Relict pine forests dominated by *Pinus sylvestris* and/or *Larix decidua* on calcareous substrates of the *Erico-Pinetea* are characterized by a significant proportion of endemic taxa together with thermophilic, dealpine as well as pre-Alpine species. Taxa typical of boreo-continental taiga are absent, while occurrences of various light-demanding species are frequent in the herb layer (Chytrý, 2013). Eurasian open pine and spruce woods in oligotrophic mires of the *Vaccinio uliginosi-Pinetea sylvestris* include rare and endangered habitats dependent on water regime which represents well-preserved fragments of unique raised bog complexes mainly in Orava region and Podtatranská brázda Furrow (Uhlířová et al., 2016). Forest communities dominated by nonnative locust tree (*Robinia pseudocacacia*) include seral forest-clearing and anthropogenic successional woodlands on nutrient-rich soils (*Robinietea*). Azonal forests, rather driven by environmental conditions such as nutrient availability in the soils and humidity instead of climatic factors can be divided into few groups. A floodplain riparian gallery forest of the *Alno glutinosae-Populetea albae* (the *Alnion incanae*) are typical on nutrient-rich alluvial soils and includes riparian, seepage and hardwood floodplain forests (Douda et al., 2016). Willow-poplar (Figure 9.10) and low open forests of riparian habitats of the *Salicetea purpureae* occur mainly along lowland rivers (Chytrý, 2012). The *Alnetea glutinosae* encompasses mesotrophic regularly flooded alder carrs and birch wooded mires together with forested

Figure 9.10   Flood-plain forest, Vojka nad Dunajom (Photo by L. Ulrych).

wetlands dominated by *Alnus glutinosa* with a species-poor herb layer with tall sedges (Chytrý, 2012).

The last group represents anthropogenic vegetation that is conditioned and dependent on human activities. It is divided into ruderal and weed vegetation, the former growing in disturbed habitats in towns, villages, industrial zones, along roads or railways and on other sites strongly influenced by humans, the latter growing in cultures of cultivated crops (Chytrý, 2009). Vegetation of forest clearings together with nitrophilous ruderal communities around sheepfolds in the mountains belongs here, as well (Jarolímek and Šibík, 2008). In addition to native species, ruderal and weed vegetation contains many alien species that have been accidentally or deliberately introduced since the Neolithic (Chytrý, 2009). When looking at diversity, there is a general decrease in the relative richness and total cover of archaeophytes and neophytes with increasing altitude in invaded habitats. There is also an observable temporal trend in archaeophytes shifting from anthropogenic towards more natural habitats (Medvecká et al., 2013). Taking the changing environment at global scale into account, we should pay attention to this threat including spreading of nonnative species into natural communities.

9.3   Biota Diversity

9.3.1   *Diversity of Cyanobacteria and Algae*

All data about the occurrence of phototrophic microorganisms published from the territory of Slovakia during the period 1791–1996 were summarized in a complex work *Checklist of non-vascular and vascular plants of Slovakia* (Marhold and Hindák, 1998). Since then other publications with more than 1800 previously not recorded taxa were elaborated. Almost 3,400 cyanobacteria and algae are known from the territory of Slovakia. However, all recorded taxa represent only a small part of the cyanobacterial and algal richness in our country. The reasons are deficiently investigated (lack of specialists) or never studied areas up to now ("white places"), and an intense grow of new taxa, thanks to the modern revision of the classification of cyanobacteria and algae based mostly on molecular approach. In addition, global climate changes provide an expansion of invasive species, which can be caused also by agriculture, urban development and increased transportation as well.

The Tatra National park (Figures 9.3 and 9.11) is rich in different types of algologically attractive habitats – glacial lakes, brooks, waterfalls, summer snow-fields (cryoseston), wetted rocks, peat-bogs, or various soil and aerial

Figure 9.11   Iconic peak of Slovakia Kriváň (Vysoké Tatry Mts) (Photo by M. Hamarová).

biotops (Hindák and Kawecka, 2010). All found cyanobacteria and algae represent approximately a half part of all taxa recorded from Slovakia. Many species have been described as new, according to the interest of specialists (cyanobacteria, desmids, green algae, diatoms), e.g., cyanophytes *Chroococcus scherffelianus* Kol. and *Scytonematopsis starmachii* Kov. et Kom., chrysophytes *Chromulina ettlii* Hindák, *Kephyriopsis tatrica* Juriš, *Carteria tatrica* Fott, green coccoid algae *Monoraphidium fontinale* Hindák and *Choricystis tatrae* (Hindák) Hindák, desmid *Cosmarium bulliferum* Růž., etc. (Hindák and Kováčik, 1993).

Peat-bogs have been studied only occasionally (*Micrasterias denticulata,* Figure 9.12), despite of this some extremely rare taxa were observed, e.g., a filamentous cyanobacterium *Katagnymene accurata* Geitler or a glaucophyte *Chalarodora azurea* Pasher. New genus and species *Neosynechococcus sphagnicola* was described from peat-bog Klin (Orava region), where it occupies different niches such as hyaline cells of *Sphagnum*, sheaths of cyanobacteria, dead cells of desmids or carapaces of dead crustaceans (Dvořák et al., 2014). Peat-bogs as special biotops (with acidic water low in nutrient) are still a challenge for further algological investigations.

Large rivers have been usually regularly monitored, mainly the Danube as an important international river. Many investigations including floristic, taxonomic and hydrobiological data are available. The first checklist of cyanobacteria and algae from the Slovak stretch of the Danube comprised 590 infrageneric taxa, a quarter of them were recorded for the first time (Hindák and Hindáková, 2000). Current monitoring is needed due to Water Framework Directive evaluations – latest results with implications

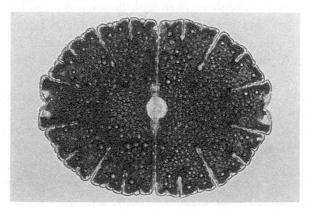

Figure 9.12 *Micrasterias denticulata Bréb. – desmid* from the peat-bog Klin (Orava, Slovakia), LM (Photo by F. Hindák).

for ecological status assessment of the benthic algae see Makovinská and Hlúbiková (2014). An important source of the microflora diversity are tributary rivers (several studies concerning the transboundary River Morava and its inundation areas exist, e.g. Marvan et al. (2004).

The phytoplankton and phytobenthos of gravel- or sand-pit lakes, mainly in Western and Eastern Slovakia, were carefully studied. As oligotrophic waters they are an important source of drinking water in our country. The microflora of these man-made biotopes can change from year to year mainly due to eutrophication. The algal diversity is large, from ca. 600 taxa recorded were several described as new for science (new genus *Granulocystopsis* was described by Hindák (1977), and species *G. pseudocoronata* (Korš.) Hindák was found by the author in the sand-pit lake at Lozorno, as well in the plankton of the river Danube in Bratislava).

Another type of man-made water bodies are fishponds or water reservoirs which serve as fishponds. Some planktonic cyanobacteria are capable of forming water blooms. Heavy water bloom occurred in water reservoir Orava (mass development of colonial species *Woronichinia naegeliana*, in castle-moat water in the town of Holíč (invasive cyanobacteria *Cylindro–spermopsis raciborskii*), and of course in the hypereutrophic water reservoir in the town Modra (strong dominance of colonial green alga *Golenkiniopsis longispina* and picoplanktic cyanobacterium *Aphanocapsa delicatissima*). In contrast to the dominance of cyanobacteria and green algae, development of centric diatoms and green flagelates is rare (cf. the water reservoir Devínske jazero in Bratislava).

Slovakia is a small country, but rich in mineral and thermal springs (ca. 1300). Many of them are used as thermal spas, one of the most known are at Piešťany and Sklené Teplice. Thermophilic microflora of both spas were studied by Hindák and Hindáková, also other mineral (thermal) springs: of the travertine pile at Sivá Brada, of the fen Močiar near Stankovany, at Gánovce, at Kováčová or cold mineral springs at Mičiná and at Kráľová pri Senci. Their biodiversity is unique (many taxa are new for Slovak flora) and very sensitive to any human activities, therefore a general protection of the whole areas is still needed.

Other biotopes are municipal fountains, city-monuments (cathedrals, sculptures at cementeries, etc.), artificial water bodies where the cyanobacteria and algae are absolutely undesired (Hindák and Hindáková, 2011). Among them, some were described as new for science, e.g., tetrasporalean alga *Chlamydocapsa mucifera* Hindák, some are new for Europe – *Makinoella tosaensis* Okada from small basin located in the campus of Slovak

Academy of Science in Bratislava. And finally, many other occasional studies exist which reveal unusual appearance of cyanobacteria and algae atypical for our country (e.g., chroococcal cyanobacterium *Synechocystis salina* in setting basin for the vast saltwater used by the crude oil production at Gbely (W Slovakia), or mass development of coccal red alga *Dixo–niella grisea* and centric diatoms *Discostella woltereckii* in greenhouse of the Botanical Garden of Comenius University in Bratislava.

The beauty of cyanobacteria and algae is amazing. In conclusion, we would like to recommend attractive books *Colour Atlas of Cyanophytes* (Hindák, 2008) and *Atlas of Euglenophytes* (Wołowski and Hindák, 2005) with cyanobacteria and euglenoids collected from different biotopes also from Slovakia.

9.3.2 *Diversity of Non-Lichenized Fungi*

Data about the occurrence and distribution of fungi in Slovakia were scarcely summarized. The first, most comprehensive checklist of fungi reported from Slovakia was compiled by Lizoň and Bacigálová (1998), who listed 2471 taxa. Later, Adamčík et al. (2003) supplemented the list with additional records, giving in total 2609 taxa of all groups of macrofungi recorded in Slovakia. A huge compendium about fungi was written by Škubla (2003) including 3723 taxa. Recent encyclopedic work by Hagara presents photographs and brief descriptions of 3230 taxa from both Czech and Slovak Republics (Hagara, 2014). Focused and intensive systematic mycological research was performed in selected orographical units and/or protected areas in Slovakia. One of the best-studied areas is part of the Eastern Carpathians, the Bukovské vrchy Mts (Kuthan et al., 1999; Červenka, 2016; Adamčík, et al., 2007, 2016a). In the Western Carpathians more thoroughly explored areas include: Protected Landscape Area Vihorlat (Ripková et al., 2007), Muránska planina National Park and adjacent areas (Kučera and Kautmanová, 2011), Cerová vrchovina Mts, Juhoslovenská kotlina basin and Laborecká vrchovina Mts (Ripková et al., 2007), Liptov region (Peiger et al., 2015), Stolické vrchy Mts and Revúcka vrchovina Mts (Mihál et al., 2011; Glejdura, 2013; Blanár and Mihál, 2002), Jelšava–Lubeník region (Mihál et al., 2015), Javorníky Mts and Strážovské vrchy Mts (Lizoň, 2006), and Malé Karpaty Mts because of its proximity to the capital Bratislava – a former center of mycology. Systematic work is further ongoing in the Západné Beskydy Mts, Biele Karpaty Mts, Poľana Mts, Liptovská kotlina Basin, Nízke Tatry Mts, Vysoké Tatry Mts (Figure 9.11) and other regions of the country. Fungi of the old-growth forests of Slovakia were studied in Havešová and Stužica in the Poloniny Mts, Vihorlat in the Vihorlatské vrchy Mts, Oblík in the Slanské vrchy Mts, Dobročský prales and Klenovský Vepor in the Veporské vrchy Mts and Badínsky prales in the Kremnické vrchy Mts (Adamčík et al., 2007, 2016a). Research proved that Slovak old grown forests belong to best preserved forest areas in Europe with occurrence of rare macrofungi (*Ceriporiopsis gilvescens, C. pannocincta, Dentipellis fragilis, Hericium coralloides, Ischnoderma resinosum, Spongipellis delectans, Pluteus umbrosus, Ionomidotis irregularis* or *Steccherinum murashkinskyi*). Research focused on grassland fungi in Slovakia resulted in several publications where a modified system of grassland ecosystems evaluation was used and new grassland fungi for Slovakia reported (Adamčík and Kautmanová, 2005; Adamčík et al., 2006; Kučera and Lizoň, 2012). Diversity of Erysiphales and Taphrinales was summarized in two volumes of the Flóra Slovenska series (Paulech, 1995; Bacigálová, 2010). Due to intensive mycofloristic investigation every year several new species of Slovak mycoflora are added. Some of them are very rare, e.g., *Cryptomyces maximus, Scutellinia colensoi, Godronia fuliginosa, Lyophyllum pseudosinuatum. Sarcodon fuligineoviolaceus* was "re-discovered" in Slovakia after 130 years (Lizoň, 2006). The world

Figure 9.13 *Microglossum griseoviride*—green earth-tongue a rare fungus described as new for science from Slovakia (Photo by V. Kunca).

mycoflora was enriched also about several fungal taxa described as new for science from our country, e.g., *Microglossum parvisporum, M. griseoviride* (Figure 9.13), *M. truncatum, M. pratense, Pseudoplectania lignicola, Pseudobaeospora mutabilis, P. basii P. terrayi, Hodophilus tenuicystidiatus, H. subfoetens, H. pallidus* (Kučera, et al., 2014a, 2014b, 2017; Glejdura et al., 2015; Adamčík and Ripková, 2004; Adamčík and Jančovičová, 2011; Adamčík et al., 2016b).

9.3.3 *Diversity of Lichenized Fungi (Lichens)*

Lichens visibly and significantly contribute to the biodiversity of Slovakia. Currently we register up to 1,688 species in Slovakia (Guttová et al., 2013; http://ibot.sav.sk/lichens/checklist.html). Major part of them includes lichenized Ascomycetes. Only three species belong to basidiolichens, namely agaricomycetes, i.e., *Lichenomphalia umbellifera, L. hudsoniana* and *Multiclavula mucida*. One of the most rich-in-species genera are lichens forming crustose thalli –, e.g., members of the genus *Verrucaria*, with up to 110 listed species; *Caloplaca* – up to 90 species, or *Lecanora* – up to 76 species. Also a group of cup lichens – genus *Cladonia*, is rich in species, with up to 65 species. On the other hand, 75 genera are represented in this territory just by one species, e.g., *Amygdalaria, Alyxoria, Anzina, Bactrospora, Cheiromycina, Dirina, Elixia, Hertelidea, Heterodermia, Hyperphyscia, Japewia, Koerberiella, Leucocarpia, Massalongia, Normandina, Peltula, Phylliscum, Polychidium, Sarcosagium, Reichlingia* and *Zahlbrucknerella*. Great part of the total of so far recorded species belongs among widely distributed species – growing as epiphytes, or on rocks, on soil, or needles of conifers, as well as on artificial substrates.

Territory of the country is a fascinating harbor of diverse lichen biota due to unique combination of geographical position, geology, geomorphology, and climate, as is true also for other organismal groups. The Western and partly Eastern Carpathians along with the Pannonian areas provide wide spectrum of habitats colonized by ubiquistic species as well as specialists to particular climatic features. Within this area we can find isolated outposts of species confined to the Mediterranean basin, like *Placolecis opaca, Leptogium hildenbrandii,* or *Leptogium ferax,* the last one only with scattered known records from Algeria, France, Hungary, Creete, Portugal and Spain (Czeika et al., 2004). Arctic-Alpine elements are well represented by the macrolichen *Nephroma arcticum,* discovered by Swedish naturalist Göran Wahlenberg (1814) in Tatry Mts, the highest part of the Carpathian arc. In the twentieth century, the species occurred scattered and sterile (Suza, 1923), and currently it is extremely rare. This circumpolar species of low and sub-arctic and the boreal-middle alpine regions is restricted in Europe only to Iceland, Scandinavia, northwest Scotland and already mentioned outpost in Tatry Mts. An example of suboceanic elements with a link to Illyrian/south-east influence is *Diploicia canescens* with placodioid thallus. Its localities are present in central and south-central parts of the country within the oro-graphical units Pohronský Inovec Mts, Štiavnické vrchy Mts, Kremnické vrchy Mts, Detvianske stredohorie Mts, and Cerová vrchovina Mts. Another suboceanic species, however with eastern limit of its distribution range in the Western Carpathians is *Xanthoparmelia mougeotii*. Its isolated locality is present in the valley of the river Váh, on steep rocky slopes covered by *Calluna vulgaris* stands, acidophilous oak woods, intermixed with birch, in the area of the picturesque ruins of the castle Strečno – Starhrad. The species *Ramalina carpatica* with shrubby, macroscopic thallus, is a boreo-montane oreophyte, growing on siliceous rocks from montane to subnival belts. Its occurrences are distributed in the Western Carpathians, Balkan (Bulgaria – Rila Mts, Vitoša Mts), and recently it was reported from the mountains in Turkey, Bozdağ Mts (Şenkardeşler and Calba, 2011). Two varieties are recognized based on secondary chemistry, nominate variety *carpatica*, and variety *teplicskaensis* Gyeln, are characterized by the presence of evernic acid in medulla. A group of rare species, with just scattered records available represents a pin-lichen *Chaenotheca servitii*, described by Nádvorník (1934) from eastern Slovakia in the area of the village Veľké Pavlovce, where it grew on the wood of *Salix* stump. Besides this site, the species was col-lected in Caucasus Mts (Titov, 1998) and in Acadian Forest in North Amer-ica (Selva, 2014). The species is red-listed extinct in Slovakia (Pišút et al., 2001). Selected lichens were included in the national red list (Table 9.1). Lichens have been legally protected in Slovakia since 1999 and the number of currently annexed legally protected species is in Table 9.2.

Unlike in the case of vascular plants, endemism is rarely reported among lichens. Current level of knowledge shows that there are no endemic lichen species restricted to the territory of the Western Carpathians, Eastern Car-pathians or Pannonia in Slovakia. It is important to mention, however, that some of lichens (e.g., Rim Lichen *Lecanora chalcophila*) are known only from the territory of the Western Carpathians, and since their discovery and description here, they have not been found anywhere else in the world. The Carpathians were known for their mineral resources too. Gold, silver or copper have been mined here since the Middle Ages. After the ores were exhausted, waste dumps with high metal content remained in the areas sur-

Table 9.1 Number of Red-Listed Species

	Total	RE+ CR/PE	Ex	CR	EN	VU	NT	LC	LR:nt	DD	NE	Resource
Fungi	309	0	5	7	39	49	0	0	51	90	32	Lizoň, 2001
Cyanobacteria and algae	283	0	0	7	80	196	0	0	0	0	0	Hindák and Hindáková, 2001
Lichens	573	0	88	140	48	169	0	0	114	14	0	Pišút et al., 2001
Bryophytes	496	0	26	95	104	112	0	0	85	74	0	Kubinská et al., 2001
Vascular plants and ferns	1218	83	0	155	171	201	347	162	0	91	8	Eliáš jun et al., 2015

Table 9.2 Overview of Number of Legally Protected Taxa (Regulation of Ministry of Environment No. 24/2003 as Amended)

Organismal group	Importance	Number
Invasive plant species	n.a.	11
Plant species (incl. subspecies)	National	668
	European	43
Bryophytes	National	14
	European	9
Lichens	National	17
Macromycetes	National	70

rounding the mines. This abandoned substratum became an empty niche, which was later colonized also by lichens. *Lecanora chalcophila* grew on rock substrates with a high concentration of copper, on old waste mine heaps. These have been gradually removed, thus opportunities to colonize more substrate significantly decreased.

There are only small differences between the lichen floras of the Western Carpathians and Eastern Carpathians in the easternmost parts of the country. Owing to diverse microclimatic conditions caused by the presence of larger areas of close-to-nature forests, some oceanic species (e.g., *Stictis urceolatum*) can be found in the Eastern Carpathians. Another evident trend is a dramatic decrease in the diversity of epiphytic lichens in close-to-nature forests during the twentieth century. In addition, the frequency of occurrence of particular species, and their abundance, is on the decrease. This has been caused by the alterations of environmental quality (mainly air) as well as fragmentation of close-to-nature habitats providing a much greater scale of various habitat types for lichens than the production forests. A rapid decline of occurrence in the second half of the twentieth century was recorded for macrolichens, e.g., *Lobaria pulmonaria* (Figure 9.14), *L. amplissima*, *Leptogium cyanescens, L. saturninum, Menegazzia terebrata, Gyalecta ulmi*, members of the genus *Nephroma*, and some species have been listed extinct for decades, e.g., *Sticta fuliginosa, S. sylvatica*, and *Usnea longissima*. Rarely recorded epiphytes are also *Alectoria sarmentosa, Bryoria bicolor, Collema nigrescens, Evernia divaricata, Heterodermia speciosa, Hypogymnia bitteri, Hypotrachyna revoluta, Parmotrema crinitum*, and *Peltigera collina*.

Grasslands are another typical landscape component of the country. Although generally, lichens do not find optimal conditions for their life

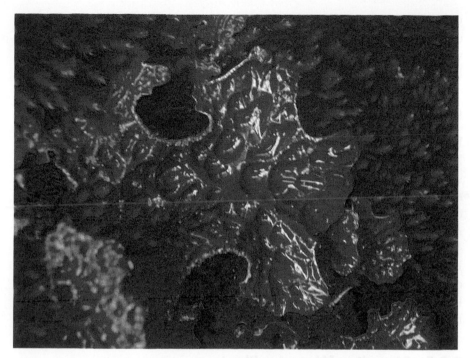

Figure 9.14 *Lobaria pulmonaria*—critically endangered epiphytic macrolichen in Slovakia, confined to well preserved, close-to-nature woodlands. Number of localities dramatically decreased during 20th century in Slovakia (Photo by A. Guttová).

here as they cannot compete with fast-growing vascular plants, locally they manage to colonize suitable microhabitats. Good example are wind-blown grasslands in (sub)alpine or siliceous grasslands (vegetation class *Caricetea curvulae*) with *Alectoria ochroleuca* as a diagnostic species, or arctic-alpine grasslands and dwarf-shrub heaths of windward ridges and edges on neutral soil (vegetation class *Carici-rupestris-Kobresietea bellardii*), with *Alectoria ochroleuca, Flavocetraria cucullata, F. nivalis,* and *Thamnolia vermicularis* as a diagnostic species plus *Cetraria islandica* and *Cladonia gracilis* as constant species (Jarolímek and Šibík, 2008). Another type of grassland habitat favorable for terrestrial lichens are thermophilic and xerophilic grasslands on rocks. From the macroscopic species, cup lichens can be found in crevices, for example, *Cladonia convoluta, Cladonia furcata, Cladonia symphycarpa,* and *Cetraria islandica.*

As for wetlands, on edges of bogs with drier peat moss hummocks which are dying away, we can find *Absconditella sphagnorum*, members of the genus

Cladonia (*Cladonia arbuscula, C. deformis, C. sulfurina*), *Cetraria islandica*, or *Icmadophila ericetorum*. In extreme habitats, such as boulders in alpine water streams or tarns covered or showered with water we can find foliose lichen *Dermatocarpon arnoldianum* or members of the genus *Verrucaria*.

9.3.4 Diversity of Bryophytes

Bryophytes are an important part of Slovak flora, currently 909 species are known, from this amount 226 species of Marchantiopsida and Anthocerotopsida and 683 species of Bryopsida. Due to a high diversity of habitats, it is a relatively high number of species not considering large area of Slovakia. The occurrence of some glacial relicts is important, e.g., *Catoscopium nigritum, Helodium blandowii, Paludella squarrosa, Meesia triquetra*. Most of this species have survived within peat-moss vegetation and/or alpine vegetation. Endemic species *Ochyraea tatrensis* was described by Váňa (1986) from the central part of Slovakia as the new taxon for science. Only two localities are known currently from Nízke Tatry Mts, where this moss occupies stones of mountain streams in steepy parts.

9.3.5 Diversity of Vascular Plants

Due to geographical position in the center of Europe, a various shape of relief, diversity of geological substrates the area of Slovakia is very important crossroads of several floristic elements, migration routes and one of the northernmost center of endemism in Europe. From that small area, a relatively high number of native species is reported: 3619 taxa (Eliáš et al., 2015) and 4713 taxa (including hybrids, common frost hardy cultivated plants, and species occasionally and for short time escaping from the culture; Marhold and Hindák, 1998). The most rich-in-species families are Rosaceae (ca. 350 taxa), Poaceae (ca. 280 taxa), Asteraceae (ca. 280 taxa), Cichoriaceae (ca. 280 taxa), Fabaceae (ca. 200 taxa), Brassicaceae (ca. 190 taxa), Ranunculaceae (ca. 150 taxa), Caryophyllaceae (ca. 150 taxa), Cyperaceae (ca. 140 taxa), Lamiaceae (ca. 120 taxa), and Apiaceae (ca. 110 taxa). As to the genera, with the highest number of species are predominantly apomictic genera: *Taraxacum* (117 taxa), *Carex* (96 taxa), *Alchemilla* (92 taxa), *Hieracium* (57 taxa), *Ranunculus* (54 taxa), *Rubus* (53 taxa), *Sorbus* (46 taxa), *Potentilla* (41 taxa), *Festuca* (37 taxa), *Veronica* (35 taxa), *Pilosella* (34 taxa), and *Trifolium* (32 taxa).

With regard to diversity of autochthonous vascular plant species, Slovakia belongs to the most rich areas in Europe (in this area ca 32% of whole

European species occurs). The numbers of species reported at regional level show, that the rich-in-species territories of the Slovak Republic are mainly in mountain regions: Veľká Fatra Mts with 1639 taxa reported (Kliment, 2008), Devínska Kobyla Mt with 1570 taxa reported (Feráková et al., 1997), Muránska planina Mts with 1150 taxa reported (http://www.npmp.sk/obsah/muranska-planina), Slovenský kras Mts. (Figure 9.4) with 1462 taxa reported (Jeník and Price, 1994), Vysoké, Západné and Belianske Tatry Mts with 1400 taxa reported (Koutná and Chovancová, 2010), Poľana Mts with 1200 taxa reported (http://www.chkopolana.eu/flora/), Záhorská nížina lowlands with 1200 taxa reported (Straka, 1998), Biele Karpaty with 1200 taxa (http://www.bielekarpaty.sk/opisBK.html).

Nowadays, about 50% of the area of Slovakia is used for agriculture purposes (Lieskovská, 2016). Over 160 plant taxa are cultivated for economic purposes (Straka, 1998). Majority part of production consists of cereals (wheat – *Triticum*, rye – *Secale cereale*, oat – *Avena sativa*, barley – *Hordeum vulgare*, maize – *Zea mays*), common sunflower – *Helianthus annuus*, rapeseed – *Brassica napus*, potato – *Solanum tuberosum*, sugar beet – *Beta vulgaris*, and legumes (pea – *Pisum sativum*, lentil – *Lens culinaris*, bean – *Phaselous vulgaris*). In lowlands there are suitable conditions for growing vegetables (tomato – *Solanum lycopersicum*, pepper – *Capsicum annuum*, cucumber – *Cucumis sativus*, melons – *Citrullus lanatus*, *Cucumis melo*, pumpkins – *Cucurbita pepo*, onion – *Allium cepa*, garlic – *Allium sativum*, lettuce – *Lactuca sativa*, etc.). From fruit species it is mainly apples – *Malus domestica*, pears – *Pyrus communis*, plums – *Prunus domestica*, cherries – *Prunus avium*, apricots – *Prunus armeniaca*, peaches – *Prunus persica*, currants – *Ribes nigrum* and *Ribes rubrum*, gooseberries – *Grossularia uva-crispa*, strawberries – *Fragaria* x *ananassa*, walnuts – *Juglans regia*, etc. The country is famous also by cultivation of grape-vine (*Vitis vinifera*).

Biodiversity of cultivated plants is the subject of investigation, collection and protection. The Research Institute of plant production (RIPP) in Piešťany realize the research of crops and crops wild relatives. Gene bank of the Slovak Republic as part of RIPP provides a tool for long-term storage of seeds of crops and their wild relatives. Very important is an investigation of the Slovak University of Agriculture in Nitra in the field of regional cultivars of fruit crops and establishing the vital collections in more orchards within Slovakia.

Very interesting is the flora of traditionally managed fields where very rare species occur today and which disappear due to mechanization and chemigation of agriculture (e.g., *Adonis flammea*, *Agrostemma githago*).

From phytogeographical point of view there are very interesting endemic species, with special importance of stenoendemics. 235 taxa are reported as endemic or subendemic from the area of the Slovakia (Kliment, 1999) with a high number of endemics in the Carpathians. Recent investigations in the Western Carpathians indicate that besides an alpine belt, the highest concentration of the endemics occurs in a mid-altitudinal zone, i.e., between 500 and 1500 m a.s.l. (Mráz et al., 2016). The highest species diversity in this zone prevails in gorges, forest cliffs, forest canyons or similar habitats, where populations of different species could survive even during the glacial periods in extra-Mediterranean northern refugia (Slovák et al., 2012; Kučera et al., 2013; Kuneš et al., 2008; Magri et al., 2006). The most iconic stenopaleoendemic species from Slovakia is *Daphne arbuscula* (Figure 9.15) distributed in Muránska planina Mts only. From category of stenoendemics we can further mention *Cyclamen purpurascens* subsp. *immaculatum* (Figure 9.16) (Veľká Fatra Mts and Nízke Tatry Mts), *Leucanthemopsis alpina* subsp. *tatrae* (Vysoké Tatry Mts), *Papaver tatricum* subsp. *tatricum* (Vysoké Tatry Mts), *Papaver tatricum* subsp. *fatraemagnae* (Veľká Fatra Mts), *Hesperis slovaca* (Nízke Tatry Mts), *Poa granitica* (Vysoké Tatry Mts), *Poa margilicola* (Veľká Fatra Mts), *Poa sejuncta* (Západné Tatry Mts), *Euphrasia*

Figure 9.15 *Daphne arbuscula*—the most iconic stenopaleoendemic plant species of Muránska planina Mts (Photo by J. Kučera).

Figure 9.16 *Cyclamen purpurascens* subsp. *immaculatum*—previously known as *Cyclamen fatrense* (Photo by J. Kučera).

stipitata (Malá Fatra Mts), *Cochlearia tatrae* (Tatry Mts), *Ranunculus altitatrensis* (Vysoké Tatry Mts), *Draba lasiocarpa* subsp. *klasterskyi* (Slovenský kras Mts), *Dendranthema zawadskii* subsp. *zawadskii* (Pieniny Mts), *Sorbus pekarovae* (Veľká Fatra Mts), etc. As a sub/endemic taxa for the whole Carpathian arc were described several others, e.g., *Aconitum firmum, Campanula carpatica, Cardaminopsis neglecta, Euphrasia tatrae, Ferula sadleriana, Leontodon pseudotaraxaci, Leucanthemum rotundifolium, Oxytropis campestris* subsp. *tatrae, Oxytropis carpatica, Platango atrata* subsp. *carpatica, Pyrola carpatica, Salix kitaibeliana, Scilla kladni*. From group of West Carpathians sub/endemics we can mention, e.g., *Aconitum firmum* subsp. *moravicum, Bromus monocladus, Campanula tatrae, Crocus discolor, Delphinium oxysepalum, Dianthus praecox* subsp. *lumnitzerii, Dianthus nitidus* subsp. *nitidus, Euphrasia stipitata, Galium fatrense, Gentianella fatrae, Hypericum carpaticum, Knautia slovaca, Pulsatilla slavica, Pulsatilla subslavica, Soldanella carpatica, Thlaspi jankae* (Figure 9.17), etc. East Carpathians sub/endemics are distributed in Slovakia mainly in NE part of Slovakia, to this category belongs, e.g., *Aconitum lasiocarpum, Campanula abietina, Festuca*

Figure 9.17 *Thlaspi jankae*—West Carpathian endemic plant species, *locus classicus* – Zoborské vrchy Mts (Photo by L. Ulrych).

saxatilis, Leucojum vernum subsp. *carpaticum, Ranunculus carpaticus, Tithymalus sojakii, Thymus alternans*, etc. Some of the Pannonian sub/endemics occur in lowlands, e.g., *Astragalus vesicarius* subsp. *albidus, Armoracia macrocarpa, Cirsium brachycephalum, Colchicum arenarium, Corispermum canescens, Dianthus serotinus, Limonium gmelinii* subsp. *hungaricum, Pulsatilla pratensis* subsp. *hungarica, Pulsatilla zimmermannii*, etc.

Many vascular plant species reach the limit of their global distribution in the area of Slovakia. Predominantly part of boundary elements create pontic-pannonian and sub-Mediterranean species which have north border of distribution in southern parts of the country, e.g., *Achillea ochroleuca, Aethionema saxtile, Althaea cannabina, Astragalus asper, Allium paniculatum, Campanula macrostachya, Carex halleriana, Colchicum arenarium, Colutea arborescens, Crambe tataria, Echium russicum, Galium tenuissimum, Hippocrepis emerus, Himantoglossum adriaticum, Ophrys sphecodes, Lathyrus pannonicus, Ononis pusilla, Quercus cerris, Salvia aethiopsis* and *Viola ambigua*. West border of distribution reach in Slovak region mainly East Carpathians taxa, e.g., *Aconitum lasiocarpum, Alyssum desertorum*,

Campanula abietina, Cirsium waldsteinii, Hierochloe odorata, Lathyrus laevigatus, Lapsana communis subsp. *intermedia, Melica altissima, Ranunculus carpaticus, Scopolia carniolica, Scorzonera rosea, Silene dubia* and *Spiraea crenata.* East border of distribution reach boreal-sub-Atlantic and sub-Atlantic floristic elements, e.g., *Asplenium adulterinum, Cirsium acaule, Dalanum angustifolium, Galium parisiense* subsp. *anglicum, Trifolium strictum, Teucrium scorodonia.* And, south border reach arctic-alpine species, e.g., *Arctous alpina, Carex rupestris, Cryptogramma crispa, Draba fladnizensis, Linnaea borealis, Pedicularis sceptrum-carolinum, Ranunculus pygmaeus, Salix starkeana, Saxifraga cernua, Sparganium angustifolium.* In Slovak region there are also very important species with disjunctive area of distribution, e.g., *Astragalus vesicarius* subsp. *albidus, Asyneuma canescens, Bellardiochloa variegata, Conioselinum vaginatum, Coronilla elegans, Crepis sibirica, Draba tomentosa, Epipactis placentina, Ligularia glauca, Loiseleuria procumbens, Orchis spitzelii, Petrocallis pyrenaica, Pedicularis comosa, Pulsatilla vernalis, Ruscus hypoglossum, Saxifraga mutata, Sibbaldia procumbens, Sparganium angustifolium.* From taxonomic-nomenclatory and historical point of view there are very valuable localities where from were described taxa new to the world (*locus classicus*). Altogether about 2400 taxa (incl. forms and varieties; Šípošová et al., 2017) were described from Slovakia, e.g., *Cardamine majovskii, Draba lasiocarpa, Festuca pseudodalmatica, Minuartia frutescens, Pulmonoria murinii, Waldsteinia teppneri.*

In total, 21.5% of the whole flora is made up of alien taxa (282 archaeophytes and 634 neophytes). The majority of alien taxa are casuals (57.6%), 39.1% are naturalized and 3.3% invasive (Medvecká et al., 2012). Invasive alien species are very significant menace for biodiversity of Slovakia. They are very adaptable, and able to compete with native species, which results into loss of biodiversity and to disruption of relationships in ecosystems. In present time there are registered in the area of Slovakia 126 invasive and potentially invasive vascular plant species (Medvecká et al., 2012). Prohibition of their cultivation, removal of individuals from nature and special monitoring of 11 species (*Ailanthus altissima, Ambrosia artemisiifolia, Amorpha fruticosa, Asclepias syriaca, Fallopia* spp., *Heracleum mantegazzianum, Impatiens glandulifera, Lycium barbarum, Negundo aceroides, Solidago canadensis,* and S*olidago gigantea*) is enshrined in law on nature and landscape protection (Table 9.2).

According to the recent published Red list of ferns and flowering plants of Slovakia (Eliáš et al., 2015), nearly 34% of native plants are at some kind of level of threat. Currently, in the Red List of vascular plants, there are 1218 taxa listed. More than 80 taxa (7%) were not confirmed in the flora (66 Regionally

Extinct, 17 Probably Regionally Extinct), 155 taxa are evaluated as Critically Endangered (13%), 171 Endangered (14%) and 201 taxa are regarded as Vulnerable (17%) (Table 9.1). Most critically endangered species of Slovak flora are distributed in habitats which are globally threatened in the whole Central Europe (peat bogs, wetlands, flooded meadows, salt marshes, sands).

Legal adjustments led to an increase in the number of taxa under state protection in 2014, thus increasing the protection of the most threatened plant taxa. Now the number of the state protected vascular plant species is 668 (Table 9.2), which is 18.5% of total number of vascular plant species of Slovakia. However, despite several measures taken in the last period there is an increase in vulnerability of plant taxa and a decrease in biodiversity of vascular plant species. While in several cases it is still unknown where is the reason and real cause of loss of their localities of occurrence.

9.4 Faunal Diversity

Due to a great surface diversity and natural environment, Slovakia is typical by highly variable fauna. Fauna composition has varying characteristic and is based on geographical (including climatic) conditions.

In Slovakia there has been ca. 38,000 animal species described, including invertebrates (but there still is an insufficient knowledge of those groups which are difficult to monitor and classify such as invertebrates and the number of animal species is estimated to be 40,000), whereby the state of fauna endangerment is still more important (it is recorded in the Red Lists). Currently, the problem of species loss is significant on the global scale. In case of all animal species, the critical requirement is to ensure the protection of their habitats – sufficiently large and preserved territories where the animals can live on their own and reproduce.

In Slovakia, there are more than 100 Carpathian endemic animal species. Most of them are invertebrates. In the mountain lakes of Tatra Mts *Tatriella slovenica* and *Trichondrilus tatricus (microdrile oligochaetes - Annelida, Oligochaeta, Lumbriculata)* can be found. There are many endemic mollusc, including *Bielzia coerulans* (Figure 9.18), *Carpathica calophara, Iphigenala testria, Acicula paarcelineata, Chondrina tatrica*, etc. The Slovak karst, the largest karst area in the Western Capathians, is a good habitat for a large number of endemic invertebrates (e. g. terrestrial crustacea Nifargus tatrensis, Mesoniscus graniger, or molluscs Alopia clathara, Sadleriana pannonica, etc.)." (Straka, 1998).

From mammal species, there are three Tatra mountain endemic subspecies – Tatra chamois (*Rupicapra rupicapra tatrica*) (Figure 9.19), alpine

Figure 9.18 *Bielzia coerulans*—terrestrial pulmonate gastropod, is an endemic species to Carpathian mountains (Photo by P. Urban).

Figure 9.19 *Rupicapra rupicapra tatrica*—endangered subspecies living exclusively in the Tatra Mts (Photo by P. Urban).

marmot (*Marmota marmota latirostris*), snow vole (*Chionomys nivalis mirhanraini*) and one – Tatra vole (*Microtus tatricus*) – is a Tatra mountain endemic species.

The root vole (*Microtus oeconomus mehelyi*) is a Danube plain endemic and belongs to glacial relics. Other mammal relict species are alpine shrew *(Sorex alpinus),* northern birch mouse *(Sicista betulina)* and snow vole *(Chionomys nivalis).*

Red Lists, that are effective today were published in 2001 and contained 466 endangered species of invertebrates and 153 species of vertebrates (Baláž et al., 2001). Only Red Lists of selected groups, e.g., molluscs (Šteffek and Vavrová, 2006), fish and lampreys (Koščo and Holčík, 2008), birds (Demko et al., 2013, 2014) and mammals (Urban et al., 2017) were updated later. According to these lists, 1,636 species of invertebrates (6.6%) and 100 species of vertebrates (24.2%) are endangered. The most threatened are cockroaches (44.4%), Ephemers (34.2%), Odonata (33.3%) and also Mollusca and Aranea (to 30%); in group of Vertebrates, the most threatened are Lampreys (100%) and Amphibians with Reptiles (over 40%). Increased numbers may not mean an increased threat to animals, they rather point to the fact that the species are better documented and subsequently added to the lists.

9.4.1 Vertebrates

The number of vertebrate species in Slovakia is shown in Table 9.3, and an overview of the actual threatened vertebrates groups is listed in Table 9.4.

Table 9.3 Number of Vertebrate Species in Slovakia

	No. of species			% of European sp. occurring in Slovakia
	World (global estimates)	Europe	Slovakia	
Mammals (Mammalia)	5,416	233	97	41.6
Birds (Aves)	9,026	530	359	67.7
Reptiles (Reptilia)	9,232	140	12	8.6
Amphibians (Lissamphibia)	6,802	83	18	9.6
Freshwater fishes (Osteichtyes)	15,000	531	79	14.9
Lampreys (Petromyzontes)	50	16	4	25

Table 9.4 Overview of the Actual Threatened Vertebrates Groups

	No of species	Threatened categories (CR, EN, VU)(No of species)	Threatened (%)	Source
Mammals (Mammalia)	97	9	9,3	Urban et al., 2018
Birds (Aves)	359 (226 breeding)	51	14,2 (22,6)	Demko et al., 2013, 2014
Reptiles (Reptilia)	12	5	41,7	Kautman et al., 2001a
Amphibians (Lissamphibia)	18	8	44,4	Kautman et al., 2001b
Freshwater fishes (Osteichtyes)	79 (54 authochtounous)	12	15,2 (22,3)	Koščo and Holčík, 2008
Lampreys (Petromyzontes)	4	4	100	Koščo and Holčík, 2008

9.4.1.1 Mammalia

In Slovakia, there were 97 terrestrial species of mammals recorded (Krištofík and Danko, 2012). The majority of Slovak mammal species are small volant and nonvolant mammals belonging to the orders Chiroptera (bats), Rodentia (rodents), and Soricomorpha (shrews and moles). At present, 27 species of bats were recorded on the territory of Slovakia (including recently distinguished sibling species, soprano pipistrelle (*Pipistrellus pygmaeus*) and Alcathoe's bat (*Myotis alcathoe*) (Hanák et al., 2010; Krištofík and Danko, 2012). Two subspecies of chamois occur in Slovakia – the autochthonous endemic Tatra chamois (*Rupicapra rupicapra tatrica*) and the introduced Alpine chamois (*Rupicapra r. rupicapra*). Slovakia is one of a few European countries with natural habitats, where for example, large carnivores, such as brown bear (*Ursus arctos*), gray wolf (*Canis lupus*) and lynx (*Lynx lynx*) (Figures 9.20–9.22), lives. Altogether 9 mammal species are threatened by extinction (categories CR, EN, VU), which represents 9% of all Slovak mammals (Urban et al., 2017). The major threats, which affect mammals in Slovakia, are habitat loss, agricultural and forestry effluents and road mortality. Hunting and trapping, also pose serious threats to mammals in the country.

Figure 9.20   *Ursus arctos* (Photo by P. Urban).

Figure 9.21   *Lynx lynx* (Photo by P. Urban).

Figure 9.22 *Canis lupus* (Photo by P. Urban).

9.4.1.2 Aves

The avifauna of Slovakia includes 359 wild-living bird species altogether (feral pigeon not including), 226 of them are breeding species. 33 diurnal raptor species have been recorded in Slovakia, 18 of which are regular breeders. Only eight species, i.e., the Red Kite (*Milvus milvus*), White-tailed Eagle (*Haliaeetus albicilla*), Goshawk (*Accipiter gentilis*), Sparrowhawk (*A. nisus*), Buzzard (*Buteo buteo*), Imperial Eagle (*Aquila heliaca*), Golden Eagle (*A. chrysaetos*) and Kestrel (*Falco tinnunculus*), are breeding residents. Most of the observed raptors are migrants, either visiting Slovakia from spring to autumn to breed, or to winter or just occurring during migration (Dravecký and Guziová, 2012).

Some bird species are threatened at national or even global level. Altogether 51 species (from all the breeding species) are threatened by extinction (categories CR, EN, VU), which represents 22% of all Slovak breeder species. Ten species (*Anas acuta, Circaetus gallicus, Aquila pennata, Falco vespertinus, Limosa limosa, Numenius arquata, Coracias garrulous, Monticola saxatilis, Tichodroma muraria, Emberiza hortulana*) face extreme risk of extinction in Slovakia they were included among critically endangered species (CR) (Demko et al., 2013, 2014).

9.4.1.3   Reptiles (Reptilia)

In Slovakia 13 species of reptiles were recorded. Altogether 5 species are threatened by extinction (categories CR, EN, VU) and they cover 42% of all Slovak reptiles. One species – turtle (*Emys orbicularis*) is critically endangered (Kautman et al., 2001b). The members of the genus *Anguis* (a small group of lizards from Anguidae family called slow worms) are widely distributed but hidden living reptile species in Slovakia. Recently, this genus was under a very intensive study, which resulted in its taxonomic revisions (Gvoždík et al., 2010, 2013). Due to their slight morphological characters among species of the genus, presence of two out of five species in the study area has only recently been confirmed. The results confirmed the dominant species *Anguis colchica* for the Slovak Republic (Jablonski et al., 2016). Habitat loss, fragmentation and degradation especially due to agricultural intensification and urbanization are the main threats to this group in Slovakia. It is also interesting to note that at least 33% of the reptile species in Slovakia may be threatened by human persecution and control, especially snakes (Anonymus, 2013).

9.4.1.4   Amphibians (Lissamphibia)

In Slovakia, 18 species of amphibians were recorded. Altogether 8 species are threatened by extinction (categories CR, EN, VU) and they cover 45% of all Slovak reptiles (Kautman et al., 2001a).

The decrease in abundance (an increase of threat) of most of the species is significant. For example, it is estimated, that a population of optically most common frog *Rana temporaria* has decreased by more than 50% (even more at places) in the last decades. Destruction of reproduction sites for agricultural, forestry and water management reasons, extensive usage of pesticides and consequential destruction of food offer (aphids, bark-beetles, agricultural pests) for insectivorous species, growing of coniferous monocultures, building of migration barriers are negatively limiting factors of the amphibian population.

9.4.1.5   Freshwater Fish (Osteichthyes) and Lampreys (Petromyzontes)

Freshwater fish recorded in the territory of Slovakia include 95 (Koščo et al., 2010), respectively 79 fish species (Hensel and Mužík, 2001). As many as one-third of these are allochthon fish species belonging to 14 families, among which several have not occurred in Slovakia recently. Analyzing the data on fish species recorded in Slovakia, only 22 nonnative species (of the total 36 nonnative species) were recorded in the wild. It seems that only 18 species established self-sustaining populations in Slovakia, and four species require continual restocking. Introduction of diseases and parasites is also a problem intimately connected with the introduction of fish in Slovakia. The origins of the exotic species seen in Slovakia are the four continents – Africa (3), North America (7), Central America (3), and Asia (13) and ten of them are from different regions in Europe (Koščo et al., 2010).

The most important threat to this group is the modification of physical and chemical characteristics of freshwater rivers and lakes due to dam construction. Water pollution caused by industrial effluent is also a main threat to this group.

In Slovakia, there were four species of lampreys recorded (*Eudontomyzon danfordi, E. mariae, E. vladykovi, Lampetra planeri*). Two are threatened by extinction (Koščo and Holčík, 2008). The impact of river regulation (channelization), obstructions (dam and small hydropons), destruction of lamprey spawning and larval habitats through dredging, drastic river level fluctuations and the decline of water quality are the most common reasons for lampreys decline.

9.4.2 Invertebrates

Fauna of Invertebrates is very rich. Among this group, the Insects (Insecta) is the richest group of animals consisting of 17,955 species, 472 families, and 4,799 genera (Table 9.5).

9.4.2.1 Beetles (Coleoptera)

In Slovakia, more than 6,500 species of beetles were recorded. The Red List encompasses 718 species. Altogether, 643 species are threatened by extinction (categories CR – 15, EN – 128, VU – 500), which represents 11% of all Slovak beetles (Holecová and Franc, 2001). More endangered are saproxylic beetles, because this group is very dependent on the dynamics of tree-aging and wood decay processes. The major threat to this group is logging and wood harvesting.

9.4.2.2 Butterflies (Lepidoptera)

Butterflies of Slovakia consist of both the butterflies and moths. Red List developed in 2013 consists of 3598 Lepidoptera species occurring in

Table 9.5 Threat of Invertebrates in Slovakia in 2015

	Percentage of endangered species (%)	Number of taxa in Slovakia	CR	EN	VU
Molluscs (Mollusca)	29.2	277	26	22	33
Spiders (Araneae)	28.3	934	72	91	101
Crustaceae	10.4	386	3	6	31
Diplopoda	18.7	75	0	7	7
Chilopoda	16.7	60	0	4	6
Ephemeroptera	34.2	120	8	17	16
Dragonfly (Odonata)	33.3	75	0	14	11
Plecoptera	18	100	5	8	5
Blattaria	44.4	9	0	0	4
Orthoptera	7.6	118	0	5	4
Heteroptera	3.4	801	14	7	6
Neuroptera	6.5	93	0	3	3
Beetles (Coleoptera)	9.7	6 498	15	128	490
Hymenoptera	4.9	5 779	23	59	203
Butterflies (Lepidoptera)	2.2	3 500	21	15	41
Mecoptera	12.5	8	0	0	1
Diptera	1.4	5 975	5	10	71

Source: Kapusta (2016).

Slovakia (Pastoralis et al., 2013), the current country list includes ca. 3,800 species (http://www.lepidoptera.sk/). Altogether, 40 species are threatened by extinction (categories CR – 21, EN – 15, VU – 41) (Kulfan and Kulfan, 2001). The main threats to butterflies are degradation and loss of habitats, which causes loss of optimal conditions for existence.

9.4.2.3 Dragonflies (Odonata)

From dragonflies there are 70 species from 9 families, which belong to Slovak fauna, 5 species were mistaken in the past (Šácha et al., 2017). Three species are considered extinct and another 41 are listed in Red List. Altogether 25 species are threatened by extinction (categories EN – 14, VU – 11) (David, 2001). This group is adversely affected by desiccation caused by dry weather, fires and increased water extraction for irrigation and human consumption. River species are also affected by ecosystem modifications such as the construction of dams and reservoirs and water quality deterioration.

9.4.2.4 Grasshoppers, Bush-Crickets and Crickets (Orthoptera)

Fauna of Orthoptera consists of 126 species, 61 families and 8 genera (Ensifera: 58 species and Caelifera 68 species) (http://www.orthoptera.sk/sk/). Altogether 9 species are threatened by extinction (categories EN – 5, VU – 4) (Krištín, 2001). The major threats for this group are habitat loss, or degradation, abandonment of traditional management, of grass cutting and grazing and following succession leading to overgrowing by shrubs and trees and human settlement expansion.

9.4.2.5 Hymenoptera

In Slovakia, there were more than 5800 species of Hymenoptera recorded. Altogether 285 species are threatened by extinction (categories CR – 23, EN – 59, VU – 203) (Lukáš, 2001).

Molluscs, spiders, or Crustaceae are interesting from other taxonomic groups.

9.4.2.6 Molluscs (Mollusca)

The list of Slovak molluscs comprises 247 native and naturalized species (excluding nonnative species restricted to greenhouses and species living in thermal waters), including 219 (88.5%) gastropods (51 aquatic and 168 terrestrial) and 28 (11.5%) bivalves (Čejka et al., 2007; Horsák et al., 2010). The native and naturalized molluscs of Slovakia were divided into 10 main groups and 33 subgroups on the basis of their ecological requirements (Lisický, 1991). The majority of terrestrial gastropods are woodland species *sensu lato* (55%), followed by euryoecious and steppe species *s. l.* (both 10%), the remaining species represent other ecological groups (Čejka et al., 2007). Altogether, 81 species are threatened by extinction (categories CR – 26, EN – 22, VU – 33) (Šteffek and Vavrová, 2006).

9.4.2.7 Spiders (Araneae)

Spiders form an important component of fauna and belong among significant predators and their presence contributes to maintaining a balance between the individual components of zoocenosis and to optimal evolution of natural conditions. This group has previously been ranked among the less popular groups of animals and this situation has not fundamentally changed until now. On the other hand, wingless spiders often have a very high bio-indicative value, because they are usually more strongly tied to a biotope than flying insects (Franc, 2000). In Slovakia, there were nearly one thousand spiders (969 species) belonging to 37 families registered (Gajdoš et al., 1999, 2018). Altogether, 264 species are threatened by extinction (categories CR – 73, EN – 90, VU – 101) (Gajdoš and Svatoň, 2001).

9.5 Nature Protection

Nature protection in Slovakia is based on species and area protection, both of which are regulated by national (Act No. 543/2002 on Nature and Landscape Protection, as amended) and European Union legislation. The national level distinguishes five levels of nature protection.

The first level of protection covers the whole area of Slovakia, where no special area protection is established. This level is the least strict. The second level of protection is mainly used in area of Protected Landscape Area – these are excellent examples of natural landscape originated on sustainable use of man, usually larger than 1,000 ha. The second level of protection is also used in buffer zones of National parks, zones D, where National Parks or Protected Landscape Areas have zones. The third level of protection is used in National Parks mainly, in buffer zones of small spaced protected areas or in some small spaced protected areas where maintenance of man is necessary. The fourth level of protection is used mainly in small spaced protected areas, where maintenance by man is useful. The highest fifth level with the most strict protection is used in some small spaced protected areas, where only wild nature rules are allowed. Overview of area protected in these levels is given in Table 9.6.

From the view of protected area types, two main groups of areas are in Slovakia – large and small spaced protected areas. Large spaced protected areas, usually greater than 1,000 ha, are National Parks and Protected Landscape Areas (Table 9.7). Only about two or three National Parks fulfill criteria of IUCN II category (IUCN Protected Areas Categories System – www. iucn.org) outstanding are in IUCN V category with vision for future to be National Park of IUCN II category. Protected Landscape Areas are examples of natural landscape originated on sustainable use of man IUCN category V (Table 9.8).

Table 9.6 Overview of Area Protected in Levels of Protection

The level of protection	Area in ha	% from the territory of Slovakia
Level 1	3,756,441	76.61
Level 2	744,564	15.19
Level 3	289,878	5,91
Level 4	26,568	0.54
Level 5	86,047	1.75

(Based on Faško, 2017)

Table 9.7 National Parks in Slovakia

Name (original Slovak)	Area (ha)	Area of buffer zone (ha)	The year of establishment, actualization
National parks			
Tatranský NP (TANAP)	738,000,000	307,030,000	1948, 1987, 2003
Pieninský NP (PIENAP)	37,496,226	224,441,676	1967, 1997
NP Nízke Tatry (NAPANT)	728,420,000	1,101,620,000	1978, 1997
NP Poloniny	298,050,514	109,732,893	1997
National parks first established as Protected landscape areas			
NP Slovenský raj	194,136,700	130,110,000	1964 as PLA, 1988, 2016
NP Malá Fatra	226,300,000	232,620,000	1967 as PLA, 1988
NP Muránska planina	203,178,021	216,979,644	1977 as PLA, 1997
NP Slovenský kras	346,110,832	117,415,677	1973 as PLA, 2002
NP Veľká Fatra	40,371,343,326	1,325,817	1974 as PLA, 2002
Together all 9 NPs	**3,175,405,726**	**2,701,275,707**	

(Based on Faško, 2017)

Small spaced protected areas, usually smaller than 1,000 ha are Nature Reserves (IUCN I), Natural Monuments (IUCN I), Protected areas (IUCN V), these are the main categories. Protected landscape elements, Private protected areas and Community protected areas are also possibilities of protection, but these types are only rarely used. 1108 of small spaced protected areas are in Slovakia, sometimes overlapping with large spaced protected areas.

Table 9.8   Protected Landscape Areas in Slovakia

Name (original slovak)	Area (ha)	The year of establishment, actualization
Vihorlat	174,852,428	1973, 1999
Malé Karpaty	646,101,202	1976, 2001
Východné Karpaty	253,071,072	1977, 2001
Biele Karpaty	445,680,000	1979, 1989, 2003
Horná Orava	587,380,000	1979, 2003
Štiavnické vrchy	776,300,000	1979
Poľana	203,604,804	1981, 2001
Kysuce	654,620,000	1984
Ponitrie	376,654,100	1985
Záhorie	275,220,000	1988
Strážovské vrchy	309,790,000	1989
Cerová vrchovina	167,712,273	1989, 2001
Latorica	231,984,602	1990, 2004
Dunajské luhy	122,844,609	1998
Together all 14 PLAs	**5,225,815,090**	

(Based on Faško, 2017).

NATURA 2000 network is the European Union system of nature protection under the two directives (Council Directive 92/43/EEC referred to as the Habitat directive and Council Directive 2009/147/EC known as the Birds Directive). This system is built from Special Protection Areas (under the Birds Directive) and Special Areas of Conservation (under the Habitat Directive). Establishing of the Special Protection Areas system is finished, special conditions of protection are established in according to subjects of protection, the system of levels of protection is not used. System of Special Areas of Conservation (SAC) is still not established, because of great overlaps with national net of protected areas. SACs will be harmonized with this national net and outstanding, not overlapping areas will be established in the form of national types of protected areas, mainly as Protected areas (IUCN V).

Keywords

• faunal diversity   • nature protection

References

Adamčík, S., & Jančovičová, S., (2011). *Pseudobaeospora terrayi*, a new species from Slovakia. *Sydowia, 63*, 131–140.

Adamčík, S., & Kautmanová, I., (2005). *Hygrocybe* species as indicators of natural value of grasslands in Slovakia. *Cathatelasma, 6*, 25–34.

Adamčík, S., & Ripková, S., (2004). *Pseudobaeospora basii*, a new species described from Slovakia. *Sydowia, 56*, 1–7.

Adamčík, S., Aude, E., Bässler, C., Christiansem, M., Dort van, K., Fritz, Ö., et al., (2016a). Fungi and lichens recorded during the Cryptogam Symposium on Natural Beech Forests, *Slovakia, 68*, 1–40.

Adamčík, S., Christensen, M., Heilmann-Clausen, J., & Walleyn, R., (2007). Fungal diversity in the Poloniny National Park with emphasis on indicator species of conservation value of beech forests in Europe. *Czech Mycol., 59*, 69–81.

Adamčík, S., Jančovičová, S., Looney, B. P., Adamčíková, K., Birkebak, J. M., Moreau, P. A., et al., (2016b). Circumscription of species in the *Hodophilus foetens* complex (Clavariaceae, Agaricales) in Europe. *Mycol. Progress, 16*, 47–62.

Adamčík, S., Kučera, V., Lizoň, P., Ripka, J., & Ripková, S., (2003). State of the biodiversity research on macrofungi in Slovakia. *Czech Mycol., 55*, 201–213.

Adamčík, S., Ripková, S., & Kučera, V., (2006). Contribution to the knowledge of macrofungi in the Biele Karpaty Mts. *Catathelasma, 8*, 17–28.

Anonymus, (2013). *Slovakia's Biodiversity at Risk a Call for Action, IUCN*, Available at: https://cmsdata.iucn.org/downloads/slovakia_s_biodiversity_at_risk_fact_sheet_may_2013.pdf.

Bacigálová, K., (2010). Flóra Slovenska, X/2: Mycota (Huby). Ascomycota (Vreckaté huby). Taphrinomycetes: Taphrinales (Grmanníkotvaré) čeľ. Protomycetaceae, čeľ. Taphrinaceae. *Veda, Bratislava*, 1–184.

Baláž, D., Marhold, K., & Urban, P., (2001). Red List of plants and animals of Slovakia. *Ochrana Prírody, 20*, 1–160.

Blanár, D., & Mihál, I., (2002). Mykoflóra okolia Revúcej I (Slovenské rudohorie – Revúcka vrchovina) In: *Výskum A Ochrana Prírody Muránskej Planiny*, Uhrín, M., (ed.). *3*, 33–52.

Čejka, T., Dvořák, L., Horsák, M., & Šteffek, J., (2007). Checklist of the molluscs (Mollusca) of the Slovak republic. *Folia Malacologica, 15*(2), 49–58.

Červenka, J., (2016). Makromycéty zaznamenané počas 11. mykologických dní v Snine. *Sprav. Slov. Mycol. Spol., 45*, 20–34.

Chapin, III. F. S., & Körner, C., (1996). Arctic and alpine biodiversity: Its patterns, causes and ecosystem consequences. In: *Functional Roles of Biodiversity – A Global Perspective*, Mooney, et al., (eds.). John Wiley & Sons, Chichester, New York, Brisbane, pp. 7–32.

Chytrý, M., (2009). Vegetace České republiky. 2. Ruderální, plevelová, skalní a suťová vegetace. *Academia, Praha*.

Chytrý, M., (2012). Vegetation of the Czech republic: Diversity, ecology, history and dynamics. *Preslia, 84*, 427–504.

Chytrý, M., (2013). Vegetace České republiky. 4. Lesní a křovinná vegetace. *Academia, Praha*.

Chytrý, M., Drazil, T., Hajek, M., & Kalnikovs, V., (2015). The most species-rich plant communities of the Czech republic and Slovakia (with new world records). *Preslia, 87,* 217–278.

Comes, H. P., & Kadereit, J. W., (2003). Spatial and temporal patterns in the evolution of the flora of the European Alpine System. *Taxon, 52,* 451–462.

Czeika, H., Czeika, G., Guttová, A., Farkas, E., Lőkös, L., & Halda, J., (2004). Phytogeographic and taxonomical remarks on 11e species of cyanophilic lichens from Central Europe. *Preslia, 76,* 183–192.

David, S., (2001). Red (cosmological) list of dragonflies (*Insecta: Odonata*) of Slovakia. In: *Red List of Plants and Animals of Slovakia,* Baláž, D., Marhold, K., & Urban, P., (eds.). *Ochrana prírody, 20,* 96–99.

Demko, M., Krištín, A., & Pačenovský, S., (2014). Red list of birds in Slovakia. *SOS/Birdlife Slovakia,* 1–52.

Demko, M., Krištín, A., & Puchala, P., (2013). Red list of birds in Slovakia. *Tichodroma, 25,* 69–78.

Douda, J., Boublik, K., Slezak, M., et al., (2016). Vegetation classification of European floodplain forests and alder carrs. *Appl. Veg. Sci., 19,* 147–163.

Dravecký, M., & Guziová, Z., (2012). A preliminary overview of monitoring for raptors in the Slovak Republic. *Acrocephalus, 33*(154/155), 261–269.

Dvořák, P., Hindák, F., Hašler, P., Hindáková, A., & Poulíčková, A., (2014). Morphological and molecular studies of *Neosynechococcus sphagnicola,* gen., sp. nov. (Cyanobacteria, Synechococcales). *Phytotaxa, 170*(1), 24–34.

Eliáš, P., Dítě, D., Kliment, J., Hrivnák, R., & Feráková, V., (2015). Red list of ferns and flowering plants of Slovakia, 5th edition (October 2014). *Biológia (Bratislava), 70*(2), 218–228.

Faško, B., (2017). Prehľad chránených území národnej sústavy, stav k 31. 12. 2016. *Chránené Územia Slovenska, 88,* 2–4.

Feráková, V., (1997). *Flóra, Geológia a Paleontológia Devínskej Kobyly.*

Franc, V., (2000). Spiders (Araneae) on the Red Lists of European countries. *Ekológia (Bratislava), 19*(4), 23–28.

Gajdoš, P., Černecká, Ľ., Franc, V., & Šestáková, A., (2018). *Pavúky Slovenska.* Veda, Bratislava.

Gajdoš, P., & Svatoň, J., (2001). Red (Cosmological) list of spiders (Araneae) of Slovakia. In: *Red List of Plants and Animals of Slovakia,* Baláž, D., Marhold, K., & Urban, P., (eds.). *Ochrana prírody, 20,* 80–86.

Gajdoš, P., Svatoň, J., & Sloboda, K., (1999). Katalóg pavúkov Slovenska. *Ústav Krajinnej Ekológie SAV, Bratislava,* 1–339.

Glejdura, S., (2013). Nové nálezy bazídiových a vreckatých húb v Stolických vrchoch (Slovensko). *Mykologické Listy, 124,* 15–39.

Glejdura, S., Kučera, V., Lizoň, P., & Kunca, V., (2015). *Pseudoplectania lignicola* sp. nov. described from central Europe. *Mycotaxon, 130,* 1–10

Gubka, A., Nikolov, C., Gubka, K., Galko, J., Vakula, J., Kunca, A., & Leontovyč, R., (2013). History, present and expected future of forests in Slovakia. *American J. Plant Sci., 4,* 711–716.

Guttová, A., Lackovičová, A., & Pišút, I., (2013). Revised and updated checklist of lichens of Slovakia. *Biológia (Bratislava), 68,* pp. 845–850. + 50 electronic appendix.

Gvoždík, V., Benkovský, N., Crottini, A., Bellati, A., Moravec, J., Romano, A., et al., (2013). An ancient lineage of slow worms, genus *Anguis* (Squamata: Anguidae), survived in the Italian Peninsula. *Molecular Phylogenetics and Evolution, 69*(3), 1077–1092.

Gvoždík, V., Jandzík, D., Lymberakis, P., Jablonski, D., & Moravec, J., (2010). Slow worm, *Anguis fragilis* (Reptilia: Anguidae) as a species complex: Genetic structure reveals deep divergences. *Molecular Phylogenetics and Evolution, 55*(2), 460–472.

Háberová, I., & Hájek, M., (2001). Scheuchzerio-Caricetea fuscae R. Tx. 1937. In: *Rastlinné Spoločenstvá Slovenska 3 Vegetácia Mokradí,* (Valachovič, M., edt.), Veda, Bratislava, 187–296.

Hagara, L., (2014). *Ottova encyklopédia húb,* Ottovo Nakladatelství, Praha, 1–1152.

Hájková, P., Navrátilová, J., & Hájek, M., (2011). Vegetace vrchovišť (OxycoccoSphagnetea). In: *Vegetace České Republiky 3 Vodní A Mokřadní Vegetace* (Chytrý, M., edt.), Academia, Praha, 705–736.

Hanák, V., Anděra, M., Uhrin, M., Danko, Š., & Horáček, I., (2010). Bats of the Czech Republic and Slovakia: Distributional status of individual species. In: *A Tribute to Bats, A Collection of Contributions on Selected Topics of Bat Research and Bat Conservation,* Horáček, I., & Uhrin, M., (eds.). Publishing House Lesnická práce s. r. o., Kostelec nad Černými lesy, 143–393.

Havašová, M., Ferenčík, J., & Jakuš, R., (2017). Interactions between windthrow, bark beetles and forest management in the Tatra National Parks. *For. Ecol. Manag., 391,* 349–361.

Hegedüšová-Vantarová, K., & Škodová, I., (2014). *Rastlinné spoločenstvá Slovenska. 5. Travinno-bylinná vegetácia.* Veda, Bratislava.

Hejcman, M., Hejcmanová, P., Pavlů, V., & Beneš, J., (2013). Origin and history of grasslands in Central Europe – A review. *Grass Forage Sci., 68,* 345–363.

Hendrych, R., (1984). *Fytogeografie.* – SPN, Praha, 1–224.

Hensel, K., & Mužík, V., (2001). Red (Ecosozological) list of lampreys (Petromyzontes) and fishes (Osteichtyes) of Slovakia. In: *Red List of Plants and Animals of Slovakia,* Báláž, D., Marhold, K., & Urban, P., (eds.). Ochrana prírody, 20, 143–145.

Hindák, F., & Hindáková, A., (2000). Checklist of the cyanophytes/ cyanobacteria and algae of the Slovak stretch of the Danube river (1926–1999). *Biológia (Bratislava), 55*(1), 7–34.

Hindák, F., & Hindáková, A., (2011). Cyanophytes and algae. In: *Plants and Habitats of European Cities,* Kelcey, J. G., & Müller, B., (eds.). Springer, pp. 100–102.

Hindák, F., & Kawecka, B., (2010). Sinice a riasy (Cyanobacteria and algae), In: *Tatry– Príroda,* Koutná, A., & Chovancová, B., (eds.). Nakladatelství Miloš Uhlíř–Baset, Praha, 313–318.

Hindák, F., & Kováčik, Ľ., (1993). Súpis siníc a rias Tatranského národného parku (Checklist of cyanophytes and algae of the Tatra National Park). – *Zbor. Prác o TANAP-e, Martin, 33,* 235–279.

Hindák, F., (1977). Studies on the chlorococcal algae (Chlorophyceae). I.-Treatise on biology (Bratislava) *Biol. Práce. Veda, Bratislava, 23*(4), 1–190.

Hindák, F., (2008). Colour Atlas of Cyanophytes. *VEDA, Bratislava,* 1–256.

Holecová, M., & Franc, V., (2001). Red (Ecosozological) list of beetles (Coleoptera) of Slovakia. In: *Red List of Plants and Animals of Slovakia,* Baláž, D., Marhold, K., & Urban, P., (eds.). Ochrana prírody, 20, 111–128.

Global Biodiversity, Volume 2

Horsák, M., Juřičková, L., Beran, L., Čejka, T., & Dvořák, L., (2010). Komentovaný seznam měkkýšů zjištěných ve volné přírode České a Sloven- ské republiky. *Malacologica Bohemoslovaca, Suppl.*, *1*, 1–37.

Jablonski, D., Jandzík, D., Mikulíček, P., Džikić, G., Ljubisavljević, K., Tzankov, N., et al., (2016). Contrasting evolutionary histories of the legless lizards slow worms (*Anguis*) shaped by the topography of the Balkan Peninsula. *BMC Evolutionary Biology*, *16*(99), 1–18.

Janda, P., Trotsuk, I., Mikolas, M., & Svoboda, M., (2017). The historical disturbance regime of mountain Norway spruce forests in the Western Carpathians and its influence on current forest structure and composition. *For. Ecol. Manag.*, *388*, 67–78.

Jarolímek, I., & Šibík, J., (2008). *Vegetation of Slovakia. Diagnostic, constant and dominant species of the higher vegetation units of Slovakia.* VEDA, Bratislava.

Jeník, J., & Price, M., (1994). *Biosphere Reserves on the Crossroads of Central Europe.* Czech Republic National Committee for UNESCO's Man and Biosphere Program.

Kapusta, P., (2016). *Threat of Animal Species in Slovakia.* Enviroportál. https://www. Enviroportal. sk/indicator/detail?id=1204.

Kautman, J., Bartík, I., & Urban, P., (2001a). Red (Ecosozological) list of amphibian (Amphibia) of Slovakia. In: *Red List of Plants and Animals of Slovakia*, Baláž, D., Marhold, K., & Urban, P., (eds.). Ochrana prírody, 20, 146–147.

Kautman, J., Bartík, I., & Urban, P., (2001b). Red (Ecosozological) list of reptiles (Reptilia) of Slovakia. In: *Red List of Plants and Animals of Slovakia*, Baláž D., Marhold, K. & Urban, P. (eds.). Ochrana prírody, 20, 148–149.

Kliment, J., & Valachovič, M., (2007). *Rastlinné spoločenstvá Slovenska. 4. Vysokohorská vegetácia.* Veda, Bratislava.

Kliment, J., (1999). Komentovaný prehľad vyšších rastlín flóry Slovenska, uvádzaných v literatúre ako endemické raxóny. *Bull. Slov. Bot. Spoločn.*, *21*(4).

Kliment, J., (2008). *Príroda Veľkej Fatry (Nature of the Veľká Fatra Mts).* Vydavateľstvo Univerzity Komenského, Bratislava.

Kliment, J., Šibík, J., Šibíková, I., Jarolímek, I., Dúbravcová, Z., & Uhlířová, J., (2010). High-altitude vegetation of the Western Carpathians – A syntaxonomical review. *Biológia (Bratislava)*, *65*, 965–989.

Kliment, J., Šibíková, I., & Šibík, J., (2011). On the occurrence of the arctic-alpine and endemic species in the high-altitude vegetation of the Western Carpathians. *Thaiszia. J. Bot.*, *21*, 45–61.

Körner, C., (2003). *Alpine Plant Life, Functional Plant Ecology of High Mountain Ecosystems.* 2nd edition. Springer, Verlag, Berlin, Heidelberg, New York.

Koščo, J., & Holčík, J., (2008). Anotovaný Červený zoznam mihúľ a rýb Slovenska – verzia 2007. *Biodiverzita Ichtyofauny ČR*, *VII*, 119–132.

Koščo, J., Košuthová, L., Košuth, P., & Pekárik, L., (2010). Non-native fish species in Slovak waters: Origins and present status. *Biológia (Bratislava)*, *Section Zoology*, *65*(6), 1057–1063. DOI: 10.2478/s11756–010–0114–7.

Koutná, A., & Chovancová, B., (2010). *Tatry – Príroda.* Baset, Příbram.

Krištín, A., (2001). Red (Ecosozological) list of orthoptera of Slovakia. In: *Red List of Plants and Animals of Slovakia*, Baláž, D., Marhold, K., & Urban, P., (eds.). Ochrana prírody, 20, 103–104.

Krištofík, J., & Danko, Š., (eds.), (2012). *Mammals of Slovakia, Distribution, bionomy and protection.* Veda, Bratislava, 1–712.

Kubinská, A., Janovicová, K., & Šoltés, R., (2001). Červený zoznam machorastov Slovenska. In: *Červený Zoznam Rastlín a Živočíchov Slovenska,* Baláž D., Marhold, K., & Urban, P., (eds.). *Ochrana Prírody, 20,* 31–43.

Kučera, J., Turis, P., Zozomová-Lihová, J., & Slovák, M., (2013). *Cyclamen fatrense* (Primulaceae) only a myth or a true West Carpathian endemic? Genetic and morphological evidence. *Preslia, 85,* 133–158.

Kučera, V., & Kautmanová, I., (2011). Contribution to the knowledge of macrofungi of the Muránska planina National Park and adjacent areas. *Reussia, 6,* 87–96.

Kučera, V., & Lizoň, P., (2012). Geoglossaceous fungi in Slovakia III. The genus *Geoglossum. Biológia (Bratislava), 67,* 654–658.

Kučera, V., Lizoň, P., & Tomšovský, M., (2014a). A new green earthtongue *Microglossum parvisporum* sp. nov. *Sydowia, 66,* 335–343.

Kučera, V., Lizoň, P., & Tomšovský, M., (2017). Taxonomic divergence of the green nakedstipe members of the genus *Microglossum* (Helotiales). *Mycologia, 107,* 46–54.

Kučera, V., Lizoň, P., Tomšovský, M., Kučera, J., & Gaisler, J., (2014b). Re-evaluation of the morphological variability of *Microglossum viride* and *M. griseoviride* sp. nov. *Mycologia, 106,* 282–290.

Kulfan, M., & Kulfan, J., (2001). Red (Ecosozological) list of butterflies (Lepidoptrea) of Slovakia. In: *Red List of Plants and Animals of Slovakia,* Baláž, D., Marhold, K., & Urban, P., (eds.). Ochrana Prírody, 20, 134–137.

Kuneš, P., Pelánková, B., Chytrý, M., Jankovská, V., Pokorný, P., & Petr, L., (2008). Interpretation of the last-glacial vegetation of eastern-central Europe using modern analogs from southern Siberia. – *J. Biogeogr., 35,* 2223–2236.

Kuthan, J., Adamčík, S., Antonín, V., & Terray, J., (1999). Huby Národného parku Poloniny [Fungi of Poloniny National Park]. *Správa Národných Parkov SR, Liptovský Mikuláš, Správa Národného Parku Poloniny, Snina, Košice,* 1–198.

Lieskovská, Z., (2016). *Životné prostredie Slovenskej republiky v kocke (Environment of the Slovak republic in focus).* Ministry of Environment of the SR, Slovak Environment Agency (SEA), Bratislava.

Lisický, M. J., (1991). *Mollusca Slovenska.* Veda, Bratislava, 1–341.

Lizoň, P., & Bacigálová, K., (1998). Huby. In: *Zoznam Nižších a Vyšších Rastlín Slovenska,* Marhold, K., & Hindák, F., (eds.). Veda, Bratislava, pp. 101–227.

Lizoň, P., (2006). Makromycéty zbierané počas 9. mykologických dní na Slovensku. *Catathelasma, 7,* 17–33.

Lukáš, J., (2001). Red (Ecosozological) list of hymenoptera of Slovakia. In: *Red List of Plants and Animals of Slovakia.* Ochrana Prírody, 20, 129–133.

Magri, D., Vendramin, G. G., Comps, B., Dupanloup, I., Geburek, T., Gömöry, D., et al., (2006). A new scenario for the quaternary history of European beech populations: Palaeobotanical evidence and genetic consequences. *New Phytol., 171*(1), 199–221.

Makovinská, J., & Hlúbiková, D., (2014). Phytobenthos of the River Danube. In: *The Danube River Basin,* Liška, I., (ed.). Springer Verlag.

Marhold, K., & Hindák, F., (1998). *Zoznam nižších a vyšších rastlín Slovenska (Checklist of non-vascular and vascular plants of Slovakia).* Veda, Bratislava.

Marvan, P., Heteša, J., Hindák, F., & Hindáková, A., (2004). Phytoplankton of the Morava river (Czech Republic, Slovakia): Past and present. *Oceanological and Hydrobiological Studies, Gdansk, 33*(4), 42–60.

Medvecká, J., Jarolímek, I., Senko, D., & Svitok, M., (2013). Fifty years of plant invasion dynamics in Slovakia along a 2,500 m altitudinal gradient. *Biol Invasions,* 1–12.

Medvecká, J., Kliment, J., Májeková, J., Halada, Ľ., Zaliberová, M., Gojdičová, E., Feráková, V., & Jarolímek, I., (2012). Inventory of the alien flora of Slovakia. *Preslia, 84,* 257–309.

Mihál, I., Blanár, D., & Glejdura, S., (2015). Enhancing knowledge of mycoflora (Myxomycota, Zygomycota, Ascomycota, Basidiomycota) in oak-hornbeam forests in the vicinity of the magnesite plants at Lubeník and Jelšava (central Slovakia). *Thaiszia, 25,* 121–142.

Mihál, I., Glejdura, S., & Blanár, D., (2011). Makromycéty (Zygomycota, Ascomycota, Basidiomycota) v masíve Kohúta (Stolické vrchy). *Reussia, 1–2,* 1–44.

Mráz, P., Barabas, D., Lengyelová, L., Turis, P., Schmotzer, A., Janišová, M., & Ronikier, M., (2016). Vascular plant endemism in the Western Carpathians: Spatial patterns, environmental correlates and taxon traits. *Biol. J. Linn. Soc.,* DOI: 10.1111/bij.12792.

Mucina, L., Bultmann, H., Diersen, K., Theurillat, J. P., et al., (2016). Vegetation of Europe: Hierarchical floristic classification system of vascular plant, bryophyte, lichen, and algal communities. *Appl. Veg. Sci., 19,* 3–264.

Nádvorník, J., (1934). Calicieae-Studien aus der Tschechoslowakei. *Feddes Repert. Spec. Nov. Regn. Veg., 36,* 307–310.

Oťaheľová, H., Hrivnák, R., & Valachovič, M., (2001). Phragmito-Magnocaricetea Klika in Klika et Novák 1941. In: *Rastlinné Spoločenstvá Slovenska 3 Vegetácia Mokradí* (Valachovič, M., ed.). Veda, Bratislava, 51–183.

Pastoralis, G., Kalivoda, H., & Panigaj, Ľ., (2013). Zoznam motýľov (Lepidoptera) zistených na Slovensku. *Folia Faunistica Slovaca, 18*(2), 101–232.

Paulech, C., (1995). *Flóra Slovenska X/1: Huby múčnatkotvaré (Erysiphales).* Veda, Bratislava, 1–297.

Peiger, M., Tomka, P., & Paulíny, M., (2015). *Huby Liptova.* Liptovské Múzeum, Ružomberok.

Pišút, I., Guttová, A., Lackovičová, A., & Lisická, E., (2001). Červený zoznam lišajníkov Slovenska. In: *Červený Zoznam Rastlín a Živočíchov Slovenska,* Baláž, D., Marhold, K., & Urban, P., (eds.). Ochrana Prírody, 20, 22–34.

Puşcaş, M., Choler, P., Tribsch, A., Gielly, L., Rioux, D., Gaudeul, M., & Taberlet, P., (2008). Post-glacial history of the dominant alpine sedge *Carex curvula* in the European Alpine system inferred from nuclear and chloroplast markers. *Mol. Ecol., 17,* 2417–2429.

Ripková, S., Adamčík, S., Kučera, V., & Palko, L., (2007). *Fungi of the Protected Landscape Area of Vihorlat.* Institute of Botany SAS, Bratislava, 1–149.

Šácha, D., David, S., Bulánková, E., Jakab, I., & Konvit, I., (2011). *Vážky Slovenskej republiky.* Available on: http://www.vazky.sk.

Selva, S. B., (2014). The calicioid lichens and fungi of the Acadian Forest ecoregion of northeastern North America, II. The rest of the story. *Bryologist, 117,* 336–367.

Şenkardeşler, A., & Calba, O. F., (2011). New lichen records from Turkey – 2, *Aspicilia, Protoparmeliopsis,* and *Ramalina. Mycotaxon, 115,* 263–270.

Šibík, J., (2007). Fyziognómia a štruktúra ako dôležitý faktor pri vytváraní univerzálneho fytocenologického systému. *Bull. Slov. Bot. Spoločn., 29,* 147–157.

Šibík, J., Šibíková, I., & Kliment, J., (2010). The sub-Alpine *Pinus mugo*-communities of the Carpathians with a European perspective. *Phytocoenologia, 40,* 155–188.

Šibíková, I., Šibík, J., Hájek, M., & Kliment, J., (2010). The distribution of arctic-alpine elements within high-altitude vegetation of the Western Carpathians in relation to environmental factors, life forms and phytogeography. *Phytocoenologia, 40*(2&3), 189–203.

Šípošová, H., Kliment, J., & Mráz, P., (2017). *Databáza – Rastliny opísané zo Slovenska (Vyššie rastliny).* Depon in Botanický ústav SAV.

Škubla, P., (2003). Mycoflora Slovaca, *Mycelium,* 1–1103.

Slovák, M., Kučera, J., Turis, P., & Zozomová, L. J., (2012). Multiple glacial refugia and postglacial colonization routes inferred for a woodland geophyte, *Cyclamen purpurascens*: Patterns concordant with the Pleistocene history of broad-leaved and coniferous tree species. *Biol. J. Linn. Soc., 105,* 741–760.

Šoltés, R., Hájek, M., & Valachovič, M., (2001). Oxycocco-Sphagnetea Br.-Bl. et Tx. ex Westhoff et al. (1946). In: *Rastlinné Spoločenstvá Slovenska 3 Vegetácia Mokradí* (Valachovič, M, ed.), Veda, Bratislava, 275–296.

Šteffek, J., & Vavrová, Ľ., (2006). Current ecosozological status of molluscs (Mollusca) of Slovakia in accordance with categories and criterion of IUCN – Version 3. 1. In: *Molluscs: Perspective of Development and Investigation*, Kyrychuk, G. Y., (ed.). Zhytomyr, 266–276.

Straka, P., (1998). *Národná správa o stave a ochrane biodiversity slovenska. National report on the Satus and Protection of Biodiversity in Slovakia.* Ministerstvo životného prostredia Slovenskej republiky, Bratislava.

Suza, J., (1923). Nový zástupce arktické vegetace lišejníkové na Vysokých Tatrách. *Čas. Mor. Mus. Brno., 22–23,* 122–131.

Thom, D., & Seidl, R., (2015). Natural disturbance impacts on ecosystem services and biodiversity in temperate and boreal forests. *Biol Rev Camb Philos Soc.* 91(3):760-81. doi: 10.1111/brv.12193.

Titov, A. N., (1998). New and rare calicioid lichens and fungi from relict tertiary forests of Caucasus and the Crimea. *Folia Cryptogamica Estonica, 32,* 127–133.

Udvardy, M., (1975). *A Classification of the Biogeographical Provinces of the World.* IUCN Occasional Paper No. 18.

Uhlířová, J., Bernátová, D., & Šibík, J., (2016). Fenomén vrchoviskových komplexov hornej Oravy. *Acta. Rer. Natur. Mus. Nat. Slov., 62,* 34–65.

Urban, P., Uhrin, M., & Ambros, M., (2017). *Red List and Red List Index of Mammals of Slovakia.* Biológia (Bratislava) (in press.).

Valachovič, M., Oťaheľová, H., & Hrivnák, R., (2001). Isoëto-Nanojuncetea Br.-Bl. et Tx. ex Westhoff et al., (1946). In: *Rastlinné Spoločenstvá Slovenska 3 Vegetácia Mokradí* (Valachovič, M., ed.), Veda, Bratislava, 347–373.

Váňa, J., (1986). *Ochyraea tatrensis* gen. et spec. nov., A remarkable pleurocarpous moss from Czechoslovakia. *J. Bryol., 14,* 261–267.

Wahlenberg, G., (1814). Cryptogamia (lichenes). In: *Flora Carpatorum Principalium Exhibens Plantas In Montibus Carpaticis Inter Flumina Waagum At Dunajetz Eorumque Ramos Arvam Et Popradum Crescentes,* Göttingen, pp. 370–395.

Willner, W., Jimenez-Alfaro, B., Agrillo, E., & Chytry, M., (2017). Classification of European beech forests: a Gordian Knot? *Appl. Veg. Sci., 20.*

Wołowski, K., & Hindák, F., (2005). *Atlas of Euglenophytes.* Veda, SAV, Bratislava, 1–136.

Zemanová, B., Hájková, P., Hájek, B., Martínková, N., Mikulíček, P., Zima J., & Bryja J., (2015). Extremely low genetic variation in endangered Tatra chamois and evidence for hybridization with an introduced Alpine population. *Conservation Genetics., 16*(3), 729–741.

Additional References

Act No. 543/2002 on Nature and Landscape Protection as amended.

http://www.bielekarpaty.sk/opisBK.html.

http://www.chkopolana.eu/flora/.

http://www.lepidoptera.sk/.

http://www.npmp.sk/obsah/muranska-planina.

http://www.orthoptera.sk/sk/.

http://www.shmu.sk.

https://www.iucn.org/theme/protected-areas/about/protected-areas-categories.

Regulation of Ministry of Environment No. 24/2003 as amended.

CHAPTER 10

Biodiversity in Sweden

ULF GÄRDENFORS

Professor of Conservation Biology and Deputy Director of the Swedish Species Information Centre, University of Agricultural Sciences, Uppsala, Sweden, E-mail: Ulf.Gardenfors@slu.se

10.1 Introduction

The land of Sweden, as for the entire Fenoscandian shield, is currently in an interglacial phase. During Pleistocene, i.e., over the last 1.8 million years, the area has been covered six times with sometimes more than a 3 km thick ice sheet. The southernmost parts of the area began to uncover from the latest glaciation about 15,000 years before present and the latest ice age ceased by 11,600 years ago when the temperature had risen almost to the levels of today. Species, such as Reindeer *Rangifer tarandus*, Moose *Alces alces*, Irish elk *Megaloceros giganteus,* and Wolly mammoth *Mammuthus primigenius,* alpine plants and soon Man *Homo sapiens*, colonized the emerging land and followed the rim of the melting ice, and when the temperature rose were followed by other species. Still, the flora and fauna of Sweden today is virtually restricted to species that have managed to colonize the area over the Holocene, i.e., the last 10,000–15,000 years. This is one explanation why the country is comparatively species poor and, in particular, holds extremely few endemic species.

What is today the kingdom of Sweden constitutes the eastern part of the Scandinavian peninsula in Northern Europe, situated between 55.3° in south and 69.1° in north (reaching well beyond the northern polar circle at 66.3°), and 11.0° in west and 24.2° in east. As the crow flies, Sweden is 1570 km long and 500 km wide. In the west, it borders to Norway and in northeast to Finland. These three countries together with Denmark in southwest (separated from Sweden by a narrow strait) and Iceland in the northern Atlantic Ocean are usually referred to as the Nordic countries, while Sweden, Norway, and Denmark are called Scandinavia, and Fennoscandia when Finland is included.

The long extension means that both climate and day-length varies a lot from south to north. Beyond the polar circle, there is a period in the summertime when the sun never sets (midnight sun), while there is a corresponding polar night in winter time when it never rises.

In spite of a comparatively northern situation, Sweden has a rather mild winter climate due to the so-called gulf stream transporting warm water from the Mexican Gulf, over the northern Atlantic Ocean to the west and north coast of Norway. Even though the summer temperatures are similar to those of America and Asia at corresponding latitudes, the winter temperatures are 20–25°C higher than, e.g., in Canada. In summer (July) the average 24 h day temperature is c. 17°C in south and 10°C in the northernmost parts. Corresponding temperatures in January is 0°C in south and −15°C in the interior parts of the north. The highest temperature ever recorded is 38.0°C and the lowest −52.6°C. The precipitation varies between 500 and 1,000 mm/year of which certain parts come as snow, particularly in the north.

The bedrock of Sweden is quite stable and almost all if it is of Precambrian age and dominated by hard weathered rocks such as granite and gneiss. In certain parts, sedimentary rocks such as limestone is present which is directly detectable through the more diverse and demanding flora and fauna.

10.2   Ecosystems and Threats

The country is dominated by forests, wetlands, and lakes, but also holds agricultural areas, urban environments and – in the northwest – mountains (the Scandes) with an alpine vegetation. In southwest, south and east it is surrounded by sea or brackish water.

Biogeographically, Sweden is part of the Palearctic region and can be subdivided – from south to northwest – in a *Nemoral* (originally broad-leaved forests), *Hemiboreal* (mixture of deciduous and conifer trees), *Boreal* (western taiga dominated by spruce and pine trees), *Sub-Alpine* and an *Alpine* zone (Figure 10.1). In the European Union (EU) terminology, the first is denominated the *Continental*, while the Hemiboreal and Boreal are clustered as *Boreal*, and the Sub-Alpine and Alpine are named the *Alpine* region.

10.2.1   Forests

The natural forest in southernmost Sweden is dominated by broad-leaved species such as Beech *Fagus sylvatica*, oaks *Quercus robur* and *Q. petraea*, Elm *Ulmus glabra*, Lime *Tilia cordata* and Norway maple *Acer platanoides*. These forests are comparatively species rich. The majority of Swedish forests, in particular towards north, naturally constitute of Norway spruce *Picea abies* and Scots pine *Pinus sylvestris* together with birches *Betula pendula* and *B. pubescens*, Aspen *Populus tremula*, Rowan *Sorbus aucuparia* and *Salix* species. However, modern forestry, with start in the nineteenth

Vegetationszoner i Sverige

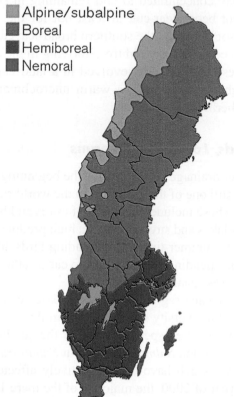

- ■ Alpine/subalpine
- ■ Boreal
- ■ Hemiboreal
- ■ Nemoral

Figure 10.1   Vegetation Zones of Sweden.

century, has changed the extension, structures and tree species composition in a way that has led to an impoverished fauna and flora (Bernes, 2011). The impingement includes felling trees at premature age, clear-cutting large areas, removal of dead wood and deciduous trees, introduction of nonindigenous species (such as Lodgepole pine *Pinus contorta*, larches *Larix* spp. and Sycamore maple *Acer pseudoplatanus*), fire-fighting, fragmentation by roads, drainage, manuring and sometimes plowing.

Due to, sometimes intense, debate connected to the high number of forest-living red-listed species, but also legislation and economic incentives following certification of the forestry (such as by the Forest Stewardship Council), the exploitation of the forests has been somewhat less intense during the last two-three decades. Still, much of the really ecologically demanding, dispersion restricted and sensitive species connected to the spruce and pine biome are today concentrated to and remaining in small areas of old forests that have not been clear-cut, at least not for the last centuries. In contrast, species associated with the southern broad-leaved forests are today mainly surviving in connection to old trees thriving in the agricultural landscape. These species have probably evolved in a landscape with scattered large trees and high insulation with a warm microclimate formed by the browsing of mega-herbivores.

10.2.2 Wetlands, Lakes and Streams

In spite of vast water drainage projects since the beginning of the nineteenth century, Sweden is still one of the countries in the world richest in wetlands, lakes, and streams. These include vast areas of mires and bogs, and also wet shores of numerous lakes and streams. With a high production of insects and other invertebrates at summer time, many wading birds and ducks breed in these habitats, while spending the remaining year in other parts of Europe, Africa or in some cases Asia.

The trenching and water regulation have decreased considerably over the last decades, but the biodiversity still suffers from the effects. This includes anadromous fish species, i.e., species living in the sea but reproducing in fresh water, such as Salmon *Salmo salar*, Trout *Salmo trutta* and European eal *Anguilla anguilla* which have been negatively affected also by power plant dams. In the year of 2000, the majority of the mere 13 amphibian species occurring in Sweden were considered threatened or near threatened, mainly due to former draining and destruction of small waters. However, due to large scale digging of new ponds, rewatering of some wetlands and active reintroduction, many of these species are today classified as Least Concern.

In the 1960s to 1980s, thousands of lakes were heavily acidified by sulfur polluted rain, emanating from sulfurous oil being burned by power plants and industries and brought by winds from southwest to Sweden. It resulted in more or less total extinction of invertebrates and fish in thousands of lakes. Extensive liming of the lakes for decades, but also changing to oil with low sulfur content, has now resulted in that most of the affected lakes have recovered.

10.2.3 Agricultural Land

Man colonized the country more or less as the ice-shield melted away and subsisted on hunting, fishing and gathering. Around 6000 years before present, people originating from the Middle East reached what is now Sweden and brought the tradition of agriculture. Since then, and in particular from c. 1500 year b.p., the landscape and its organisms has been affected by agricultural activities.

Cultivation, and grazing and browsing by cattle, sheep and goats, created new types of environments. Fields, meadows, pastures and open groves provided habitats for an array of species, earlier restricted to shores, impediments or storm and fires glades, or in most cases new to the country, brought by man and its undertakings. In particular, a large proportion of the vascular plants, and insects associated with these plants, occurring in Sweden are of this category.

A quite distinctive ecosystem type, the *alvar*, is found mainly on the Baltic islands Öland and Gotland. It is formed on limestone with very thin soil, high insulation and low precipitation. In combination with grazing by domesticated animals, a very particular steppe-like flora and fauna have thrived and developed over millenia. Several endemic subspecies and varieties are found here, while very few have had the time to evolve to distinctive species.

Since the beginning of the nineteenth century, accelerating in the 20th, the general agricultural landscape has undergone large changes. These include trenching (drainage), ceased haymaking, ceased grazing, fertilization, pesticide spraying, overgrowth, afforestation and creation of large, mainly monocultural fields. As a result, species richness and habitat diversity of the former agricultural landscape has drastically decreased. Remnants are today concentrated to small areas maintained either by farmers receiving subsidies from the government to uphold a traditional cultivation or by active nature conservation management.

10.2.4 Urban Environments

The human population in Sweden amounts to 10 M inhabitants, while the area of the country is 447,435 km^2 (double the area of UK and 25% larger than Germany), leaving the population density to 22 inhabitants/km^2. The cities are in general small, and many have good amount of parks with old and sometimes hollow trees. Therefore, today many species earlier found in agricultural habitats or even broad-leaved forests are found in urban areas.

10.2.5 Mountains

The Scandinavian mountains form a natural geophysical boarder between Sweden and Norway. The highest peaks (Galdhøpiggen 2,469 m and Glittertind 2,465 m above m.s.l.) are situated in central Norway, while the highest peak of Sweden (mount Kebenekajse 2,106 m) is located in the north.

Due to a harsh climate (latitude and altitude), the flora and fauna of the mountains are rather species poor. On the other hand, there are relatively many species only occurring here in Sweden.

The tree border, i.e., the upper limit of the subalpine zone, is normally formed by Downy birch (*Betula pubescens*) and runs at c. 900 m a.s.l. in the south and c. 500 m in the north. The subalpine vegetation is, besides Downy birch, dominated by *Salix* shrubs and often tall herbs.

The lower alpine zone has a rich variety of *Salix* bushes and a number of flowers, sedges and birds that does not occur elsewhere in the country. The middle and high alpine zones are generally poor in vegetation and animal species and are often dominated by lichens, grasses and *Carex* species together with boulders and snow patches that melt late in the season and even hold a few glaciers.

Both the subalpine forest and, in particular, the alpine meadows have been grazed by domesticated rain-deer for centuries which has affected both the tree limit and the amount of vegetation. Today, in some regions there are fewer rain-deer, which in combination with a warmer climate and nitrogenous fall-out leads to denser vegetation and a tree limit that wanders upwards. The temperature rise of the past century has also led to severe melting of the glaciers.

10.2.6 The Sea Including the Baltic

The seawater along the Swedish west coast is almost as saline as the rest of the Atlantic, while the salinity of the Baltic Sea decreases from c. 1% in the south to almost fresh water in the upper parts of Gulf of Bothnia in northeast. The richness of the flora and fauna is paralleled with the salinity. Consequently, the Baltic Sea is rather species poor, but on the other hand in several cases quite distinctive.

There used to be very good fish stocks of Atlantic cod *Gadus morhua*, Atlantic herring *Clupea harengus*, European eel *Anguillla* and several other species. However, heavy overfishing has resulted in many fish populations today quite depleted.

Also, beginning around 1970, increased eutrophication was observed through nutrient supply from land in large parts of the country's coastal waters and marine areas with increased algae bloom, altered benthic fauna and oxygen deficiency in the Baltic Sea. This may, together with effects of the introduced American mink *Mustela vison*, be the cause of the decrease of several bird-species breeding in coastal areas.

10.3 Species in Sweden

10.3.1 Species Richness

There is a general gradient in species richness from south to north, associated to both climate and ecosystems. The highest numbers are found in the broad-leaved forests, old-fashioned agricultural landscapes with old trees and calcareous dry grounds (in particular in the *alvar* on the islands Öland and Gotland) in the south. Considerably lower numbers dwell in the northern boreal pine and spruce forests, and in the Scandinavian mountains, even though, in particular, the latter harbor quite a number of species found nowhere else in the country.

Roughly 60,000 species are currently known to be resident and reproducing in Sweden (Table 10.1; www.dyntaxa.se). The number includes both indigenous species and introduced species that have established and naturalized in the country. In addition, more than 2,600 apomictic (reproducing as asexual entities) vascular plant species of the genera *Taraxacum, Hieracium, Ranunculus, Rubus* and *Alchemilla* are indigenous to Sweden. Furthermore, 2,240 vascular plant species, 260 birds, and a number of fish, insects and marine invertebrates are known as occasional, regular migrants or vagrants.

No doubt, Sweden is one of the best-surveyed countries of the world when it comes to biodiversity. Still, much remains to be discovered or resolved. The numbers of Eubacteria and Archaea 'species' refer to taxonomic described entities and are a very rough estimate since there is no reliable account. The real number could be several million. Also concerning fungi and insects, the actual number can be assumed to be considerably larger.

Taking insects (hexapodes) as an example, Linnaeus (1766–68) knew and described some 1,500 species from Sweden. Lindroth (1967) listed >18,000 species, and Gärdenfors et al. (2003) 24,475. Today we know 26,590 described species (Dyntaxa), but the real number might be even 5,000 species higher (of which quite a few already are found but not formally described).

10.3.2 The Swedish Taxonomy Initiative

In 2002, the Swedish government commissioned The Swedish Species Information Centre (ArtDatabanken) to lead and conduct a Swedish Taxonomy Initiative (STI). The overall goal was to chart and describe all species occurring in Sweden (Ronquist and Gärdenfors, 2003; Miller, 2005, ArtDatabanken). The STI includes (1) inventories and taxonomic research and

Table 10.1 Number of Resident Species Known From Sweden and Percentage
Evaluated for the Red List*

The numbers include introduced and naturalized species. Non-breeding migrants,
visitors, vagrants and other occasional species are not included, nor apomictic
vascular plant microspecies.

Organism Group	2017	2015
	Known resident species	Evaluated for the Red List (% of all)
EUBACTERIA (BACTERIA)	3,000	1
ARCHAEA	30	0
PROTOZOA	1,200	0
CHROMISTA	5,000	4
PLANTAE (PLANTS)	4,400	80
Glaucophyta	2	0
Rhodophyta (red algae)	193	90
Chlorophyta (green algae)	757	6
Charophyta (stoneworts)	355	10
Marchantiophyta (liverworts)	248	100
Anthocerophyta	2	100
Bryophyta (mosses)	781	100
Tracheophyta (vascular plants)**	2,055	100
Lycophyta (club mosses)	8	100
Monilophyta (ferns and horsetails)	40	100
Spermatophyta (phanerogams)	2,025	100
FUNGI	11,300	40
Microsporidia	80	0
Chytridiomycota	625	0
Glomeromycota	18	0
Zygomycota	590	0
Ascomycota (incl. most lichens)	5,900	30
Basidiomycota (mushrooms et al.)	4,100	70
ANIMALIA (ANIMALS)	35,480	40
Mesozoa	4	0
Porifera (sponges)	160	0
Cnidaria	205	18

Table 10.1 *(Continued)*

Organism Group	2017	2015
	Known resident species	Evaluated for the Red List (% of all)
Ctenophora (comb-jellies)	4	0
Xenacoelomorpha	75	0
Rotifera	680	0
Acanthocephala	40	0
Cycliophora	1	0
Entoprocta	26	0
Platyhelminthes (flatworms)	600	1
Nemertea ribbon worms)	60	0
Mollusca (molluscs)	675	75
Solenogastres	5	0
Caudofoveata	6	0
Polyplacophora (chitons)	12	85
Gastropoda (snails et al.)	460	60
Cephalopoda (squids and octopods)	14	60
Bivalvia (bivalve molluscs)	175	90
Scaphopoda (tusk shells)	4	100
Sipuncula (peanut worms)	9	0
Annelida (bristle-worms)	800	10
Polychaeta (marine annelids)	506	13
Clitellata (earthworms, leeches et al.)	300	4
Gnathostomulida	14	0
Bryozoa (sea-mats)	142	0
Brachiopoda (lamp-shells)	4	100
Phoronida	5	0
Chaetognatha (arrow worms)	6	0
Gastrotricha	99	0
Tardigrada (water bears)	101	0
Arthropoda	30,100	42
Myriapoda (centipedes, millipedes et al.)	94	100
Hexapoda (*insects*)	26,590	45

Table 10.1   *(Continued)*

Organism Group	2017	2015
	Known resident species	Evaluated for the Red List (% of all)
Protura	4	0
Collembola (springtails)	299	0
Diplura (two-tailed bristletails)	5	0
Archaeognatha (bristletails)	3	0
Thysanura (silverfish)	3	0
Ephemeroptera (mayflies)	59	100
Odonata (dragonflies)	58	100
Blattodea (cockroaches)	5	100
Orthoptera (grasshoppers, crickets et al.)	37	100
Dermaptera (earwigs)	5	0
Plecoptera (stoneflies)	37	100
Psocoptera (book-lice)	67	0
Phthiraptera (biting lice)	267	0
Thysanoptera (thrips)	143	0
Hemiptera (bugs)	1,805	60
Megaloptera (alder-flies)	5	100
Raphidioptera (snake-flies)	4	100
Neuroptera (lacewings and antlions)	65	12
Coleoptera (beetles)	4,425	100
Mecoptera (scorpion flies)	6	0
Siphonaptera (fleas)	54	0
Strepsiptera (stylops)	7	0
Diptera (two-winged flies)	7,880	25
Trichoptera (caddies-flies)	225	100
Lepidoptera (butteflies and moths)	2,682	100
Hymenoptera (wasp et al.)	8,440	10
Crustacea	1,570	8
Chelicerata	1,840	40
Araneae (spiders)	741	100
Opiliones (harvestmen)	23	100

Table 10.1 *(Continued)*

Organism Group	2017	2015
	Known resident species	Evaluated for the Red List (% of all)
Pseudoscorpiones (false scorpions)	20	100
Acari (mites and ticks)	1,140	0
Pycnogonida (sea-spiders)	15	0
Nematoda (roundworms)	1,030	0
Nematomorpha (hair-worms)	7	0
Kinorhyncha	18	0
Loricifera	1	0
Priapulida	2	0
Echinodermata	73	92
Hemichordata	4	0
Chordata	534	99
Tunicata	53	95
Cephalochordata (lancets)	1	0
Craniata (vertebrates)	480	100
Myxinomorphi (hagfish)	1	100
Petromyzontomorphi (lampreys)	3	100
Chondrichthyomorphi (cartilaginous fish)	9	100
Actinopterygii (bony fish)	120	100
Amphibia (amphibians)	13	100
Reptilia (reptiles)	6	100
Aves (birds)***	250	100
Mammalia (mammals)	78	100
Total	**60,400**	**36**

*The numbers include introduced and naturalized species. Non-breeding migrants, visitors, vagrants and other occasional species are not included, nor apomictic vascular plant microspecies.

**In addition, 2622 apomictic species and 2250 occasional species.

*** In addition, 260 nonbreeding migrants and vagrants.

revisions, (2) financial support to Swedish biodiversity museums in order to strengthen their curation, digitalization, taxonomic competence and capacity to support STI with reference material and storing new material, (3) a taxonomic database (Dyntaxa), and (4) description of the fauna and flora on the web and in the book series *Encyclopedia of the Swedish Flora and Fauna*. The annual budget 2017 is 70 MSEK (c. 8 M US$).

Hitherto, the project has resulted in the discovery of more than 3,000 species earlier not known from Sweden. More than 1,000 species of these are believed to be new to science and will successively be formally described. The majority of the latter belongs to the dipteran families Phoridae (scuttle flies) and Cecidiomyidae (fungus gnats). But also among other insect groups, marine invertebrates and fungi a considerable number of new species have been found.

10.3.3   Endemicity

The approximately 15,000 years since recolonized after the latest ice-age started has in most cases not been sufficient for the evolution of entirely new species, and very few plants and animals are therefore unique to Sweden. Examples among the very few endemic vascular plants are *Corydalis gotlandica, Artemisia oelandica* and *Arenaria gothica*, all closely related to other more widespread species. To these can be added a substantial number of apomictic species of the genera *Taraxacum, Hieracium*, etc., which have evolved within the country. Many insects and other invertebrates are hitherto only known from Sweden, but the majority of these no doubt occurs also elsewhere in the world.

If the scope is enlarged to the Fennoscandian shield there are some more true endemics. Examples are *Primula scandinavica, Deschampsia bottnica*, the Norway lemming *Lemmus lemmus* and a number of well-studied insects.

10.3.4   Threatened Species

The first Red List of threatened species (vertebrates) in Sweden was produced in 1975 (Ahlén 1975). In 2000, a Red List including evaluation of more than 20,000 species from most major organism groups and based on the new IUCN Categories and Criteria was published (Gärdenfors, 2000; Table 10.1). That process has been repeated at the Swedish Species Information Centre every five years (2005, 2010, 2015), leading to a good understanding of threats, needed measures and distribution of species at risk of extinction in Sweden.

Two hundred and two species were classified as Regionally Extinct in 2015 (ArtDatabanken 2015), but hitherto none of these are considered globally extinct. Examples of RE species are Black stork *Ciconia nigra*, European roller *Coracius garrulous*, Kentish plover *Charadrius alexandrinus*, Black rat *Rattus rattus*, Water caltrop *Trapa natans,* the lichen *Erioderma pedicellatum*, Rosalia longicorn *Rosalia alpina*, and three bumble bees *Bombus cullumanus, B. pomorum* and *B. ruderatus.*

Among the 215 species classified as *Critically endangered* can be mentioned Pollack *Pollachius pollachius*, European eel *Anguilla anguilla*, Spiny dogfish *Squalus acanthias*, Lesser white-fronted goose *Anser erythropus*, White-backed woodpecker *Dendrocopos leucotos*, Assmann's fritillary *Melitea britomartis*, Great capricorn beetle *Cerambyx cerdo*, Scots elm *Ulmus glabra*, and Field elm *Ulmus minor*. The latter two are trees that have been severely smitten by the Dutch elm disease over the last decades.

The two most important factors affecting the extinction risk of the red-listed species are felling of old trees (affecting c. 33% of all red-listed species) and overgrowth of pastures and other open areas (almost as many). These are followed by plantation of agricultural land (c. 12%), eutrophication (almost as many) and drainage of the landscape (c. 7% of all red-listed species).

Sweden has also calculated a Red List Index (RLI) for the years 2000, 2005, 2010 and 2015 (Sandström et al., 2015). The RLI measures trends in extinction risk over time, where 1 means that every species are classified as Least Concern, while 0 means that every species are Extinct or Regionally Extinct (Butchart et al., 2007). In 2015, the overall RLI for assessed Swedish species (n = 7069 in 2015) was 0.88, and it has been very stable since 2000. Broken down to organism groups, bryophytes had the highest index (0.92) while bees (Apoidea) had the lowest (0.84). The most remarkable change has been for Amphibians + Reptiles that had a RLI of 0.77 in 2000, but had risen to 0.85 in 2015 due to active and successful conservation actions.

10.3.5 Invasive Species

Sweden is comparatively spared from invasive alien species, even though this can be anticipated to become a growing problem with a changing climate. Still, examples of species that already cause considerable problems are not lacking. American mink *Mustela vison* and Raccoon dog *Nyctereutes procyonoides* predate on ground-breeding birds and the former may be one important reason to the sharp population decrease of species like Common eider *Somateria mollissima* and Velvet scoter *Melanitta fusca*. Spanish slug *Arion vulgaris (lucitanicus)* causes much problems in gardens and vegetable cultivations. Japanese knotweed *Reynoutria japonica* and Beach rose *Rosa rugosa* are examples of species outcompeting and overgrowing other species. Giant hogweed *Heracleum mantegazzianum* can cause skin irritation to people touching it. Canadian waterweeds (*Elodea canadensis* and *E. nuttallii*) are literally filling ponds and piscicultures. Pacific oyster *Crassostrea gigas* outcompetes the indigenous European flat oyster *Ostrea edulis*. The Signal crayfish *Pacifastacus leniusculus* with its Crayfish plague *Aphanomyces astaci* has caused the extinction of the highly prized for food European crayfish *Astacus astacus* over large areas of Sweden. Ash dieback fungus *Hymenoscyphus fraxineus* and Dutch elm disease *Ophiostoma ulmi* have severely reduced the populations of Ash *Fraxinus excelsior* and Elms (*Ulmus* spp.) in a quite short time.

10.3.6 Charting Species Distribution Using Citizens

Knowledge of species distribution and occurrence in Sweden is probably among the better in the world. There is a long tradition of inventorying and observing species in the country. NGOs like botanical and ornithological societies are charting the country in a systematic way, producing good atlases and knowledge of species distribution. While the number of professional taxonomists is shrinking the quantity of skilled amateur biologists observing all kinds of species is growing. There are at least two reasons for the latter. With the publication of *Encyclopedia of the Swedish flora and fauna* and other books, as well as the Internet, presenting a growing number of well-illustrated identification guides, many groups are now available for studies and field observations. The other important factor is the Swedish Species Observation System—*Artportalen,* https://artportalen.se/—a website open for anybody to report and study other people's observations. Exposition of names of the reporters and competition lists contribute to trigger people to observe and report, including uploading confirming photos. Artportalen is used not only by amateurs, but also by governmental and professional bodies archiving results from inventories, monitoring and even research. Currently, Artportalen contains some 60 M species observations of more than 30,000 species and 10,000–15,000 new observations are submitted every day around the year. With the exception of a few species sensitive to revealing exact location, all data is freely available to anybody to take part of.

10.4 Nature Conservation and Protected Areas

Some 13% of Sweden is currently protected as National parks (1.5%), nature reserves (9.5%) or Natura 2000 designations. The majority of the reserves are located in sparsely populated areas in northern parts of the country. Thus, more than 40% of the mountains (the Scandes), and forests close to those are protected by law. In contrast, in several parts of Southern Sweden less than 1% of the area is protected. Calculated on forested areas all over the country some 3.8% is protected. To this comes a certain amount of voluntarily set-asides by forest companies and private landowners. The pace of protecting new areas is currently (2017) comparatively high, including a growing number of marine reserves.

Keywords

- nature conservation
- protected areas
- threats
- species

References

Ahlén, I., (1975). Hotade ryggradsdjur (exkl. fiskar) i Sverige. Preliminär lista med kategorier delvis baserade på "Red Data Book." *Sverige Natur,* Årsbok, 126–129.

ArtDatabanken, (2015). *Rödlistade arter i Sverige.* ArtDatabanken SLU, Uppsala [in Swedish]. http://www.artdatabanken.se/publikationer/bestall-publikationer/rodlistan2015/.

ArtDatabanken, (The Swedish Species Information Centre). http://www.artdatabanken.se/en/.

Artportalen, (The Swedish Species Observation System): https://artportalen.se/.

Bernes, C., (2011). *Biologisk Mångfald i Sverige.* Monitor 22. Naturvårdsverket.

Butchart, S. H. M., Akçakaya, H. R., Chanson, J., Baillie, J. E. M., Collen, B., Quader, S., Turner, W. R., Amin, R., Stuart, S. N., & Hilton, T. C., (2007). Improvements to the Red List Index. *PLoS One, 2*(1), e140. DOI: 10.1371/journal.pone.0000140.

Dyntaxa. *Swedish Taxonomic Database.* Swedish Species Information Centre, Swedish University of Agricultural Sciences, Uppsala. https://www.dyntaxa.se.

Encyclopedia of the Swedish Flora and Fauna (Nationalnyckeln till Sveriges flora och fauna). Hitherto, 17 hard copy volumes published, and successively material is also published at https://artfakta.artdatabanken.se/.

Gärdenfors, U., (2000). *The 2000 Red List of Swedish Species.* ArtDatabanken, SLU, Uppsala.

Gärdenfors, U., Hall, R., Hallingbäck, T., Hansson, H. G., & Hedström, L., (2003). Animals, fungi and plants in Sweden. Catalogue of number of species per family. *ArtDatabanken Rapporterar, 5.*

Lindroth, C. H., (1967). *Entomologi. Biologi 7.* Almqvist & Wiksell, Stockholm.

Linnaeus, C., (1766–1768). *Systema Naturae.* 12th ed.

Miller, G., (2005). Linnaeus's Legacy Carries on. *Science, 307,* 1038–1039.

Ronquist, F., & Gärdenfors, U., (2003). Taxonomy and biodiversity inventories: Time to deliver. *Trends in Ecology and Evolution, 18*(6), 269–270.

Sandström, J., Bjelke, U., Carlberg, T., & Sundberg, S., (2015). Tillstånd och trender för arter och deras livsmiljöer – Rödlistade arter i Sverige. ArtDatabanken Rapporterar 17. *ArtDatabanken, SLU.* Uppsala.

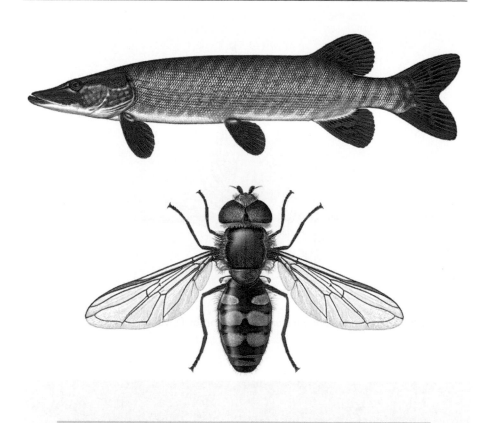

Plate 10.1 (from top to botom) *Esox lucius* (illustration by Linda Nyman), *Eupeodes corollae* (illustration by M. Eric Nasibov), *Stauropus fagi* (illustration by Torbjörn Östman)

Plate 10.2   *Racomitrium lanuginosum* (illustration by Polyanna von Knorring.)

Plate 10.3   The sun does not rise over the horizon in mid-winters of northernmost Sweden. Still, the snow-cover lights up the nature. Here a mire at night of Central Sweden. Photo by Johan Carlberg.

Plate 10.4 The taiga of Northern and Central Sweden is dominated by conifer (spruce and pine) forest with a mosaic of extensive mires and lakes. Photo by Håkan Berglund.

Plate 10.5 Forests of Southern Sweden are dominated by broadleaved and deciduous forest, such as beech, oakes, elmes, ash, maples and lime. Here a beech (*Fagus sylvatica*) in autumn which has droped most of its leaves that have turned reddish brown. Photo by Michael Krikorev.

Plate 10.6 Sweden is very rich in all types of water and wetlands. Here a typical river of the taiga. Photo by Samantha Dawson.

Plate 10.7 In Lapland, at the foothills of the Scandes, extensive mires spread, dominated by *Carex* and other sedges. Here, among others, the cottongras *Eriophorum scheuchzeri*. Photo by Sebastian Sundberg.

Plate 10.8 Oligotrophic mire-ponds often harbour the Goldeneye *Bucephala clangula*. Photo by Tomas Carlberg.

Plate 10.9 In Southern Sweden, the agricultural fields are vast with little or no intermixed other habitats. Photo by Tomas Carlberg.

Plate 10.10   In Central Sweden, the agricultural landscape is typically intermixed with forests and lakes. Photo by Sebastian Sundberg.

Plate 10.11   The Scandes, a mountain ridge in north-western Sweden and Norway, hold a subarctic biodiversity. Here the Glacier buttercup *Ranunculus glacialis*. Photo by Hjalmar Croneborg.

CHAPTER 11

Biodiversity in Turkey

ALİ A. DÖNMEZ and **SEDAT V. YERLİ**

Department of Biology, Faculty of Science, Hacettepe University, 06800 Beytepe, Ankara, Turkey
Email: donmez@hacettepe.edu.tr

11.1 Introduction

Turkey is a country occupying a meeting point of the Asian and European continents between the 35–42°N latitudes and 25–45°E longitudes, and it covers an area of 783,562 km². Having extensive variety of climate, topography, main rocks, soil and aquatic habitats with running water systems, the country has magnificent biodiversity. Furthermore, glacial periods had only a limited destructive effect on the country at higher altitudes (Sarıkaya et al., 2011), thus its biodiversity was kept from natural destruction, unlike European biodiversity. Hence, in respect of biodiversity, Turkey is an important center worldwide and especially in the Mediterranean phytogeographical region and it has plenty of biodiversity hotspots for conservation priorities (Medail and Diadema, 2009). Moreover, not only the Mediterranean region but also the Irano-Turanian part of Turkey is important biodiversity centers and new taxa are continuously described (e.g., Kuru et al., 2014; Dönmez et al., 2015; Mutlu and Karakuş 2015; Binzet, 2016; Daşkın and Bağçıvan, 2017). In respect of zoogeography, Turkey is in the Palearctic region and it contains many animal species belonging to the Afro-Ethiopia, Iran-Caspian, East Asia and Angora regions of Central Asia.

Turkey's seas contain many different habitats because of their geological and geomorphological structures, hydrological conditions and special locations. Along the shores of Turkey, there are thousands of sea caves with many different geological structures and they inhabit numerous fish species as well as other sea creatures.

Based on the floristic studies carried out in Turkey, approximately 10,150 vascular plant species are known to grow in the country. Among them, approximately 3100 species are endemic. Turkey has a high fauna biological diversity with 197 mammals, more than 420 birds, approximately 130 reptiles, 28 amphibians (Yerli et al., 2012), approximately 371 freshwater fishes (Kuru et al., 2014) and nearly 449 marine fish species (Evirgen, 2007). Unlike more accurate data about vascular plants and higher animal groups, due to insufficient data about algae, fungi, invertebrates, and microorganisms, we are currently far away to give accurate numbers of taxa composing species biodiversity (Table 11.1).

11.2 Phytogeography of Turkey

In respect of phytogeography, three different phytogeographical regions are represented in the country according to the climax vegetation and characteristic taxa, namely the Mediterranean, Irano-Turanian, and Euro-Siberian phytogeographical regions (Davis, 1965). This phytogeographical evaluation of Turkey is based on the climatic, topographic and other environmental conditions. Hence, they also affect the distribution of animal taxa.

11.2.1 Euro-Siberian Phytogeographical Region

This region stretches alongside the Black Sea and covers nearly the whole region, from seashores to peaks of mountains, as a narrow strip. The eastern part of the Euro-Siberian region has higher precipitation and it is accepted as a subprovince of the region, called as the Colchic sector. Melet stream (in Ordu) lies at the west most border of the Colchic sector in Turkey (Davis, 1971). The area is predominantly covered by mesophytic vegetation of deciduous and conifer forests or mixed forests of various taxa. The area has

Table 11.1 Species Number of Living Organisms in Turkey*

Organisms	Known Species	Endemism (%)
Seed plants	10 150	31
Mosses	942	1
Ferns	85	–
Algae	~5000	?
Fungi	2158	?
Lichenes	1588	–
Mammalia	197	3
Birds	420	0.5
Reptiles	130	13
Amphibians	28	57
Freshwater fishes	371	16
Marine fishes	449	–

*References are given in the text. Species numbers are lowests.

sclerophylic macquis vegetation at the slopes facing the Black Sea and the vegetation is called as pseudomacquis (Akman, 1995). The eastern part of the region is more humid than the middle one and the highest precipitation is recorded in Rize. The Karadeniz mountain range has important peaks of Turkey, and the Kaçkar peak (3932 m) is the highest among them. The Küre Mountains are one of the most important biodiversity reservoirs in the Central Black Sea region, and the area is covered with both conifer and deciduous forests with openings as pastures.

11.2.2 Mediterranean Phytogeographical Region

The Mediterranean region includes a small part of the Thracian region, Aegean area and Akdeniz region of Turkey. In respect of phytogeography, all of these areas belong to the Eastern Mediterranean province. The Taurus Mountain range is the most prominent topographical structure in the area. Unlike, low altitudes and extensive agricultural areas characterize the Taurus Mountain range, Thracian part, and the Aegean area.

The Mediterranean region is characterized by macquis and red pine forests from seashore to 1000 m elevations. After these elevations, cedar, fir, juniper, and black pine forests cover the areas up to upper zones of the forest. The most common main rock of the area is limestone followed by outcrops of serpentine that has many endemic plant taxa. Kaz Mountain (1774 m) is the most prominent elevation of the Aegean area and it hosts endemic *Abies nordmanniana* subspecies *equatrojanii, Galanthus trojanii*, etc. The Taurus Mountain range has several important peaks with plenty of endemic plant species, for instance, Babadağ (2308 m) which has charismatic plant species of *Sternbergia candida*. The other peaks are Beydağları (3069 m), Akdağ (3014 m), Tahtalıdağ (2375 m), Bolkar Mountain (Medetsiz 3524 m), and Aladağlar (Demirkazık 3688 m). The Amanus Mountain range, as part of the Meditteranean area, has its own characteristic plants and enclaves of the Euro-Siberian flora, including *Fagus orientalis*.

Alongside the high peaks of the Taurus mountain range, the area has several important plains harboring many endemic taxa. In addition to that, wetlands and agricultural areas support the biodiversity, both native and agricultural. One of the largest of them is the Çukurova plain situated in the area of Mersin and Adana where cotton, wheat and other crops are extensively grown. The Antalya plain is another important area that is famous for citrus plantation and greenhouse cultivation of tropical plants. The other plains lying west of İzmir, Aydın and Muğla are important agricultural areas of grape, fig, citrus and other fruits of horticulture.

11.2.3   Irano-Turanian Phytogeographical Region

The Irano-Turanian area is the largest of the three regions and it is composed of mostly plateaus and high mountains, especially eastwards. The boundary of the area is mostly covered by forests, both within the Euro-Siberian and Mediterranean regions, while various types of steppe vegetation cover the inner parts and the eastern parts. The two largest lakes of Turkey lie in the region, one of them is Tuz Lake which harbors many endemic halophytic plant taxa. Several well known volcanic mountains can be found in the area (Erciyes Mountain 3917 m, Hasandağı 3268 m, Nemrut Mountain 3050 m and Ağrı Mountain 5122 m). Kızılırmak, Tigris, Euphrates and several other river systems contribute to habitat diversity in the area.

Extensive elevation range, various types of main rocks, such as gypsum, serpentine, lakes, ponds and river systems in the area compose together a very rich habitat diversity. Consequently, various vegetation types such as steppe, scrub or forests of deciduous, conifer, alpine grasses and many others survive in the area. Dry areas, especially in the Iğdır and Harran plains, are covered by steppe or halophytic vegetation. Beside this, recently established dams are slightly changing the climate of the Harran plain and the other areas. Extensive irrigations create salinity problem in the fertile soil, especially in the Harran plain.

11.3   Historical Background and General Features of Turkey

11.3.1   Geology

The Tethys Sea is an important area of geologic history for Turkey as well for the world. The sea lies between the Laurasia and the Gondwana at the end of Paleozoic epoch (Erinç, 1982). While some parts of the land around İstanbul belong to Silurien and Devonien epoch, Karadeniz Ereğlisi, Zonguldak, parts of Amanos Mountain and Diyarbakır belong to Paleozoic epoch (Ketin, 1966). Fossil specimens of the genera *Lepidodendron, Sigillaria, Annullaria,* and *Sphenophyllum* were determined from the cool beds belonging to Carbonifer epoch (Meriç, 1982). Approximately 125 million years ago, the Tethys Sea started to regress and this regression resulted in the formation of a land area (Şengör and Yılmaz 1981). Alongside the land formation of Turkey, flora and fauna were evolved during the same course of time.

The Central Anatolia is a wide plateau, which was originally formed during the Miosen period, and due to volcanic activities, its formation continued during the Pliosen period (Atalay, 1987). The Miosen flora of Turkey involved *Subtriporopollenites simplex, Intratriporopollenites instruetus,* and various species of *Castanea, Quercus* and Fagaceae. The members of Poaceae, Apiaceae, Ephedraceae and Asteraceae families were also found abundantly (Kayseri and Akgün, 2010).

Glaciations severely affected and destroyed flora of northern countries. In Turkey, only high mountain altitudes were affected which is regarded as another reason for floristic richness of Turkey in comparison with European countries. For example, the existence of *Sequoiadendron giganteum* and *Pseudodioon akyoli* is only known from their fossils found in Turkey (Erdei et al., 2010). However, distribution areas of some relict endemic taxa such as *Liquidambar orientalis, Flueggea anatolica* and *Eryngium thorifolium* declined by the time.

The Asian elephant (*Elephas maximus asurus*) is known to have lived near lakes and wetlands around Hatay by the first half of the 1st century BC. In the valley of the Fırat (Euphrates) and Dicle (Tigris) Rivers, the wild ox (*Bos primigenius*) has been known to live around the same period as well as various other parts of Anatolia. The Anatolian wild ass (*Equus hemionus anatoliensis*) has been reported to have lived around Fırat and Karasu until the end of the 12th century. The Asiatic lion (*Panthera leo persica*) was seen in Birecik around the Fırat valley in the 19th century. Cheetah (*Acinonyx jubatus*) also lived in Southeastern Anatolia by the 19th century. According to a report in 1970 a Hazar tiger (*Panthera tigris virgata*) was hunted in Southeastern Anatolia for the last time. Anatolian leopard, *Panthera pardus* is another big cat, which has not been seen for a long time (UN Report of the World summit on sustainable development, Johannesburg, 2002 in Tekeli et al., 2006).

11.3.2 Fertile Crescent and Turkey

The Fertile Crescent is known as the cradle of civilization and the land is regarded as the birthplace of agriculture, urbanization, writing, trade, science, history and organized religion in the Middle East. The Tigris-Euphrates river system supports all kinds of life activities. Both of the rivers have many tributaries that originate in the high mountains of Turkey and empty into the Persian Gulf. The largest plane of Turkey, the Harran Plain, is a part of the Fertile Crescent in Turkey and it is an important genetic diversity center of wheat. Wild wheat and its parents, *Aegilops* taxa, naturally grow in this area.

11.3.3 Topography

The Taurus Mountain range creates high rate differentiation in Turkey in the Mediterranean Region. Alongside extreme peaks, it has two important plains called Antalya plain and Çukurova plain. Endemism rate of plant species in the Mediterranean region is very high due to habitat richness based on topography. Similarly, the Black Sea Region also has mountain ranges, creating extremely rich habitat variations. Especially, the northeast part of the country hosts important peaks and ice lakes. Several mountains lie in the east; the highest one is Ararat Mountain (5122 m) representing the region. As an extension of the Zagros Mountains into Turkey, several important peaks represent SE Turkey with ice lakes. Additionally, several volcanic mountains stand in the central part of the country (e.g., Erciyes Mountain 3917 m). The ratio of the field that is higher than 1000 m is 56%.

11.3.4 Climate

A large part of Turkey is dominated by the Mediterranean climate regime. According to the Koppen-Geiger climate classification system (Kottek et al., 2006), six different climates influence the country. The main one is warm temperate climate with hot and dry summers (Csb) in the western and southern parts of Turkey. Warm temperate with dry and cool summer (Csc) in Central Turkey, warm temperate with fully humid and cool summers (Cfc) in the Black Sea region, snowy, fully humid and warm summer (Dfb) in the Northern East Turkey and snowy, dry and warm summer (Dsb) in the Central East part and the East, and arid, dry summer and cold arid (Bsk) in Central Turkey. In Turkey, the last 35 years (1981–2016) average rainfall is 574.0 mm (TSMS, 2017), but based on diverse climate pattern, precipitation varies greatly by location from nearly 240 mm in eastern Iğdır (in 2008) to 1920 mm in Rize (in 2001).

Based on various reasons such as topography, effects of the surrounding sea and wind, Turkey has rather complex climate systems and its subgroups. However, the following three main groups can be classified as basic climate types in Turkey.

11.3.4.1 Terrestrial Climate

It is characterized by dry and hot summers and cold and snowy winters. Most of the Irano-Turanian area falls under this climate regime. The least annual precipitation per square meter is observed in this area, namely Aralık

(197 mm/m^2), lying in Iğdır province in Turkey. Vegetation period in the areas with terrestrial clime is very short.

11.3.4.2 Mediterranean Climate

It is characterized by hot and dry summer and warm and rainy winters. Akdeniz, Aegean, and partially Marmara regions are under this climate. A wildfire of red pine forests in the Mediterranean area is an important problem threatening the biodiversity. The second location with low annual precipitation can be found in Taşburun (225 mm/m^2) situated near Mersin in Turkey.

11.3.4.3 Oceanic Climate (Locally the Black Sea Climate)

It is characterized by heavy rains without a drought period throughout the year. The climate in the area is cool in summers and warm, cold and rainy in winter. Rize situated in the Black Sea region is the rainiest province in Turkey, with 1920 mm/m^2 recorded in 2001.

11.4 An Overview of the Ecosystems in Turkey

Based on extensive topography and climate, natural ecosystems in Turkey are categorized under 11 titles. These vegetation groups are characterized by dominant plants and climate of the regions.

11.4.1 Steppe

Nearly 70% of Anatolia was covered by forests 4,000 years ago; steppes only covered a small area near Salt Lake (Çalışkan & Boydak, 2017). Today, the steppe fauna has taken over Anatolia due to increasing deforestation. The steppe vegetation is characterized by mostly herbaceous plants composed of chamaephytes, cryptophyte, hemicryptophyte, and terophyte that are adapted to dry areas. It is located in Turkey in Inner Aegean, Central and East Anatolia. Altitudes in these areas vary from 350 m (in Harran) to 1750 m (in Erzurum). While physiognomy of the vegetation during spring and early summer is green, then they fastly turn into grey due to drying of the plants in the beginning of the summer (Figure 11.1). Summer drought is an important characteristic of this vegetation type. Plant biodiversity in the steppe is quite high due to less competition among plants.

Habitat variation in steppe areas in Turkey is very rich due to the main rocks, water and climate variations. The area surrounding Salt Lake in Cen-

Figure 11.1   A view from steppe with wheat fields.

tral Anatolia is characterized by halophytic plants (Figure 11.2). It supports several endemic plant taxa, such as *Kalidium wagenitzii, Ferula halophila, Gladiolus halophilus, Acantholimon halophilum, Cousinia birandiana, Allium scabriflorum* and *Sphaerophysa kotschyana*, and the area has 42 endemic plant species. Among the endemic taxa, two of them are monotypic endemic genera, namely *Vuralia* (Uysal et al., 2014) and *Pseudodelphinium* (Vural et al., 2012).

A small size of halophytic vegetation can be found in various steppe areas across Turkey. Among them, the steppe area of Tuzluca in Turkey has many halophytic plants such as *Halanthium rarifolium, Salsola dendroides, Suaeda altissima, Camphorosma monspeliaca, Halothamnus glaucus, Halostachys belangeriana* and *Kalidium capsicum,* (Adıgüzel, 2005).

Steppe vegetation faces three major threats in Turkey, namely land clearing, overgrazing, and plantation. An extensive irrigation project in the Harran Plain has decreased steppe area in the region. Recent developments in rural life and intensive urbanization in Turkey especially around the cities, are decreasing natural steppe areas.

Steppe vegetation in Turkey has a rich biodiversity and especially the genera mentioned below are generously represented by many species in Turkey: *Astragalus, Verbascum, Salvia, Trigonella, Stipa, Bromus,* and *Thymus*.

Figure 11.2 *Salicornia* association at Salt Lake.

The steppe ecosystem of Turkey is habitat for many important species such as the Anatolian mouflon (*Ovis orientalis anatolica*) which is an endemic sub-species, caracal (*Felis caracal*), wolf (*Canis lupus*), Caucasian birch mouse (*Sicista caucasia*), mole (*Talpa europea*), European ground squirrel (*Spermophilus citellus*), and bird species under global threat in Europe like the great bustard (*Otis tarda*), lesser kestrel (*Falco naumanni*) and many other important species such as the short-toed snake eagle (*Circaetus gallicus*), buteo (*Buteo* spp.), falcon (*Falco* spp), Eurasian hobby (*Circus aeruginosus*), little bustard (*Otis tetrax*), hoopoe (*Upupa epops*) and quail (*Coturnix coturnix*) (Öztürk and Yerli, 2002; Tekeli et al., 2006).

11.4.2 Dry Deciduous Forests

Alongside steppe, dry deciduous forests are present in areas from Inner Aegean to East Anatolia at various altitudes. Dominant plant of this deciduous forest is at least one species of *Quercus*, but mostly two or more species grow together. This oak forest and environs of it include many different species of woody rosaceous plants which belong to *Prunus, Crataegus, Pyrus, Amygdalus, Cotoneaster, Sorbus, Amelanchier, Rosa,* and *Cerasus*. Moreover, the genera *Cistus, Rhamnus, Lonicera* and *Cornus* are also represented here with at least single species within the deciduous forest. Among them, *Pyrus eleagrifolia* and *Crataegus azarolus* rarely covers certain areas as pure stand within the forest. Around openings various species of *Trigonella, Centaura, Salvia, Poa, Minuartia, Astragalus, Silene, Bromus, Phlomis, Teucrium, Dianthus*, etc., can be found.

11.4.3 Dry Evergreen or Mixed Forests

These forests are composed of pine, juniper, mixed taxa of gymnosperms or angiosperms in the area of Inner Aegean and throughout the Irano-Turanian phytogeographical region. Native *Pinus nigra* forests are the most common in the area. Traditional planting practice of black pine has been carried out for the last 50 years. Some of the plantation areas have a native forest physiognomy. *Juniperus excelsa* is another important taxon in the dry evergreen or mixed forest, and it is possible to see its pure stands in Afyon and inside the Taurus Mountain range. These forests are rather poor in terms of biodiversity.

Most large mammals in Turkey live in forest ecosystems. Carnivorous mammals like bear (*Ursus arctos*), fox (*Vulpes vulpes*), jackal (*Canis aureus*), lynx (*Lynx lynx*) and hyena (*Hyena hyena*), herbivorous mammals like deer (*Cervus elaphus* and *Capriolus capriolus*), rupicapra (*Rupicapra rupicapra*), wild goat (*Capra aegaprus*) and wild boar (*Sus scrofa*), small mammals like wild cat (*Felis silvestris*), European badger (*Meles meles*), marten (*Martes foina*), porcupine (*Erinaceus concolor*), hare (*Lepus capensis*), weasel (*Mustela nivalis*) and squirrel (*Sciurus vulgaris*), reptiles like some snake species, chameleon (*Chameleo chameleon*), some lizards and turtles (*Testudo graeca, Testudo hermanni*) and also birds like pheasent (*Phasianus colchicus*), Caspian snowcock (*Tetraogallus caspius*), caucasian grouse (*Tetrao mlokosiewiczi*), woodpecker (*Dendrocopus major*), birds of prey (*Accipiter* spp.), cinereous vulture (*Aegyphius monachus*), eastern imperial eagle (*Aquila heliaca*), greater spotted eagle (*Clanga clanga*) and lesser spotted eagle (*Clanga pomarina*), *Circus* spp, *Buteo* spp, *Pandion haliaetus, Falco* spp, *Pernis apivorus*, various owls and many songbirds (ÇB, 2001; Tekeli et al., 2006).

11.4.4 Steppe of Mountains and High Plateaus

This vegetation type starts at the end of the forest zone (approximately 1,500 m) of the Irano-Turanian and Mediterranean phytogeographical regions (Figure 11.3). Vegetation period is shorter due to low temperatures and annual

Figure 11.3 High mountain steppe around Kars.

rainfalls. These areas are mostly covered by herbaceous steppe vegetation of chamaephytes, cryptophyte and hemicryptophytes. *Onobrychis cornuta, Astragalus angustifolius, Acantholimon acerosum, Festuca* spp. and *Bromus* sp. are widely distributed species in these areas. Alongside the herbaceous vegetation, various shrubs compose a patchy vegetation dominated by *Daphne* spp. and *Cerasus prostrata.* On the other hand, various plant species of chasmophytic vegetation are common in the alpinic zones, including *Silene odontopetala, Draba acaulis, Arenaria* and *Dianthus.* Due to limited vegetation period, topography and the other ecological factors, these areas have not been used for agricultural purposes (Figure 11.4).

11.4.5 Humid Deciduous Forests

These forests are composed of various angiosperm taxa and they grow throughout the Black Sea Region, which forms a part of the Euro-Siberian phytogeographical region (Figure 11.5). The forests range from sea level to 1500 m in various parts of the region. Beech (*Fagus orientalis*) is one of the most dominant plant species in this kind of forest and it occupies pure stands in the area. It also makes mixed forest together with various

Figure 11.4   *Fritillaria imperialis* from Hakkari where composed of high mountain.

Figure 11.5   Humid mixed forest from Artvin.

plant taxa, both deciduous and evergreen. Among the deciduous plants, oak (*Quercus*), hornbeam (*Carpinus*), wild apples (*Malus*), azalea (*Rhododendron*), maple tree (*Acer*), willow (*Salix*), hawthorn (*Crataegus*), rowan (*Sorbus aucuparia*), sweet cherry (*Cerasus avium*) and holly (*Ilex*) are the most common taxa with single or several species. Within the forest, various climbers such as *Hedera helix, Smilax aspera, Dioscorea communis* and *Clematis cirrhosa* are distributed throughout the region. These forest areas host various herbaceous plant species of *Ajuga, Doronicum, Arctium, Equisetum, Helleborus orientalis, Primula, Trifolium, Typha,* etc. Alongside the small streams, the region is covered by *Alnus glutinosa, Salix, Rubus* and the other taxa.

11.4.6 Humid Conifer Forests

Humid conifer forests grow in the Black Sea region and they are generously represented in the eastern part of the area. Fir (*Abies nordmanniana*) and spruce (*Picea orientalis*) are the dominant species of these forests. Black pine (*Pinus nigra*) and yellow pine (*Pinus sylvestris*) incorporate into the forests at various altitudes. While fir forest prevails from Giresun to Artvin at high altitudes, spruce forest creates pure stands around Kastomonu, Uludağ and the surrounding areas. Although mostly occupying high altitudes of the humid forest, it rarely grows naturally at seashores (Sinop and Bartın). These taxa compose pure stands and mixed forests with each other and with other deciduous plant taxa.

Owing to closed canopy and soil surface being covered by hardly decomposing leaves, biodiversity within the forest is rather poor. Various taxa of mushrooms and some parasitic or semiparasitic plants (*Monotropa hypopitys* and some orchid taxa) flourish the forest.

11.4.7 High Mountain Grasses

The areas above the forest zone and the large flatlands among the forests are covered by grasses composed of various herbaceous plants. These areas are floristically poorly known in Turkey and they are more sensitive to any kind of human activity and environmental threats. Higher humidity, high precipitation and underground water are basic reasons allowing these grasses to survive in Turkey. Flora of these areas is very similar to Caucasus and European alpinic grasses (Figure 11.6). Among these grasses can be found the taxa of *Carex, Poa, Sibbaldia, Aster, Alchemilla, Potentilla,* and *Geranium,* etc.

Figure 11.6 High mountain grasses at Palandöken from Erzurum.

11.4.8 Coastal Sand Vegetation

Turkey lies on a peninsula, which is surrounded by three seas in the north, west and partially south. Hence, the country has 8333 km of seashore, including islands. The coverage of sand vegetation is very weak. However, it has its own plant species adapted to sandy habitat. Largest sand vegetations can be found in the Mediterranean Region.

Sand vegetation is evaluated under two categories as herbaceous and shrubs according to their physiognomy. Herbaceous plants are represented by *Pancratium maritumum, Eryngium maritumum, Trigonella maritima, Juncus maritimus, Otanthus maritimus, Ipomoea stolonifera, Cakile maritima, Maresia nana, Phyla nodiflora, Juncus bufonius* and *Inula crithmoides*. As a shrub vegetation of sandy shores, it is possible to find small populations of *Thymelea hirsuta, Pistacia lentiscus, Rubus sanctus, Vitex agnus-castus*, and *Helianthemum stipulatum* at the sandy areas of the Mediterranean and Aegean shores.

11.4.9 Macquis

This vegetation is a characteristic component of the Mediterranean area in Turkey and is composed of evergreen or deciduous shrubs. The macquis extend alongside the whole Mediterranean basin, California, Chile, South Africa and Australia. Every region has its own name for this type of vegetation. The macquis in Turkey is dominantly composed of *Quercus coccifera, Spartium junceum, Olea europaea, Arbutus andrachne, Arbutus unedo, Euphorbia dendroides, Laurus nobilis, Pistacia lentiscus, Pistacia terebinthus, Sarcopoterium spinosum, Lavandula stoechas, Calicotome villosa, Cistus creticus, Myrtus communis, Ceratonia siliqua, Phillyrea latifolia, Sytrax officinalis* and *Erica manupuliflora* (Figure 11.7). Because the climate macquis has many geophyte species and Turkey is one of the richest countries according to the number of geophyte plants, many of them are endemic, such as *Muscari macrocarpum, Muscari muscarimi, Fritillaria forbesii, Crocus antalyensis, Galanthus peshmenii, Cyclamen persicum, Narcissus seratonius, Fritillaria hermonalis, Fritillaria latakiensis*, and *Gladiolus antakiensis*.

11.4.10 Mediterranean Conifer Forest

The Mediterranean region of Turkey is mostly covered by conifer forests composed of various taxa. *Pinus brutia* has the largest distribution area among the Mediterranean conifer taxa. Forest of the red pine in the area

Figure 11.7 Macquis at Kaş from Antalya.

starts from sea level up to 1000 m and above this altitude *Pinus nigra* or other conifer taxa grow (Figure 11.8). At lower altitudes, native *Cupressus sempervirens* stands occur very locally in Denizli, Antalya, Mersin and Adana (Anonymous, 2013). *Abies cilica* is another important conifer species in the Mediterranean region distributed at altitudes above 1000 meters and it is represented by two varieties. *Cedrus libani* is an important native species with an extensive distribution up the upper forest zone, 1800 m. Owing to be a high quality timber plant, it has been logged since antique periods for ship constructions in the region. *Juniperus drupacea* and the other juniper species (*J. oxycedrus, J. foetidissima* and *J. phoenica*) appear in various elevations among the other conifer forests.

Stone pine (*Pinus pinea*) is a native plant of Turkey and it has very a narrow distribution in the country, found at Kozak pasture (İzmir). Based on economical importance, it has been cultivated in Muğla, Balıkesir, Bursa, İstanbul in the west and in other parts of Turkey.

As an extension of the Taurus mountain range, Amanos Mountain is covered mostly by red pine and other conifer species. As an enclave of Humid Deciduous Forests in the Black Sea region, Amanos Mountain harbors *Fagus orientalis, Carpinus orientalis, Taxus baccata, Ilex colchica, Buxus sempervirens* and *Tilia argentea.*

Figure 11.8 *Cedrus libani* and *Abies cilicica* forest at Taurus Mountain.

11.4.11 Marine, Aquatic and Wetland Plants

Turkey has 8,333 km long shores excluding the islands and is surrounded by three different seas; namely the Black Sea, the Mediterranean Sea, the Marmara Sea which is an inner sea and the Aegean Sea each having different ecological properties (Figure 11.9). Turkey is also very rich in inland waters. Statistically, Turkey has 910,000 ha with 200 natural lakes; 373,000 ha dams and ponds; and 200,000 km of rivers in length. The total surface area of coastal (freshwater) lagoons is assumed to be more than 40,000 ha (Yerli, 2016).

The yearly average surface flow of Turkey's 25 catchment areas is 186 billion m³. Some rivers having lengths longer than 500 km and their names are; Kızılırmak (1,355 km), Fırat (1,263 km), Sakarya (824 km), Seyhan (560 km), Dicle (523 km), Yeşilırmak (519 km) and Ceyhan (509 km). The delta of Kızılırmak River, which discharges into the Black Sea, is very important for migratory bird species, which directly cross the Black Sea. Lagoon and deltas are important with their contribution to the biological diversity of species foremost aquatic bird species and their fertile soils. All of these rivers have fish habitats with high bio-geographical diversity (ÇB, 2001; Tekeli et al., 2006).

Figure 11.9   Regression of the Eber Lake.

Turkey has a great variation of aquatic and wetland plants (Figure 11.10). The lakes of various sizes host hundreds of plant species (Seçmen and Leblebici, 1997) such as *Nymphaea alba, Nuphar luteum, Potamogeton* spp. and *Trapa natans*. Beside the lakes, marshy areas with characteristic shallow water also host various plant species such as *Leucojum aestivum, Iris pseudoacorus, Butomus umbellatus, Juncus* spp., *Cyperus* spp., etc. Alongside rivers or smaller streams, *Tamarix* spp., *Myricaria* spp. and the other shrubs are widely distributed throughout the country. One of the species, namely *Liquidambar orientalis*, grows in humid areas in the Mediterranean region. The species is a relic endemic restricted only to Muğla and Aydın provinces very locally in Turkey.

The largest natural lake in the Eastern Anatolia region is Van Lake (374,000 ha). Tuz Lake is the second largest lake in Turkey and it is the largest salty shallow lake of the Middle Anatolian steppes (128,000 ha). This natural salt resource has a great economical significance (ÇB, 2001). Foremost, Van and Tuz Lakes and many other lakes along with them are under the threat of becoming polluted.

The wetlands of Turkey are important not only for species that breed and winter there but also vital for migratory species. Two of the three important migration routes in the world pass over Turkey (in between west Palearctic

Figure 11.10 Bays and Islands from Datça, Muğla.

and Africa). Every year more than 200,000 birds of prey fly over Çoruh River and Çoruh valley in the Eastern Black Sea region of Turkey and rest in the wetlands of the Eastern Anatolia region. This migration route is known as the largest birds of prey migration of the world in the west Palearctic region. Another important migration route is the Bosphorus migration, of over 250,000 storks in groups of 200–700, arriving from the western Black Sea and flying over Thrace and the Bosphorus advancing towards Anatolia. During these long migrations of birds, the wetlands of Turkey are vital (Tekeli et al., 2006).

On the other hand, there are some alien species threatening wetlands like the invasive fish species *Pseudorasbora parva*, *Carasssius* spp. etc.

11.5 Species Diversity in Turkey

Due to extensive habitat and climate diversity, Turkey has a rich fungal diversity. Studies on Turkish fungi have been focused on mostly Ascomycetes and Basidiomycetes due to their economic value and relatively cheaper research methods. One of the recent checklists for fungi reported (Sesli and Denchev, 2014) that 2,158 species of three main groups are naturally present in Turkey. The authors listed the taxa under the main names of myxomycetes, ascomycetes and basidiomycetes. In respect of fungal taxonomy, ascomycetes and basidiomycetes are represented in the list with their correct name and current classification. Besides this, the rest of the fungal taxa are listed under the name of myxomycete. The species number is 232 for myxomycetes, 215 for ascomycetes, and 1943 for basidiomycetes in Turkey up to the publication date of the checklist. The genera *Physarum* (25 species), *Arcyria* (18), and *Trichia* (12) are the largest genera of the myxomycetes. Many of the species of both ascomycetes and basidiomycetes have economic importance with edible species in Turkey. *Helvella* (21), *Morchella* (21), and *Tuber* (7) belonging to ascomycetes are commonly consumed by local people and *Agaricus* (37), *Amanita* (33), *Boletus* (34), *Clitocybe* (41), *Coprinellus* (16), *Cortinarius* (98), *Entoloma* (46), *Inocybe* (73), *Lactarius* (53), *Macrolepiota* (8), *Mycena* (59) and *Pleurotus* (13) are widely distributed genera in the country and they have economic importance since they are consumed as food (Figures 11.11–11.14).

The algal diversity of Turkey is known from limited researches of marine and inland ecosystems. Instead of higher taxonomic categories, descriptive names are used for various algae groups in this text for a better understanding. Macroalgae live in the marine ecosystems and 800 species live within the Turkish marine waterlands such as *Sargassum* and *Acetabularia*.

Figure 11.11 *Disciotis venosa* (Identified by E. Sesli).

Figure 11.12 *Lactarius piperatus* (Identified by E. Sesli).

Figure 11.13 *Fomes fomentarius* (Identified by E. Sesli).

Figure 11.14 *Coprinus comatus* (Identified by E. Sesli).

Alongside macroalgae, the other groups are called microalgae and they have approximately 300 species within Turkish seas. Soil algae are poorly known organisms in Turkey and it is estimated that they have 450 species in Turkey. Aerophytic algae lives on trees and aerial parts of other things and they are composed of 350 species. Freshwater algae are composed of two artificial groups: Diatoms and the other algae taxa and all of them have approximately 1,340 species in Turkey (pers. com. A. Şenkardeşler). *Pinnularia*, *Navicula* and *Cyclotella* are the predominantly known genera of Diatomes, whereas *Spirogyra*, *Scenedesmus*, *Cosmarium* and *Chara* are the most common genera of the other algae taxa in freshwater ecosystems in Turkey. On the other hand, based on the literature survey up to the date, approximately 5,000 algae naturally live in Turkey (Gönülol, 2017).

Moss flora of Turkey is extensively rich and it is composed of 942 species, 10 of them are endemic (Erdağ and Kürschner, 2017). Taxonomically they are treated under three divisions, namely Marchanthiophyta with 179 species, Bryophyta with 759 species and Anthoceratophyta with 4 species. Among them, three species of Marchanthiophyta and seven species of Bryophyta are endemic. Due to high annual precipitation, the Euro-Siberian Region is densely covered by various moss taxa, both various kinds of terrestrial habitats and aerophytic. Moreover, forest and scrub areas have many of the moss species which flourish during spring seasons.

As lower vascular plants, ferns are represented by 85 (pers. com. G. Kaynak) species in Turkey. Most of the taxa grow in the Black Sea region due to high annual precipitation. Among them, *Athyrum* is widely distributed and *Dryopteris* is represented by 13 species as the largest genus (Figures 11.15–11.16).

11.5.1 Gymnosperm Diversity in Turkey

Among 1,000 gymnosperm species (The Plant List, 2013), 22 species belonging to 8 native genera naturally grow in Turkey. They compose nearly half of the Turkish forests (Işık, 1987). Some of the taxa create pure stands of single species or mixed species, both deciduous taxa and conifers. Due to habitat diversity in Turkey, the gymnosperm taxa have extensive intraspecific diversity which leads to description of several new taxa.

- **Pinus:** The largest group of gymnosperm forests in Turkey is composed of pine species. These species represent pure stands, both deciduous and the other conifer taxa in Turkey. Alongside the native conifer forests, cedar, black pine, and yellow pine taxa have been

Figure 11.15 *Ophioglossum lusitanicum.*

Figure 11.16 *Asplenium trichomonas.*

planted throughout Turkey in recent years. The most widely distributed species in Turkey are *P. brutia, P. nigra, P. sylvestris, P. pinea* and *P. halepensis*. All of the pine species have various value and usage, for example as timber, for ornamental, obtaining resin, in honey and food (*P. pinea*) industry (Figures 11.17–11.18).

- **Abies:** Fir is represented in Turkey by two species, *A. nordmanniana* is distributed in the north while *A. cilicica* is distributed in the south. *A. nordmanniana* is divided into three subspecies, namely *A. nordmanniana* subsp. *nordmanniana*, subsp. *bornmuelleriana* and subsp. *equa-trojanii*. The latter is endemic to Kaz Dağı (Ida Mountain) and the epithet "equa-trojanii" denotes its historical name.

- **Juniperus:** The genus *Juniper* is represented by 7 species in Turkey. They mostly grow in various forests or scrubs. *Juniperus excelsa* rarely creates of pure stands in various parts of Turkey. Fruits of junipers are traditionally used for treatment purposes by local people (Baytop, 1999).

11.5.2 Angiosperms in Turkey

Asteraceae is the largest family in Turkey (Davis 1965–1985; Davis et al., 1988) based on the number of species (approximately 1400) and Fabaceae comes second with approximately 1070 species. In respect of endemism,

Figure 11.17 *Taxus baccata.*

Figure 11.18 *Abies nordmanniana.*

Turkey does not have an endemic family. Furthermore, the diversity of Brassicaceae is the highest in the USA. However, in Turkey, it holds second place with 571 species (Al-Shehbaz et al., 2007). Poaceae and Apiaceae are represented by many genera, 142 and 100 respectively, in Turkey. In respect of intrageneric diversity or economic importance, Turkey hosts many important plant genera (Table 11.2). The table prepared according to the Checklist (Güner et al., 2010) with some corrections and additional literature.

- **Triticum:** Wheat is represented by 10 species, 6 of them are cultivated in Turkey. *T. aestivum* is extensively cultivated in Turkey as well as in other parts of the world. Moreover, parents of the wheat naturally grow in Turkey. Wheat consists of a polyploidy series; diploid (*T. baeoticum, T. monococcum*), tetrapoid (*T. timopheevi, T. dicoccoides, T. dicoccon, T. durum, T. turgidum, T. carthlicum*), and hexaploid (*T. aestivum*). The ancestors of the all ploidy series have natural representatives with local names and usages in Turkey. Based on the archaeobotanical findings, one of the earliest wheat cultivation of *T. monococcum* and *T. dicoccum* in the Fertile Crescent of Western Asia started at Caferhöyük (Malatya) in eastern Anatolia about 9,000 years ago (de Moulins 1997).

Table 11.2   Vascular Plant Genera Have the Most Species Diversity in Turkey

Genera	Species/Endemic (E)	Subspecies/E	Variety/E	Taxon/E
Astragalus	444/197	17/11	13/6	474/214
Verbascum	343/186	5/0	25/10	373/200
Allium	179/70	12/3	2/2	193/75
Centaurea s.str.	166/100	22/9	13/7	211/116
Silene	145/75	15/6	7/4	165/75
Campanula	123/63	6/4	3/2	133/69
Hieracium	115/70	0/0	0/0	117/70
Trifolium	105/13	5/0	29/0	139/13
Galium	106/55	14/5	2/0	122/6
Alyssum	101/55	5/2	5/1	111/60
Salvia	100/53	4/0	3/3	107/57
Dianthus	76/44	1/1	8/8	77/53

- **Hordeum:** Barley is represented by seven species; two of them are cultivated in Turkey. Due to the economic importance, they are cultivated mostly in Central Turkey.
- **Allium:** The genus is represented by 179 species and several of them such as onion, garlic and parsley are widely cultivated. Moreover, several wild species are used by local people as spice or vegetable (Baytop, 1999).
- **Tulipa:** The genus is represented by 17 species in Turkey (Eker et al., 2014). All of the species are used as ornamental plants and they are widely cultivated in botanical gardens worldwide. Tulip cultivation was a major tradition at the Ottoman Palaces (Baytop, 1998), and, one of the Ottoman periods is called "the Tulip era". Despite of the general belief of people that tulips originated in the Netherlands, they were in fact imported there from Turkey, just three hundred years ago.
- **Galanthus:** Snowdrop is represented by 13 species in Turkey, 4 of them are endemic. Due to various conservation measurements and legislation, native populations are under conservation. Snowdrop is the most commonly exported plant from Turkey (12 tonnes in 2016 from which 7 tonnes came from nature and 5 tonnes came from cultivation; Resmi Gazete, 2016).
- **Crocus:** The genus is represented by 44 species in Turkey, flowering both in spring and autumn. Among the taxa, *Crocus biflorus* has

extremely intraspecific variation with the accepted 10 subspecies (Mathew, 1984). Saffrron (*Crocus sativus*) was extensively cultivated during the Ottoman period as a spice. Today, its cultivation is restricted to a limited area in Safranbolu (Karabük) district (Figures 11.19–11.20).

- **Ranunculus:** Buttercup plants are widely diversified with 81 species. They do not have remarkable economic value. However, in respect of biodiversity, they grow in all types of habitats in Turkey with many infraspecific taxa, including marshy places and shallow aquatic habitats.
- **Dianthus:** The genus is represented by 76 species and they naturally grow mostly in steppe, forest opening and rocky places in Turkey.
- **Hypericum:** Specimens of the genus is popular in folk medicine for the treatment of various health problems. The genus is represented by more than 70 species with several subspecific taxa, mostly growing in steppe areas.
- **Astragalus:** It is the largest genus found in Turkey, growing throughout the whole country. The species are the most common plants of steppe and high mountain openings. Some of the species are traditionally used in gum industry and the roots are used as fodder during harsh winter periods.

Figure 11.19   *Crocus sativus.*

Figure 11.20 *Crocus ancyrensis,* an endemic species.

- **Cicer:** This small legume genus is represented by 11 native species in Turkey. *Cicer arietinum* is the most important legume plant used for commercial purposes (Figure 11.21).
- **Prunus:** Wild plum is distributed in every part of the country and local people traditionally use the fruits as natural sour. *Prunus cocomilia* grows in the *Quercus* scrub and in the openings and it has extensive variation of fruit sizes, colors and tastes.
- **Amygdalus:** Wild almond is extensively cultivated for its fruits and for ornamental purposes. Among the native species, *A. orientalis* is the most common, especially in Central Turkey. One of the most cultivated forms is called *Datça Bademi,* characterized by easily broken shell.
- **Cerasus:** Wild cherry is common in the Black Sea region in deciduous forest and the epithet of *Prunus cerasus* denotes of the province of Giresun in the area. Both of the species, *Cerasus vulgaris* and *Cerasus avium*, are important crop plants with extensive cultivation.
- **Rosa:** Rose is one of the most common ornamental plants in Turkey with hundred cultivated forms in gardens. Fruits of wild species, especially *Rosa canina,* are used for the production of jam and tea and for other purposes. Oil of *Rosa damascena* is used commercially and it has a very long tradition in Isparta and around.

Figure 11.21   *Cicer uludereensis,* an endemic species.

- **Crataegus:** Turkey is one of the important diversity centers of the genus and fruits of some species are used as food. Cultivation of valuable fruits of *Crataegus azarolus* var. *aronia* is becoming more popular in Hatay province (Figure 11.22).
- **Pyrus:** Wild pear species are widely distributed all over Turkey and they have local usages. Cultivation of *Pyrus communis* for its fruit and ornamental purposes is very common. *P. elaeagrifolia* is used by local people as rootstock for grafting.
- **Alyssum:** The genus has extensive diversity in Turkey and it is represented by 99 species, out of which 55 are endemic. Some of the endemic species are very local and some are subject to nickel accumulation on serpentine soils (Reeves et al., 2001).
- **Cyclamen:** All of the native species are used for ornamental purposes and their collection from nature is under control. The genus is represented by 11 in Turkey. The species are distributed in the Mediterranean and Euro-Siberian regions of the country.
- **Salvia:** Sage is one of the most important medicinal plants in Turkey (Figure 11.23). It is especially consumed as tea. Beside sage, various species of *Sideritis* and *Origanum* are also used as spice and tea.

Figure 11.22 *Crataegus tanacetifolia*, an endemic species.

Figure 11.23 *Salvia hedgeiana*, an endemic species.

- **Thymus:** Thyme is one of the most prominent species in steppe areas in Turkey. Due to essential oil content, nearly all of the species have been collected by local people for medicinal purposes. It is also used as spice.
- **Verbascum:** It is the second largest genus in Turkey, represented by 343 species, out of which 200 are endemic. Specimens of the genus has not economic value for local people, with the exception of occasional medicinal usage (Baytop, 1999). The species of the genus are commonly hybridizing with each other according to Huber-Morath (1978), and he reported 114 hybrid species from Turkey. These proposals for hybridization among *Verbascum* species require further researches.

Generic endemism is an important criterion for evaluation of the biodiversity in any area. Turkey has 15 endemic plant genera (Table 11.3) and most of them are monotypic (Figure 11.24). Nearly all of these endemic genera are local endemic and they require conservation measures. Asteraceae is the largest family in respect of species richness, both in Turkey and worldwide. However, it does not have any endemic genera in Turkey.

Table 11.3 Endemic Plant Genera of Turkey

Taxon	Family	Distribution
Nephelochloa orientalis	Poaceae	Muğla, Uşak, Denizli, Kütahya, Afyon
Pseudophleum gibbum	Poaceae	Manisa, Uşak, Denizli, Isparta
Oreopoa anatolica	Poaceae	Antalya
Pseudodelphinium	Ranunculaceae	Konya
Tchihatchewia isaetida	Brassicaceae	East part of Turkey
Physocardamum davisii	Brassicaceae	Ağri, Erzurum
Phryna ortegioides	Caryophyllaceae	East part of Turkey
Thurya capitata	Caryophyllaceae	Niğde, Amanus Mt.
Cyathobasis fruticosa	Amaranthaceae	Central Turkey
Sartoria hedysariodes	Fabaceae	Antalya
Vuralia turcica	Fabaceae	Konya
Dorystoechas hastata	Lamiaceae	Antalya
Aegokeras caespitosa	Apiaceae	Western Turkey
Microsciadium minutum	Apiaceae	Western Turkey (Lesvos and Kos Islands)
Microsciadium tenuifolium	Apiaceae	Western Turkey
Ekimia bornmuelleri	Apiaceae	Burdur

Figure 11.24   *Tchihatchewia isaetida*, an endemic genus.

Only genera were published and considered as endemic in Turkey in the past. However, with further taxonomic studies, they lost their endemic genera status. For example, *Necranthus orobanchoides* (Orobanchaceae) was a taxon known as endemic in the past. Today, the genus is reduced to synonym of *Orobanche gamosepala* (Teryokhin, 2001). A recent study of the genus by Zare (PhD thesis-unpublished) (2012) supports its taxonomic evaluation under the genus *Orobanche*. *Leucocyclus formosus* (Asteraceae) was a narrowly distributed endemic genus and it is now accepted as a species of *Achillea* with the name of *A. formosa*. *Crenosciadium siifolium* (Apiaceae) was an endemic genus confined to Central Anatolia is accepted as a species of *Opopanax* with the name of *O. siifolia*. *Kalidopsis wagenitzii* (Amaranthaceae) which was formerly known as an endemic genus is now accepted as a species of the genus *Kalidum, K. wagenitzii* (Kadereit et al., 2006). *Olymposciadium caespitosum* was an endemic taxon at a generic level. Evaluation of the taxon with the name of *Aegokeras caespitosa* has not been changed its generic endemism status and it is given in Table 11.3.

11.5.3 Fish

Fish biodiversity in inland waters of Turkey consists of 371 species that belong to 27 families. More than half of the existing species were the members of the Cyprinidae family. Nemacheilidae, Gobiidae, Cobitidae, and Salmonidae were also rich families by species (Kuru et al., 2014). The Cyprinidae family is represented by 193 species (52% of the total fish fauna). In addition, Nemacheilidae (11%), Gobiidae (7%), Cobitidae (5%) and Salmonidae (5%) have rich species in number. In terms of protection status of fish fauna as EN (Endangered), VU (Vulnerable), CR (Critically endangered), 100 species are in the IUCN Red List. The listed species within 29 species are in the Bern List (in Appendix III) and six species are in both the Bonn (for migratory species) and CITES (Appendix II) lists (for the endangered species).

Three species (Tarek, common carp and sand smelt) dominate the catches (26% of total inland production in 2015 (TUIK, 2016). Tarek (*Alburnus tarichii*) are primarily caught from Lake Van in eastern Anatolia where there are soda lakes. Common carp (*Cyprinus carpio*) and sand smelt (*Atherina boyeri)* are found in inland waters. Sturgeons (*Acipencer* spp.) are among the most endangered species living in the Black Sea and rivers surroundings on it. *Salmo trutta labrax* is an endemic subspecies. Striker, *Cyprinion macrostomus*, and licker, *Garra rufa*, are used for the treatment of psoriasis and skin diseases. These species live in the hot springs and pools in Kangal (Sivas) and in the basins of Eurphrates, Tigris and Orontis Rivers.

Turkey seas have rich fish diversity. Turkey seas have 449 fish species, roughly (Evirgen, 2007). Anchovy, mackerel, European pilchard and bonito have been represented by large populations. Golden grouper, *Epinephelus alexandrines*, and dusky grouper, *Epinephelus guaza*, are the species whose populations are declining due to over and illegal fishing.

11.5.4 Amphibia

There are 14 species belonging to Urodela (frog with a tail) and a family of the Salamandridae and 14 species of five families from Anura (frog without a tail) taxon in Turkey. In terms of protection status of amphibian fauna as EN (Endangered), VU (Vulnerable), CR (Critically endangered), 10 species are in the IUCN Red List, respectively (Yerli, 2012). *Rana holtzi* (the Taurus frog) is a mountain frog and its length is 7–8 cm. It is an endemic and protected species that only lives in Turkey, Karagöl, which is 2500 m in altitude on the top of the Bolkar Mountains (Öztürk and Yerli, 2002). *Bombina bombina, Lissotrion vulgaris, Neurergus strauchii, Pelophylax ridibundus* and *Rana tavasensis* are some endemic amphibians (Baran, 2005; Yerli et al., 2012).

11.5.5   Reptilia

There is a rich diversity of reptiles in Turkey with 130 species. Five species of two families live in freshwater, three species of one family in the sea, and three species of one family on the ground from the taxa of turtles (Ordo Testudinata). There are two taxa of squamous (Ordo Squamata), as lizards (Subordo Lacertilia) and snakes (Subordo Ophidia) (Öztürk and Yerli, 2002; Yerli et al., 2012).

The two species of snakes that inhabit the regions of the Black Sea and southeast Anatolia are found only in Turkey. *Pelias barani* has the length of 55 cm and it is entirely black. This species lives in shrubs and is endemic in Turkey. *Montivipera wagneri* has the length of 90 cm. It lives among rocks with little vegetation, the rocky sides of mountains and plateaus (Öztürk and Yerli, 2002; Yerli et al., 2012).

Suphan lizard, *Eremias suphani*, which lives in sandy and stony places, is an endemic species. It is distributed only around the cities of Van and Bitlis. *Blanus strauchi, Montivipera xanthina, Apathya cappadocica, Anatololacerta danfordi Crocodura arispa* and *Eirenis barani* are some endemic species of Reptilia (Öztürk and Yerli, 2002; Baran, 2005).

The loggerhead turtle *Caretta caretta* and the green turtle, *Chelonia mydas* regularly nest along Mediterranean coasts of Turkey, *Dermochelysis coriacea* very rarely nests in the Turkish coasts. There are 21 major nesting beaches on the Aegean and Mediterranean coasts.

11.5.6   Birds

Turkey located on the big migration routes between Asia, Europe and Africa is a key area for many bird species. Every year, depending on the season, millions of migrants fly through the country. Thus, species diversity is high as a number of 420 species (Yerli et al., 2012).

In the eastern Anatolia, on the north-south migration route, birds of prey can be observed. Every year, more than 200,000 of prey fly over Çoruh River and Çoruh valley in the Eastern Black Sea region of Turkey and rest in the wetlands of the Eastern Anatolia region (Tekeli et al., 2006).

In the west, the main migration route of storks and birds of prey is Bosphorus in İstanbul. Another important migration route is the Bosphorus migration, of over 250,000 storks in groups of 200–700, arriving from the western Black Sea region and flying over Thrace and Bosphorus advancing towards Anatolia. During these long migrations of birds, the wetlands of Turkey are vital (Tekeli et al., 2006). In the middle part of the country, a part of the Siberia-Africa migrants can be observed on the North-South direction. The Kızılırmak Delta shelters more than 320 species of birds, The Göksu Delta in the Mediterranean coast shelters nearly 332 species of birds in the Black Sea Coast (Öztürk and Yerli, 2002).

A total of 420 species of birds can be found in Turkey (Yerli et al., 2012). All of these species are protected by national laws and international agreements (Figures 11.25–11.26). Some of these species are globally endangered (e.g., the great bustard *Otis tarda*). Stork, flamingo, spoonbill, avocet, heron and ducks are the most common birds found in the wetlands of Turkey (Öztürk ve Yerli, 2002).

Many endangered species (Dalmatian pelican, marbled duck, pygmy cormorant, white-headed duck, slender-billed curlew, bittern, lesser white-fronted goose, red-breasted goose and ferruginous duck species etc.) breed in Turkey. The western swamphen (*Porphyro porphyro*) populations habituated in the Mediterranean region, are decreasing in number and they mostly breed in the Göksu delta in Turkey with a high percentage (Tekeli et al., 2006).

The night heron, *Nycticorax nycticorax*, is now very rare in some regions. Cattle egret, *Bubulcus ibis*, colonies use the islands of Lake Van as nesting areas. The grifton vulture, *Gyps fulvus*, lives in the Taurus Mountains and

Figure 11.25   Common buzzard, *Buteo buteo* (photo by Cem Orkun Kıraç).

Figure 11.26 *Cinereous vulture, Aegypius monachus* (photo by Cem Orkun Kıraç).

Middle Anatolia. Their number has now decreased significantly. The black vulture, *Aigrupus monachus*, is rarely seen in Anatolia. The white stork, *Ciconia ciconia*, migrates through Turkey and they can be seen in the hundreds of thousands while they use the İstanbul Strait as a migration route. (Öztürk and Yerli, 2002).

There are 357 bird species that overwinter and breed in Turkey. Threatened species number is 35 with nearly 10%. Some of them are *Ketupa zeylonensis* (CR), *Ammomanes deserti* (CR), *Aquila nipalensis* (CR), *Oxyura leucocephala* (EN), *Geronticus eremita* (RE), *Haliaeetus albicilla* (VU), *Ceryle rudis* (EN), *Aythya marila* (VU), *Vanellus vanellus* (VU), *Alcedo atthis* (VU) and *Neophron percnopterus* (EN) (Birdlife International, 2015).

11.5.7 Mammals

Many species of terrestrial mammals that originated in Europe, Asia and Africa entered into Anatolia in the past 50,000 years. A total of 197 species of mammals can be found in Turkey (Yerli et al., 2012). Some species of insectivorous like mole (*Talpa europaea*), Caucasian shrew (*Sorex caucasicus*) and toothed shrew (*Crocidura leucodon*) can be seen everywhere. Bats (Chiroptera) usually live in caves. The Egyptian fruit bat (*Rousettus aegyptiacis*) lives on the Mediterranean. Bechstein bat (*Myotis bechsteinii*) and Leisler bat (*Nyctalus leisleri*) are rare species. Caucasian pine vole (*Microtus majori*), one of the rodents, forest dormice (*Dryomys laniger*) and spiny mouse (*Acomys cilicicus*) are endemic species (Öztürk and Yerli, 2002).

The fox (*Vulpes vulpes*) and beach marten (*Marten foina*), brown bear (*Ursus arcos*), lynx (*Felis lynx*), caracal (*Felis caracal*), wildcat (*Felis silvestris*), Wolf (*Canis lupus*), hyena (*Hyaena hyaena*), badger (*Meles meles*), jackal (*Canis auereus*), big poppy (*Mustela erminea*) and small poppy (*Mustela nivalis*) are some mammal species of Turkey. Otter (*Lutra lutra*), a semi aquatic species is threatened species. A type of wild sheep (*Ovis gmelinii anatolica*) is under protection in Central Anatolia. This species is a typical member of steppe ecosystems (Öztürk and Yerli, 2002).

The seas of Turkey are rich in marine mammal (Cetacean) diversity, especially in whales and dolphins, namely *Delphinus delphis*, *Tursiops truncates*, *Phocoena phocoena*, *Stenella coeruleoalba*, *Globicephala melas*, *Grampus griseus*, *Pseudorca crassidens*, *Physeter catodon*, *Ziphius cavirostris*, *Balaenoptera physalus* (Öztürk and Yerli, 2002). The Aegean Sea is also very important for the endangered Mediterranean Seal (*Monachus monachus*).

11.5.8 Invertebrata

Anatolian fauna also catches the eye with insect diversity of over 90,000 species (roughly estimation) Kuru (1987) predicted 1500 invertebrate species for inlands. There are 5,000 and 1,700 known animal species in the Mediterranean and Black Sea, respectively (Kocataş et al., 1987). Despite the lack of data, invertebrates constitute the largest group among the identified living species. Eken et al. (2006) reported 345 species of butterfly, 85 species of Odonata. The total number of invertebrate species in Turkey is about 19,000 of which about 4,000 species/subspecies are endemic (www. iucn.org/content/biodiversity-turkey).

11.6 Conclusion

High biodiversity in Turkey would be attributed to topography, climate, geographical location and evolutionary processes of million years. Similarly, high biodiversity in the terrestrial areas, marine systems, and shores in Turkey host extensive biodiversity with various threats to them. Similar to the world biodiversity crisis, Turkey has many important threats to biodiversity (Şekercioğlu et al., 2011). The following threats to the Turkish biodiversity and conservation measurements are highlighted here to educate the next generations and to save the nature in the country.

1   Uncontrolled or inadequately controlled chemicals and fertilizers used for agricultural and horticultural purposes. By the food chain in habitats, the natural biota has been decreasing in terms of the species richness.
2   Habitat loss in steppe, wetland, aquatic and the other areas for urbanization, farming and the other purposes. Despite both national and international legislations, all kinds of habitats in Turkey are under extensive habitat loss pressure (Figures 11.27–11.28).
3   Dams and hydropower plants are extensively put in action during the recent 20 years.
4   Industrial pollution in the coastal and marine areas.
5   Residential and touristic activities, both throughout coastal areas and pastures, especially in the Black Sea Region.
6   Erosion due to agricultural and overgrazing activities.

In spite of these major threats to the Turkish biodiversity, there are many attempts for conservation of biodiversity, at both species and ecosystem levels. Tuz Lake (Kurt et al., 2010) is an important conservation area at eco-

Figure 11.27 Urbanization at Marmaris.

Figure 11.28   Invasion of the pasture at Sis Dağı (Giresun).

system level in respect of unique biodiversity. On the other hand, various projects for species based conservations, (such as *Cyanus tchihatcheffii, Liquidambar orientalis* (Figure 11.29) and *Phoenix theophrasti*) are underway by governmental and non-govermental organizations.

There are some studies/projects/programs that can be referred to as flag species, which set an example within species conservation practices in Turkey (Tekeli et al., 2006). The projects about the protection of the endangered sea turtles and the Mediterranean monk seal, which were started by non-governmental organizations and supported by companies, created an important historical background. Thus, species conservation practices will be divided into three as; sea turtles, Mediterranean monk seal and other conservation projects. Mountain goat (*Capra aegagrus*), Anatolian wild sheep (*Ovis gmelinii anatolica*), rupicapra (*Rupicapra rupicapra*), deer (*Cervus elaphus*), Eurasian otter (*Lutra lutra*), bear (*Ursus arctos*), roe deer (*Capreolus capreolus capreolus*), gazelle (*Gazella subgutturosa*) and northern bald ibis (*Geronticus eremita*) are under special protection programs as endangered species. The sandbar shark *Carcharinus plumbeus*, which is in vulnerable status breeds only in the Boncuk Bay in the Mediterranean Sea. *In-situ* protection measures have been considered for the breeding sites of this vulnerable species. There are numerous projects carried out by both governmental

Figure 11.29 *Liquidambar orientalis*, a relict endemic species in Marmaris.

and non-governmental organizations to protect fields with rich biological diversity.

Field protection programs as national parks (also marine and submarine park like Gökçeada and Fenike), habitat parks, nature protection fields, wildlife protection fields, private habitat protection fields, natural sites, natural properties gene protection and management fields and ex-situ programs as seed collection gardens, arboretums, botanical gardens and gene banks carry out. There are 25 different protection positions in Turkey. The national law covers 13 of them and 5 of them are included in international agreements (Table 11.4).

On the other hand, Eken et al. (2006) reported 305 important natural areas with 20,280,149 ha for Turkey.

The Black Sea is the world's most isolated sea from oceans. This marine habitat having no life forms under the depth of 100–200 meters is under threat due to overfishing, marine transportation, pollution, alien species and eutrophication. From the 26 commercially significant species only 6 of them are important in terms of fishing (Tekeli et al., 2006). The Bosphorus and the Dardanels is a biological corridor for the Mediterranean and Black Sea. The protection of this corridor is vital for the protection of biodiversity. Marine

Table 11.4 On Site Protection Areas in Turkey (in-situ)

Protection status	Related law	Number of protected areas	Total surface area (hectare)
Field protection positions in Turkey and their coverage			
National park	National Park Law 2873	42	845,814
Natural park	National Park Law 2873	209	99,378
Natural protection area	National Park Law 2873	30	47,244
Natural monument	National Park Law 2873	111	7,142
Wildlife development site	Hunting Law 4915	81	1,189.293
Wildlife conservation site	Hunting Law 4915	1	8000
RAMSAR	RAMSAR agreement, Regulations for protection of wetlands 2873	14	184,487
Wetlands with national importance	Regulations for protection of wetlands 2873	38	469,830
Wetlands with local importance	Regulations for protection of wetlands 2873	6	1,602
Protection forest	Forest Law 6831	55	251,548
City forest	Forest Law 6831	145	10,550
Gene conservation forest	Forest Law 6831	295	39,732
Seed orchard	Forest Law 6831	187	1,442
Seed stands	Forest Law 6831	330	43,858
In forest recreation	Forest Law 6831	86	3,826
Natural site	Conservation of Cultural and Natural Property, 2863	2434	1,991,700
Natural asset (tree)	Conservation of Cultural and Natural Property, 2863	8724	
Natural asset (cave)	Conservation of Cultural and Natural Property, 2863	249	
Biosphere reserve	UNESCO Man and Biosphere Programme	1	27,752
World cultural and natural heritage site	Agreement of world cultural and natural heritage protection	2	10,961
Areas of Special Conservation Interest (ASCIs), EMERALD Network	Agreement of Europe's natural life and habitat protection – Bern Convention	9	716.529

Table 11.4 *(Continued)*

Protection status	Related law	Number of protected areas	Total surface area (hectare)
Specially protected area	Protocol of establishing private protection areas and biodiversity, Barcelona Convention	16	2,459,749
Important plant area	Convention on Biodiversity	122	
Aquaculture production site	Aquaculture Law	–	Where water living organisms available
Natura 2000 sites (SPA and SAC sites)	European Union protection of birds regulation 9/409/EEC, European Union regulation of protection of habitats and species 92/43/EEC	In progress	–
Important natural area*	Related convention and rules	305	20,280,149

*Macro protection aimed.

Sources:

www.milliparklar.gov.tr/Anasayfa/istatistik.aspx?sflang=tr.

www.whc.unesco.org/en/list/357Cappadocia.

www.whc.unesco.org/en/list/485Pamukkale-Hierapolis.

www.unesco.org/new/en/natural-sciences/environment/ecological-sciences/biosphere-reserves/europe-north-america/turkey/camili/.

transport, pollution and alien species are the main threats to these marine ecosystems.

Acknowledgments

The authors thank the following experts for their valuable comments on the text Zübeyde Uğurlu Aydın, and Emel Oybak Dönmez (archaeobotanical data and English revision of the text) and Adnan Erdağ (bryophyte), Gönül Kaynak (ferns), Ayhan Şenkardeşler (algae and lichenes), Evren Cabi (Poaceae), A. Emre Yaprak (Amaranthaceae), Birol Mutlu (Brassicaceae).

Keywords

- endemic species
- invertebrata

- mammals
- species diversity

References

Adıgüzel, N., (2005). *Iğdır Ovası.* In: *Türkiye'nin 122 Önemli Bitki Alanı,* Özhatay, N., Byfield, A., & Atay, S., ed., İstanbul, pp. 338–339.

Akman, Y., (1995). *Türkiye Orman Vejetasyonu,* Ankara.

Al-Shehbaz, I., Mutlu, B., & Dönmez, A. A., (2007). The Brassicaceae (Cruciferae) of Turkey, Updated, *Turk. J. Bot., 31,* 327–336.

Anonymous, (2013). *Orman Atlası. T. C. Orman ve Su işleri Bakanlığı. Orman Genel Müdürlüğü, 107s.* Ankara.

Atalay, İ., (1987). *Türkiye Jeomorfolojisine Giriş,* İzmir.

Baran, İ., (2005). *Turkiye Amfibi ve Sürüngenleri,* Turkiye Bilimsel ve Teknik Araştırma Kurumu (TUBİTAK) Popüler Bilim Kitapları: 207, Başvuru Kitaplığı: *21,* Ankara.

Baytop, T., (1998). *İstanbul Lalesi,* Ankara.

Baytop, T., (1999). Türkiye'de Bitkiler ile Tedavi, Geçmişte ve Bugün, Ankara.

Binzet, R., (2016). A new species of *Onosma* L. (Boraginaceae) from Anatolia. *Turk. J. Bot., 40,* 194–200.

BirdLife International, (2015). *Avrupa Kuşları Kırmızı Listesi.* Lüksemburg Avrupa Toplulukları Resmi Yayın Ofisi. *82.*

Çalışkan, S., & Boydak, M., (2017). Afforestation of arid and semiarid ecosystems in Turkey. *Turkish Journal of Agriculture and Forestry., 41,* 317–330.

Çevre Bakanlığı, (ÇB), (2001). *Ulusal Biyolojik Çeşitlilik Stratejisi ve Eylem Planı,* Çevre Bakanlığı.

Daşkın, R., & Bağçıvan, G., (2017). A new species of *Galium* (Rubiaceae) from Southwest Anatolia, Turkey. *Phytotaxa, 308*(2), 267–274.

Davis, P. H., (1965–1985). *Flora of Turkey and the East Aegean Islands,* vol. 1–9, Edinburgh University Press, Edinburgh.

Davis, P. H., (1971). Distribution patterns in Anatolia with particular reference to endemism. In: *Plant Life of Southwest Asia,* Davis, P. H., Harper, P. C., & Hedge, I. C., (eds.). *Bot. Soc. Edinburgh,* pp. 15–27.

Davis, P. H., Mill, R. R., & Tan, K., (1988). *Flora of Turkey and the East Aegean Islands, vol. 10,* Edinburgh University Press, Edinburgh.

De Moulins, D., (1997). Agricultural Changes at Euphrates and Steppe Sites in the Mid-8th to the 6th Millennium BC. In: *Paléorient, 24*(1), 113–114.

Dönmez, A. A., Uğurlu, Z., & Işık, S., (2015). *Polygala turcica* (Polygalaceae), A new species from E Turkey, and a new identification key to Turkish *Polygala. Willdenowia., 45,* 429–434. DOI: http://dx.doi.org/10.3372/wi.45.45309.2015.

Eken, G., Bozdoğan, M., İsfendiyaoğlu, İ., Kılıç, D. T., & Lise, Y., (2006). *Türkiye'nin önemli doğa alanları,* Doğa Derneği, Ankara.

Eker, İ., Babaç, M. T., & Koyuncu, M., (2014). Revision of the genus *Tulipa*, L. (Liliaceae) in Turkey. *Phytotaxa, 57(1)*, 1–112. DOI: 10.11646/phytotaxa.157.1.1, ISSN: 1179–3163.

Erdağ, A., & Kürschner, H., (2017). *Türkiye Bitkileri Listesi (Karayosunları)*, İstanbul.

Erdei, B., Akgün, F., & Lumaga, M. R. B., (2010). *Pseudodioon akyoli* gen. et sp. nov., an extinct member of Cycadales from the Turkish Miocene. *Plant Syst. Evol., 285*, 33–49.

Erinç, S., (1982). *Jeomorfoloji I*, İstanbul.

Evirgen, A., (2007). *Fotoğraflarla Türkiye Deniz Balıkları*, İstanbul.

Gönülol, A., (2017). *Turkishalgae electronic publication*, Samsun, Turkey. http://turkiyealgleri.omu.edu.tr.

Güner, A., Ekim, T., Vural, M., Babaç, M. T., & Aslan, S., (2012). *Türkiye Bitkileri Listesi*. NGBB Press. İstanbul.

Huber-Morath, A., (1978). *Verbascum* L. In: *Flora of Turkey and The East Aegean Islands*, Davis, P. H., (ed.). Edinburgh University Press, Edinburgh, *vol. 6*, pp. 461–602.

Işık, K., (1987). Ormanlar ve Milli Parklar, *Türkiye'nin Biyolojik Zenginlikleri* (Koor. A. Kence), Türkiye Çevre Vakfı Yayını Önder Matbaa, Ankara, 93–116.

Kadereit, G., Ladislav, M., & Freitag, H., (2006). Phylogeny of Salicornioideae (Chenopodiaceae): diversification, biogeography, and evolutionary trends in leaf and flower morphology. *Taxon., 55*(3), 617–642.

Kayseri, M. S., & Akgün, F., (2010). The Late Burdigalian–Langhian time interval in Turkey and the palaeoenvironment and palaeoclimatic implications and correlation of Europe and Turkey: Late Burdigalian–Langhian palynofloras and palaeoclimatic properties of the Muğla–Milas (Kultak). *Geol. Bull. Turkey, 53*(1), 1–44.

Ketin, İ., (1966). Anadolu'nun tektonik birlikleri. *Maden Tetkik ve Arama Dergisi., 66*, 20–34.

Kocataş, A., Ergen, E., Mater, S., Özel, İ., Katağan, T., Koray, T., Önen, M., & Kaya, M., (1987). Deniz faunası, *Türkiye'nin Biyolojik Zenginlikleri* (Koor. A. Kence), Türkiye Çevre Vakfı Yayını Önder Matbaa, Ankara, 149–168.

Kottek, M., Grieser, J., Beck, C., Rudolf, B., & Rubel, F., (2006). World map of the Koppen-Geiger climate classification updated. *Meteorol. Z., 15*, 259–263.

Kurt, L., Yiğit, N., & Kaya, M., (2010). *Tuz Gölü Özel Çevre Koruma Bölgesi Habitat İzleme Projesi Sonuç Raporu*. T. C. Çevre ve Orman Bakanlığı, Özel Çevre Koruma Kurumu Başkanlığı (In Turkish), pp. 181.

Kuru, M., (1987). İçsu faunası, *Türkiye'nin Biyolojik Zenginlikleri* (Koor A. Kence), Türkiye Çevre Vakfı Yayını Önder Matbaa, Ankara, 133–148.

Kuru, M., Yerli, S. V., Mangıt, F., Ünlü, E., & Alp, A., (2014). Fish biodiversity of Turkey. *J. Academic Documents for Fisheries and Aquaculture, 3*, 93–120.

Mathew, B., (1984). *Crocus*. In: *Flora of Turkey and the East Aegean Islands*, Davis, P. H., Mill, R. R., & Tan, K., (eds.). Edinburgh University Press, Edinburgh, *8*, pp. 413–438.

Medail, F., & Diadema, K., (2009). Glacial refugia influence plant diversity patterns in the Mediterranean Basin. *Biogeography, 36*(7), 1333–1345.

Meriç, E., (1982). *Tarihsel Jeoloji*, Konya.

Mutlu, B., & Karakuş, Ş., (2015). A new species of *Campanula* (Campanulaceae) from Turkey. *Phytotaxa, 234*(3), 287–293.

Öztürk, B., & Yerli, S., (2002). *Natural Heritage of Turkey* (contributors: Ertan, A., Baran, İ., Albayrak, İ., Öztürk, B., Yerli, S., Öztürk, A.) Turkish Marine Research Foundation, İstanbul.

Reeves, R. D., Kruckeberg, A. R., Adıgüzel, N., & Kramer, U., (2001). Studies on the Flora of Serpentine and Other Metalliferous Areas of Western Turkey. *S. Afr. J. Sci.*, *97*, 53–507.

Resmi, G., (2016). Doğal Çiçek Soğanlarının 2016 Yılı İhracat Listesi Hakkında Tebliğ (Tebliğ No: 2015/47).

Sarıkaya, M., Çiner, A., & Zreda, M., (2011). Quaternary glaciations of Turkey. In: *Quaternary Glaciations – Extent and Chronology*, Ehlers, J., Gibbard, P., & Hughes, P., (eds.). Oxford, UK, Jordan Hill, pp. 393–403.

Sesli, E., & Denchev, C. M., (2014). *Checklists of the Myxomycetes, Larger Ascomycetes, and Larger Basidiomycetes in Turkey*. 6th edn. Mycotaxon Checklists Online (http://www.mycotaxon.com/resources/checklists/sesli-v106-checklist.pdf), 1–136.

Seçmen, Ö., & Leblebici, E., (1997). *Türkiye Sulak Alanlarının Bitkileri ve Bitki Örtüsü*, Bornova: İzmir.

Şekercioğlu, Ç. H., Anderson, S., Akçay, E., Bilgin, R., Can, Ö. E., Semiz, G., Tavşanoğlu, Ç., Yokeş, M. B., Soyumert, A., İpekdal, K., Sağlam, İ. K., Yücel, M., & Dalfes, H. N., (2011). Turkey's globally important biodiversity in crisis. *Biological Conservation*, *144*, 2752–2769.

Şengör, A. M. C., & Yılmaz, Y., (1981). Tethyan evolution of Turkey: A plate tectonic approach. In: *Tectonophysics*, Elsevier Scientific Publishing Company, Amsterdam, pp. 181–241.

Tekeli, İ., Çağatay, G., Yerli, S. V., Algan, N., Vaizoğlu, S., Kaya, A. D., Öztürk, B., Mutlu, B., & Demirayak, F., (2006). *Dünya'da ve Türkiye'de Biyolojik Çeşitliliği Koruma*, Türkiye Bilimler Akademisi Raporları, Türkiye Bilimler Akademisi, Ankara.

Teryokhin, E. S., (2001). De genere *Necranthus* Gilli (Scrophulariaceae) notula. *Novosti Sist. Vysschich Rasteni*, *33*, 205–207.

The Plant List, (2013). Version 1.1. Published on the Internet, http://www.theplantlist.org/ (accessed 1st January).

TUİK, (2016). (Türkiye İstatistik Kurumu), Su ürünleri istatistikleri, Haziran, (www.tuik.gov.tr/PreHaberBultenleri. do?id=21720).

Turan, D., Ekmekçi, F. G., Kaya, C., & Güçlü, S. S., (2013). *Alburnoides manyasensis* (Actinopterygii, Cyprinidae), A new species of cyprinid fish from Manyas Lake basin, Turkey. *ZooKeys*, *276*, 85–02. doi:10.3897/zookeys.276.4107.

Turkish State Meteorological Service (TSMS), (2017). Devlet Meteoroloji Müdürlüğü)., Yıllık Toplam Yağış Verileri (Yearly Rainfall Data). https://www.mgm.gov.tr/veride-gerlendirme/yillik-toplam-yagis-verileri. aspx (in Turkish).

Uysal, T., Ertuğrul, K., & Bozkurt, M., (2014). A new genus segregated from *Thermopsis* (Fabaceae: Papilionoideae): *Vuralia. Plant. Syst. Evol.*, *300*, 1627–1637.

Vural, M., Duman, H., Aytaç, Z., & Adıgüzel, N., (2012). A new genus and three new species from central Anatolia, Turkey. *Turk. J. Bot.*, *36*, 427–433.

www.iucn.org/content/biodiversity-turkey,BiodiversityinTurkey,june13,2017.

www.milliparklar.gov.tr/Anasayfa/istatistik.aspx?sflang=tr.

www.unesco.org/new/en/natural-sciences/environment/ecological-sciences/biosphere-reserves/europe-north-america/turkey/camili/.

www.whc.unesco.org/en/list/357Cappadocia.

www.whc.unesco.org/en/list/485Pamukkale-Hierapolis.

Yerli, S. V., (2012). *Doğal Sit Alanlari Teknik Esaslar Rehberi*, Ek 1 ÇŞB, TVK GM, Ankara.

Yerli, S. V., (2016). The ecology of inland fisheries of Turkey. In: *Freshwater Fisheries Ecology*, Craig, J. F., (ed.). Wiley and Blackwell pp. 304–310.

Yerli, S. V., Albayrak, İ., Kıraç, C. O., Ilgaz, Ç., & Küçük, F., (2012). *Doğal Sit Alanları Teknik Esaslar Rehberi (Mammalia, Aves, Amphibia, Reptilia, Pisces), Ek,* ÇŞB, TVK GM, Ankara.

Zare, G., (2012). Türkiye *Orobanche,* L. (Orobanchaceae) Cinsinin Taksonomik Revizyonu ve Moleküler Filogenisi. Hacettepe Üniversitesi Fen Bilimleri Enstitüsü, (in Turkish, unpublished PhD thesis), pp. 321.

CHAPTER 12

Biodiversity in the United Kingdom

RANEE OM PRAKASH[1] and **FRED RUMSEY**[2]

[1]Curator-Flowering Plants, Department of Life Sciences, Natural History Museum, South Kensington, SW7 5BD, United Kingdom, E-mail: r.prakash@nhm.ac.uk

[2]Senior Curator-in Charge, British, European and Historic Herbaria, Department of Life Sciences, Natural History Museum, South Kensington, SW7 5BD, United Kingdom

12.1 Introduction

12.1.1 Geography, Geology, and Land Characteristics

The United Kingdom of Great Britain and Northern Ireland, commonly known as the United Kingdom (UK) or Great Britain is made up of England, Scotland, Wales, and Northern Ireland (Figure 12.1) and has sovereignty over 14 Overseas Territories and 3 Crown Dependencies (https://www.gov.uk/government/policies/uk-overseas-territories, http://www.justice.gov.uk/downloads/about/moj/our-responsibilities/Background_Briefing_on_the_Crown_Dependencies2.pdf) (Figure 12.2). The UK is located in northwestern Europe (55.3781°N, 3.4360°W). It lies between the North Sea and the North Atlantic Ocean. The major part of the country includes the island of Great Britain, Ireland and some smaller surrounding islands. It is bounded by the Atlantic Ocean, the North Sea to its east, and the English Channel to its south and the Celtic Sea to its southwest. The Irish Sea lies to the west between Ireland and Great Britain.

The country has an area of 93,627.8 sq mi and as of 30th June 2016 has an estimated population of 65.64 million inhabitants (Office for National Statistics, 2017). The coastline of Great Britain is 11,073 miles long and is connected to Europe by the Channel Tunnel; the longest (31 miles) underwater tunnel in the world (https://www.eurotunnel.com/uk/build/).

The country consists of lowland terrain, with mountainous terrains northwest of Tees-Exe line (an imaginary northeast–southwest line that can be drawn on a map of Great Britain roughly dividing the country into lowland and upland regions) and includes the Cumbrian Mountains of the Lake District, The Pennines, Exmoor, and Dartmoor. Some of the major rivers and estuaries are: The Thames, Severn, Tees, Tyne, Avon, Exe, Mersey, and the Humber. Scafell Pike (978 meters) in the Lake District is England's highest mountain (https://www.nationaltrust.org.uk/wasdale/features/enjoying-eng-

Figure 12.1 Map of United Kingdom (*Source:* http://www.worldatlas.com/
webimage/countrys/europe/lgcolor/ukcolor.htm; reprinted with
permission).

lands-highest-mountain---scafell-pike-in-wasdale). Scotland accounts for
just under a third of the total area of the UK, covering 30,410 sq mi and
includes nearly 800 smaller islands (however, only around 130 islands have
permanent inhabitants). The larger groups include the Hebrides, Orkney and
Shetland Islands (http://www.independent.co.uk/travel/uk/the-complete-
guide-to-the-scottish-islands-633851.html).

Scotland (Capital: Edinburgh) is the most mountainous region in the
British Isles and has a distinct topology – the Highland Boundary Fault (a

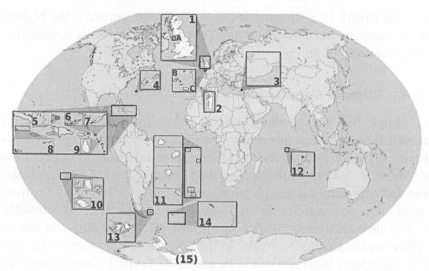

Figure 12.2 Locations of UK dependencies (crown dependencies alphabetized, overseas territories numbered): A – Isle of Man; B – Guernsey; C – Jersey; 1 – United Kingdom; 2 – Gibraltar; 3 – Akrotiri and Dhekelia; 4 – Bermuda; 5 – Turks and Caicos Islands; 6 – British Virgin Islands; 7 – Anguilla; 8 –Cayman Islands; 9 – Montserrat; 10 – Pitcairn Islands; 11 – Saint Helena, Ascension and Tristan da Cunha; 12 – British Indian Ocean Territory; 13 – Falkland Islands; 14 – South Georgia and the South Sandwich Islands; (15) – British Antarctic Territory (*Source:* Wikimedia Commons; https://en.wikipedia.org/wiki/File:United_Kingdom_(overseas%2Bcrown_dependencies),_administrative_divisions_-_Nmbrs_(multiple_zoom).svg)

geologic rock failure). This fault traverses Scotland from Arran and Helensburgh in the west to Stonehaven in the east. It divides the country into two different regions; namely the Highlands to the north and west and the lowlands to the south and east. Ben Nevis at 1,344 meters is the highest point in the British Isles (http://www.bennevisweather.co.uk/index.asp).

Wales (Capital: Cardiff) covers 8,020 sq mi and accounts for less than a tenth of the total area of the UK. North and mid-Wales is mostly mountainous, though South Wales has sandy beaches. There are extensive sand dune systems in mid and N.Wales too and the Brecon Beacons reach 886 m on the Pen y fan. Snowdonia including Snowdon (3,560 ft.) is the highest peak in Wales (http://news.bbc.co.uk/1/hi/world/europe/country_profiles/6233450.stm).

Northern Ireland (Capital: Belfast) is separated from Great Britain by the Irish Sea and North Channel and is mostly hilly. Lough Neagh is the largest lake in the British Isles by area (150 sq. mi). Slieve Donard is the highest peak in Northern Ireland (852 meters) (http://www.atlapedia.com/online/countries/unitedki.htm).

12.1.2 Climate Conditions

Great Britain has a Temperate Oceanic climate influenced by the North Atlantic Drift and the southwesterly winds. This gives cool to mild winters and cool summers with moderate variation in temperature throughout the year. One of the greatest influences on the climate of the UK is the Atlantic Ocean and especially the Gulf Stream, which carries warm water up from lower latitudes and modifies the high latitude air masses that pass across the UK. This thermohaline circulation has a powerful moderating and warming effect on the country's climate. In the absence of this warm water current, temperatures in winter would be about 10°C (18°F) lower than they are today, similar to eastern Russia or Canada which occupy the same latitude. The temperature records for England are continuous back to the mid-17th century. The Central England Temperature (CET) record is the oldest in the world, and is a compound source of cross-correlated records from several locations in central England. In England, the average annual temperature varies from 8.5°C (47.3°F) in the north to 11°C (51.8°F) in the south, but over the higher ground, this can be several degrees lower. The recorded extremes in the UK run from a maximum of 38.5°C to a minimum of −27.2°C. In summer there may be a difference of 20°C, or more, between the south of England and the islands off the northern Scottish Coast. Mild to warm weather is generally in the months of May, June, September, and October (https://www.metoffice.gov.uk/public/weather/climate).

The warm ocean currents also bring substantial amounts of humidity that contributes to the notoriously wet climate that western parts of the UK experience. Precipitation records date back to the eighteenth century; the modern England and Wales series beginning in 1766. Rainfall amounts can vary greatly across the United Kingdom and generally the further west and the higher the elevation, the greater the rainfall. In the wettest parts of the country as much as 4,577 millimeters (180.2 in) of rain can fall annually, making these locations some of the wettest in Europe. However, in some years rainfall totals in Essex and South Suffolk can be below 450 millimeters (17. 7 in), less than the average annual rainfall in some semi-arid parts of the world.

The high latitude and proximity to a large ocean to the west means that the United Kingdom experiences strong winds. The prevailing wind is from the southwest, but it may blow from any direction for sustained periods of time. Winds are strongest near westerly facing coasts and exposed headlands, with gusts to 142 mph or more recorded. Flooding is becoming increasingly common.

12.2 Habitats in the UK

As per Joint Nature Conservation Committee (JNCC, 2017), land and freshwater habitats, including fens, dunes, grassland, heathland, marsh, rivers, and woodlands are the major ecosystems which are home to many thousands of species. The requirement for the classification of habitats has only been developed in detail in recent decades. Several terrestrial and freshwater classifications have emerged as important standards for conservation in the UK and JNCC plays an important role in maintaining these standards. Some of the popular terrestrial habitat/vegetation classifications in use include:

- National Vegetation Classification (NVC);
- Phase 1 Habitat Classification;
- UK Biodiversity Action Plan broad habitat types;
- UK Biodiversity Action Plan priority habitats;
- JNCC freshwater classifications.

These have been classified with different objectives and JNCC (see http://jncc.defra.gov.uk/page–1425) is also an important resource for National Biodiversity Network Habitats Dictionary.

Some of the habitats such as coastal lagoons, inland salt meadows, fixed dunes with herbaceous vegetation, decalcified fixed dunes with *Empetrum-nigrum*, Atlantic decalcified fixed dunes, coastal dunes with *Juniperus* spp., Mediterranean temporary ponds, Turloughs, Temperate Atlantic wet heaths with *Erica ciliaris* and *Erica tetralix*, Dry Atlantic coastal heaths with *Erica vagans*, species-rich *Nardus* grassland, on siliceous substrates in mountain areas, active raised bogs, blanket bogs, calcareous fens with *Cladium mariscus* and species of the *Caricion davallianae,* petrifying springs with tufa formation (Cratoneurion), alpine pioneer formations of the Caricion-bicoloris-atrofuscae, limestone pavements, Tilio-Acerio forests of slopes, screes and ravines, Caledonian forest, bog woodland, alluvial forests with *Alnusglutinosa* and *Fraxinus excelsior* and *Taxus baccata* woods of the British Isles are classed as priority habitats under the UK Biodiversity Action Plan (UKBAP) and form habitat types on Annex I of the EU habitats. For

details, see http://jncc.defra.gov.uk/Publications/JNCC312/UK_habitat_list.
asp.

The UK has 15 National Parks (10 in England, 3 in Wales and 2 in Scotland). Some of the larger ones are Lake District, Cairngorms, and Snowdonia (see Figure 12.3).

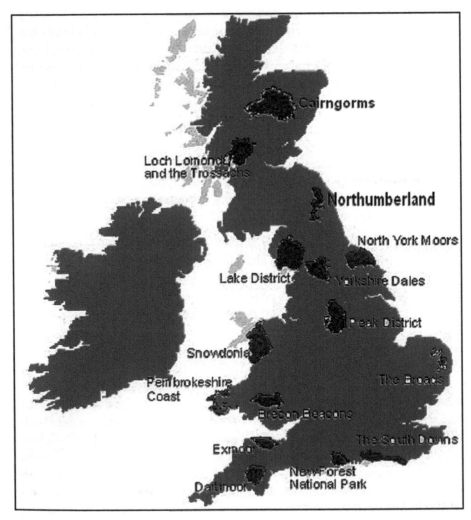

Figure 12.3 The National Parks of Britain (*Source:* http://www.walking-uk.com/natparks.htm).

12.3 Organisms and Taxonomic Groups

Although there are numerous floras, atlases and literature available on the biodiversity of UK, *The Changing Wildlife of Great Britain and Ireland* edited by Hawksworth (2003) gives a synoptic view of the national biodiversity spanning 26 chapters contributed by various authors. These chapters give neat summaries of various organism groups. More than a decade has passed since this publication and in the recent era of digital revolution, the way we capture data has significantly advanced our knowledge, especially in the less well-studied groups such as viruses, nematodes, mites, and fungi. Of recent interests are innovative projects such as microverse, a citizen science project to discover the diversity of microscopic life in urban environments using DNA technologies (see http://www.nhm.ac.uk/schools.html) and life in the air.

Crowdsourcing events on recording wildlife such as Bioblitzes and Citizen Science are now becoming immensely popular. However, one thing which is alarming and is of concern for future generations is the decline of professional taxonomists, largely as a consequence of a shift away from whole organism teaching, and this decline in general taxonomic expertise is something that must be addressed if knowledge of Britain's biodiversity is to be maintained. Information on UK Biodiversity and the expertise in identifying and recording it is increasingly to be found within the amateur naturalist community.

The British Isles has a very long tradition of biological recording, the collections, some dating back to the mid 17th century preserved in our national and regional museums and botanic gardens. A vibrant network of societies and dedicated individuals, dominated for most organism groups by volunteer naturalists are active in recording nature such as the Botanical Society of Britain and Ireland (BSBI, see http://bsbi.org/), London Natural History Society (LNHS, see http://www.lnhs.org.uk/), the South London Botanical Institute (SLBI, see http://www.slbi.org.uk/), Royal Society for the Protection of Birds (RSPB, see https://www.rspb.org.uk/), People's Trust for Endangered Species (produced Apps, e.g., Mammals on Roads; Traditional Orchard Survey), Earthworm Society of Britain, British Bryological Society (recorded around 1069 moss species in 2013; 9 moss species have been added since then) and the British Mycological Society to name a few (see list of some societies and organizations in Appendix 1). Much of this data has traditionally been maintained by the Biological Records Centre (BRC, see brc.ac.uk/irecord). To make this data more accessible the National Biodiversity Network (NBN) (see: https://nbn.org.uk/) was formed as a collaborative partnership for exchanging biodiversity information. A charitable organization, it has over 180 members with a membership including many UK wildlife conservation organizations, government, country agencies, environmental agencies, local environmental records centers, and several voluntary groups. It has over 215 million wildlife records available through the NBN Gateway and has UK's largest collection of freely available biodiversity data. NBN is also the UK node for Global Biodiversity Information Portal (GBIF).

Over 70,000 species of animals, plants, fungi, and single-celled organisms are found in the UK (Hayhow et al., 2016). The UK Species Inventory project led by Angela Marmont Centre for UK Biodiversity (see http://www.nhm.ac.uk/take-part/centre-for-uk-biodiversity.html) aims to bring all of the standard reference names for all species of flora and fauna in the UK, together in one place in order to ensure accurate recording and tracking of UK biodiversity.

The UK's Overseas Territories (OTs) are of great importance too for wildlife worldwide; over 32,000 native species have been recorded in the OTs, of which 1,557 do not occur anywhere else in the world and an estimated 70,000 species may remain undiscovered in the OTs (Hayhow et al., 2016).

The State-of-the-Nature–2016 report (Hayhow et al., 2016) summarized data from across the UK biodiversity sector produced by a partnership of more than 50 organizations involved in the recording, researching and conservation of nature in the UK and its OTs:

- Between 1970 and 2013, 56% of species declined, with 40% showing strong or moderate declines. 44% of species increased, with 29% showing strong or moderate increases. Between 2002 and 2013, 53% of species declined and 47% increased. These measures were based on quantitative trends for almost 4,000 terrestrial and freshwater species in the UK.
- Of the nearly 8,000 species assessed using modern Red List criteria, 15% are extinct or threatened with extinction from Great Britain.
- An index of species' status, based on abundance and occupancy data, has fallen by 16% since 1970. Between 2002 and 2013, the index fell by 3%. This is based on data for 2,501 terrestrial and freshwater species in the UK.
- An index describing the population trends of species of special conservation concern in the UK has fallen by 67% since 1970, and by 12% between 2002 and 2013. This is based on trend information for 213 priority species.

- A new measure that assesses how intact a country's biodiversity is, suggests that the UK has lost significantly more nature over the long term than the global average. The index suggests that we are among the most nature-depleted countries in the world.
- The loss of nature in the UK continues. Although many short-term trends suggest improvement, there was no statistical difference between our long and short-term measures of species' change, and no change in the proportion of species threatened with extinction.
- Of the 7,964 terrestrial and freshwater species that have been assessed using modern IUCN Red List criteria, 1,057 (13%) are thought to be at risk of extinction from Great Britain, and 142 (2%) are known to have gone extinct from Great Britain.
- Of the terrestrial and freshwater species that have been assessed using modern IUCN Red List criteria, 19% of plants, 11% of invertebrates and 11% of fungi are classified as being at risk of extinction from Great Britain.
- The recent Birds of Conservation concern four assessments, which uses different criteria from IUCN Red Lists, assessed 247 bird species. 67 species (27%) were red-listed, a substantial increase from 52 species back in 2009.

12.3.1 Fauna of UK

12.3.1.1 Invertebrates

12.3.1.1.1 Molluscs (Non-Marine)

As per Anderson (2005), 220 species of non-marine molluscs have been recorded as living in the wild in Britain.

12.3.1.1.2 Insects

There are over 20,000 species of insects in Great Britain (Chinnery, 1977). *The Moths and Butterflies of Great Britain and Ireland* by Emmet et al., (1999), *The Millennium Atlas of Butterflies in Britain and Ireland* by Asher et al. (2001), *Britain's Butterflies* by Tomlinson and Still (2002) and *The Butterflies of Britain & Ireland* by Thomas and Lewington (2014) are some of the useful resources listing butterflies and moths which are naturalized, extinct and are of doubtful origin (see some of the butterflies in Figures 12.4c and 12.4d).

12.3.1.1.3 Arachnids

The Checklist of British spiders compiled by the British Arachnological Society in 1991 is now obsolete but an up to date naming is now provided as per the checklist of the accepted names of British spiders by Merrett and Murphy (2000), where 668 species are listed. An illustrated field guide (Bee et al., 2017) has just been published. See an example of one of the largest British spiders, the raft spider *Dolomedes fimbriatus* in Figure 12.4f

Other works worth checking are *The Odonata* (Dragonflies and Damselflies) which are very well recorded and have good field guides, *Field Guide to the Bees of Great Britain and Ireland* by Falk & Lewington (Carreck, 2016). The Royal Entomological Society of London has produced a range of identification guides in a series called *Handbooks to the Identification of British Insects*. Likewise, the British Entomological and Natural History Society have produced several authoritative works such as on *British soldierflies and their allies* (Stubbs and Drake, 2014) and also *British Hoverflies* (Stubbs, 1996). *Britain's Hoverflies: A Field Guide* by Ball (2015) is also a handy reference.

There is an excellent *Guide to the Ladybirds (Coccinellidae) of Britain and Ireland* produced by the BRC/FSC in 2011 authored by Roy et al. and the FSC have produced other guides too such as *A Pictorial Guide to British Ephemeroptera* by Macadam and Bennett (2010) which arose out of a citizen science project developed at the NHM which evolved into the Riverfly partnership.

12.3.1.2 Vertebrates

12.3.1.2.1 Amphibians

The species of amphibian native to Britain are the great crested newt, smooth newt, palmate newt, common toad, natterjack toad, common frog and the pool frog (extinct in the wild 1995 but re-introduced using continental European stock, see https://www.arc-trust.org/pool-frog). Several other species have become naturalized. All native amphibians are protected to some degree under Schedule 5 of the Wildlife and Countryside Act, 1981. For the commoner species this just extends to prohibiting their sale, whereas the Great Crested Newt, Natterjack toad, and Pool Frog are completely protected.

Figure 12.4 Some examples of UK fauna (Images credits: Figure 3.1a (Fred Rumsey), 3.1.b by Importé sur Commons par Salix. Sur Flikr par Polandeze (Flickr: original page) [CC BY 2.0 (http://creativecommons.org/licenses/by/2.0)], via Wikimedia Commons, 3.1c, d, e, f by Fred Rumsey).

12.3.1.2.2 Reptiles

Great Britain has fewer species of snakes, with the European adder being the only venomous snake to be found here. The other snakes found in Great Britain are the grass snake and the smooth snake. Great Britain has three native breeds of lizard: slowworms, sand lizards and viviparous lizards. There are also turtles, such as leatherback turtles to be found in the Irish Sea, although these are rarely seen by the public. Other reptile species exist but are not native: aesculapian snake, wall lizard, and the green lizard. All native reptiles are protected to some degree under Schedule 5 of the Wildlife and Countryside Act, 1981. For the commoner species, this just extends to prohibiting their sale and their intentional injury, or killing. All turtles, the Sand Lizard, and Smooth snake are completely protected.

12.3.1.2.3 Birds

The avian fauna of Britain is largely composed of Palaearctic species and is similar to that of Europe. The UK has fewer breeding species and many species that cannot cope with arduous winter conditions can winter in Britain. About 250 wintering species are recorded in Great Britain. At least 146 species of British Birds are listed in Red Data Book of Birds. For details see: http://www.nhm.ac.uk/our-science/data/uk-species/checklists/NBN-SYS0000000020/index.html.

12.3.1.2.4 Mammals

The red deer is the largest native mammal species, and is found throughout England, Scotland and Wales. The Roe deer is the other native species. Other species, which have become well established in UK, include the fallow deer (originating from France), Muntjac (originating from China), and the Sika deer (originating from Japan) (Ratcliffe, 1987).

The hedgehog is the most widely known insectivore. Other insectivores include the Shrews and the pygmy shrew (one of the smallest mammals in the world, see Figure 12.4b.). The Brown rat is the most common rodent. Rabbits, brown hares and the mountain hare (found only in Scotland and in Ireland as an endemic race) are also found in UK. Many sea mammals such as the seal, dolphins, porpoises and orcas are seasonally found on British coastlines and shores.

12.3.1.2.5   Bats

These are the only flying mammal species. There are around 1300 species of bats worldwide and around 27 species of bats are listed for UK as per the checklist prepared by the Bat Conservation Trust (2004), see http://www.nhm.ac.uk/our-science/data/uk-species/checklists/NHMSYS0000527885/index.html). *Alcathoe* was recorded as a new bat species in 2010. *Pipistrelle* is the most common and smallest bat weighing between 4–7 g, with a wing span of 18–25 cm and the largest bat is the *Noctule* weighing up to 40 g with a wingspan of between 33–45 cm (http://www.bats.org.uk/data/files/Amazing_Bats_2015.pdf).

12.3.1.2.6   Fish

Many native freshwater fish are present in UK, Salmon being the largest one. Shark (saltwater fish) is also found. Trout, Perch, Pike, Roach, Salmon, Dace, and Grayling are also found in lakes and rivers. The Checklist of British freshwater fishes published by BRC lists 82 species. It details the scientific and vernacular names of all fish species that have been recorded from freshwater as well as brackish water habitats in UK. For details see: http://www.nhm.ac.uk/our-science/data/uk-species/checklists/NHMSYS0000544284/index.html.

The list of fauna is endless and we have covered only a few examples here (see Figure 12.4a–f):

a   ***Sciurus vulgaris:*** Red Squirrel, once was common across the UK, have undergone a decline mainly due to competition for food and living space by the invasive non-native, American Grey squirrel. An estimated only 160,000 now remain in the UK, maximum of which are in Scotland (https://scottishwildlifetrust.org.uk/our-work/our-projects/saving-scotlands-red-squirrels/). Conservation efforts by various bodies are ongoing.

b   ***Sorex minutus***: Pygmy Shrew is one of the smallest mammals in the world.

c   ***Phengaris arion:*** The Large Blue butterfly became extinct here c. 1979 and has been the subject of a successful reintroduction programme.

d   ***Erebia aethiops:*** The Scotch argus like several of our butterflies has distinct races one of which is endemic.

e ***Stethophyma grossum:*** The large Marsh Grasshopper – is the biggest British grasshopper species. It has a distinct greenish-yellow or olive-brown coloration, with a red flush underneath the femurs and black and yellow bands on the tibia of the hind legs and generally found in marshy places, prime habitats being acid bogs with clumps of grass (http://sounds.bl.uk/Environment/British-wildlife-recordings/022M-W1CDR0001382-0200V0).

f ***Dolomedes fimbriatus:*** Raft Spider (derives its name from its ability to walk on water), is a widespread European spider and one of the two largest spiders found in UK.

12.3.2   Flora of UK

As per the *New Atlas of the British and Irish Flora* (2002), the numbers of species in various groups of Vascular plants (comprising of flowering plants, conifers, ferns, horsetails and club mosses) such as native species (which arrived in the UK without human intervention), archaeophytes (naturalized in the UK prior to AD 1500), neophytes (naturalized since AD 1500) and casuals (non- native and not naturalized) are: Natives (1407), Archaeophytes (149), Neophytes (1155), and Casuals (240) leading to a total of 2951 number of species.

In addition to this, there are approximately 1034 native species of bryophytes which include 746 species of Mosses, 284 species of Liverworts and 4 species of Hornworts.

There is a fully comprehensive checklist of the British and Irish Bryophytes published by the British Bryological Society, which includes common names, accepted names, synonyms by Blockeel and Long (1998) listing 1194 species.

The British Lichen Society maintains the list of British Isles List of Lichens and Lichenicolous Fungi and currently 2,390 species are listed. Prior to this, Checklist of British Lichen-Forming, Lichenicolous and Allied Fungi was published by Hawksworth et al., in 1980 listing 1701 species, distributed through 294 genera. Of these species 1471 are lichen-forming, 183 lichenicolous, and 47 allied fungi. They made considerable changes in nomenclature for the first time and listed one new genus (*Herteliana*) and 56 new combinations.

Important sources to check the number of terrestrial and freshwater algae is *A Coded List of Freshwater Algae of the British Isles* (contains the accepted scientific names, thorough synonymy and a taxonomic hierarchy for freshwater and terrestrial algae in the British Isles, see http://www.nhm.

ac.uk/our-science/data/uk-species/checklists/NHMSYS0000591449/index. html), which lists over 5000 species of algae in the UK. BSBI maintains a checklist of charophytes. The Natural History Museum (NHM) also houses one of the world's largest collections of Algae with over 300,000 specimens including around 9,810 type specimens and around 6,000 species. Of particular note are Adam Buddle's algal specimens collected from UK shores in the 1690s (for details, see http://www.nhm.ac.uk/our-science/collections/ botany-collections/algae-collections.html). See also, Seaweed collections online (data from national and regional museums) by Wilbraham and Brodie (2013) and a revised checklist of the seaweeds of Britain produced by Brodie et al., in 2016.

Approximately, 2800 species of freshwater diatoms have been recorded from Britain and Ireland (http://naturalhistory.museumwales.ac.uk/diatoms/).

There are collections at NHM (http://www.nhm.ac.uk/our-science/collections/botany-collections/diatom-collections.html) and at other Museums in the UK, see a volunteers based project recently undertaken at NHM (http://diatoms.myspecies.info/, Yesilyurt et al., 2016).

The Dictionary of UK species contains over 200 checklists of species provided by a variety of government, conservation and scientific experts and organizations (see http://www.nhm.ac.uk/our-science/data/uk-species/ checklists/index.html).

The comital flora of the British Isles (Mora Comitalis Britannicae: Fl. Com. Brit) published by Druce in 1932 gave a comprehensive account of the distribution of British (including a number of non-indigenous) plants throughout the 152 vice-counties of Great Britain, Ireland and the Channel Islands, with the place of growth, elevation, world-distribution, grade, chief synonyms and first names by which the plants were recorded as British. This book also has an original colored map showing the botanical vice-counties presented by William James Patey. Later on in 2003, Stace et al., compiled *The Vice-county Census Catalogue of the Vascular Plants of Great Britain* which published by the Botanical Society of the British Isles giving a complete picture of the vice-county distribution of vascular plant species in Great Britain, the Isle of Man, and the Channel Islands (an online version of the catalogue can be viewed through: http://www.botanicalkeys.co.uk/flora/vccc/index.html).

At the recent plenary session of the Linnean Society of London in September, it was pointed out that the first Atlas of British Flora which was published in 1962 had around 1,500,000 records, and now in 2017, there are over 40 million records.

Below are a few examples of endemic and threatened flora (Figure 12.5a–f):

a   ***Drosera x eloisiana:*** A very rare hybrid between *Drosera rotundifolia* and *D. intermedia* – almost restricted to a single bog complex, Godlingston Heath, Dorset.

b   ***Primula scotica*** Hook: One of the very few sexual flowering plant endemics found only on the coast of N. Scotland and the islands north of there (Orkneys, Shetlands).

c   ***Vandenboschia speciosa*** G. Kunkel (= ***Trichomanes speciosum***): An example of the rich Atlantic flora to be found in the British Isles, many species showing very disjunct sometimes Neotropical affinities. This species has been protected in British law since 1976 and is on the EU Habitats directive and the Bern Convention.

d   ***Ditrichum plumbicola*** Crundw: A near endemic moss restricted to lead rich soils. First described in the 1970s this has since been found in Ireland and Germany on old mine workings. It was a subject of the UK BAP. Its sites continue to be lost as no new lead-rich spoil is produced and old sites become naturally ameliorated or deliberately remediated.

e   ***Sorbus cuneifolia*** T.C.G. Rich: An example of an endemic apomictic (micro) species. These have generally formed through hybridization and most are very narrow endemics. This one is just known from a few km of limestone crags near Llangollen in N. Wales.

f   ***Tephroseris integrifolia*** **subsp.** ***maritimus*** (Syme) B. Nord: Endemic to coastal heath and maritime cliffs near South Stack, Anglesey.

12.3.3   *Invasive Non-Native Species*

The Non-Native Species Secretariat (NNSS) is responsible for helping to coordinate the approach to invasive non-native species in Great Britain (see http://www.nonnativespecies.org/home/index.cfm). This online resource is useful to find out about the various invasive species currently in UK, how to identify the main problem species, how to tackle these problems and the current projects on invasive species taking place across UK. At the moment, invasive garden ant – *Lasius neglectus*, Water Primrose – *Ludwigia grandiflora*, Quagga Mussel – *Dreissena rostriformis bugensis*, Asian hornet – *Vespa velutina,* Killer shrimps – *D. villosus, and D. haemobaphes,* Carpet Sea-squirt – *Didemnum vexillum* are listed in species alerts on its website, for details, see (http://www.nonnativespecies.org/alerts/index.cfm).

Figure 12.5   Some examples of endemic and threatened flora of UK (Images credit: Fred Rumsey).

a. ***Drosera x eloisiana***

b. **Primula scotica** Hook

c. ***Vandenboschia speciosa*** G. Kunkel

d. ***Ditrichum plumbicola*** Crundw

e. ***Sorbus cuneifolia*** T.C.G. Rich

f. ***Tephroseris integrifolia subsp. maritimus*** (Syme) B. Nord.

The EU Invasive Alien Species Regulation—Frequently Asked Questions (updated July 2017)—is a useful document addressing questions for Pet owners, Pet shops, Animal businesses, Zoos and wildlife parks, Horticulture trade, Gardeners, Landowners, Crayfish Operators and what happens on leaving the European Union (see http://www.nonnativespecies.org/index.cfm?sectionid=7).

The EU Invasive Alien Species (IAS) Regulation (1143/2014) came into force on 1 January 2015. The Regulation imposes strict restrictions on a list of species known as "species of Union concern" (these are species whose potential adverse impacts across the European Union are such that concerted action across Europe is required).

The first list of 37 species—23 animals and 14 plants—was approved at a meeting of EU Member States in December 2015. Subsequently, the European Commission published the implementing regulation (2016/1141) in an Official Journal on 14 July 2016. The list came into force on 3 August 2016.

There are currently 23 animals and 14 plants on the list of Union concern (Table 12.1).

Recently, on 13 July 2017, the European Commission published Commission Implementing Regulation 2017/1263 adding a further 12 species to the list of species of Union concern as below:

- *Animals*
  - Egyptian goose *Alopochen aegyptiacus*
  - Raccoon dog *Nyctereutes procyonoides* and
  - Muskrat *Ondatra zibethicus*
- *Plants*
  - Alligator weed *Alternanthera philoxeroides*
  - Milkweed *Asclepias syriaca*
  - Nuttall's waterweed *Elodea nuttallii*
  - Chilean rhubarb *Gunnera tinctoria*
  - Giant hogweed *Heracleum mantegazzianum*
  - Himalayan balsam *Impatiens glandulifera** (Figure 12.6) [*This plant was originally introduced from India to UK as an ornamental garden plant but has spread rapidly in some parts of UK and is considered a noxious plant (Prakash, 2016).]
  - Japanese stiltgrass *Microstegium vimineum*
  - Broadleaf watermilfoil *Myriophyllum heterophyllum*
  - Crimson fountain grass *Pennisetum setaceum*

See some of the important sources of invasive species maps in Appendix 2.

Table 12.1   Species List of Union Concern

Category	Species
Plants	American skunk cabbage *Lysichiton americanus*
	Asiatic tearthumb *Persicaria perfoliata (Polygonum perfoliatum)*
	Curly waterweed *Lagarosiphon major*
	Eastern Baccharis *Baccharis halimifolia*
	Floating pennywort *Hydrocotyle ranunculoides*
	Floating primrose willow *Ludwigia peploides*
	Green cabomba *Cabomba caroliniana*
	Kudzu vine *Pueraria lobata*
	Parrot's feather *Myriophyllum aquaticum*
	Persian hogweed *Heracleum persicum*
	Sosnowski's hogweed *Heracleum sosnowskyi*
	Water hyacinth *Eichhornia crassipes*
	Water primrose *Ludwigia grandiflora*
	Whitetop weed *Parthenium hysterophorus*
Animals	Amur sleeper *Perccottus glenii*
	Asian hornet *Vespa velutina*
	Chinese mitten crab *Eriocheir sinensis*
	Coypu *Myocastor coypus*
	Fox squirrel *Sciurus niger*
	Grey squirrel *Sciurus carolinensis*
	Indian house crow *Corvus splendens*
	Marbled crayfish *Procambarus fallax f. virginalis*
	Muntjac deer *Muntiacus reevesi*
	North American bullfrog *Lithobates (Rana) catesbeianus*
	Pallas's squirrel *Callosciurus erythraeus*
	Raccoon *Procyon lotor*
	Red swamp crayfish *Procambarus clarkia*
	Red-eared terrapin/slider *Trachemys scripta* (this includes all sub-species of *Trachemys scripta,* e.g., yellow-bellied slider, red-eared slider, Cumberland slider, slider and common slider)
	Ruddy duck *Oxyura jamaicensis*
	Sacred ibis *Threskiornis aethiopicus*
	Siberian chipmunk *Tamias sibiricus*
	Signal crayfish *Pacifastacus leniusculus*
	Small Asian mongoose *Herpestes javanicus*

Table 12.1   *(Continued)*

Category	Species
	South American coati *Nasua nasua*
	Spiny-cheek crayfish *Orconectes limosus*
	Topmouth gudgeon *Pseudorasbora parva*
	Virile crayfish *Orconectes virilis*

Source: The EU Invasive Alien Species Regulation – Frequently Asked Questions, July 2017.

Figure 12.6   *Impatiens glandulifera* Royle (Image credit: Fred Rumsey).

12.4 Crop Wild Relatives (CWR) Project Review

Climate change will be one of the main drivers predicted to cause a substantial decline of agricultural production in the next few years. Funded by the Norwegian Government for a ten-year initiative and led by the Global Crop Diversity Trust (Crop Trust), The CWR Project is a partnership with Kew's Millennium Seed Bank in collaboration with national and international agricultural research institutes aimed at collecting, conserving and utilizing the wild relatives of food crops. This is a global effort to conserve crop wild relatives. It is envisaged that these wild plants contain essential traits that could be bred into crops to make them more hardy and adaptable to different climatic conditions expected in the coming years (see an interview with Dr. Sandy Knapp at https://vimeo.com/116740177).

Although, Kew has not compiled any guides for UK Crop Wild relatives (pers. Comm. Richard Allen, Kew Gardens, 2017), Fielder et al. (2015) have published an article on Enhancing the Conservation of Wild Crop Relatives in England. They mention that at the moment there is not only any provision for long-term conservation *in situ,* but also comprehensive *ex situ* collection and storage of CWR is lacking too. Fielder et al. (2015) further identified a series of measures aimed at enhancing the conservation of English CRW and provided an inventory of 148 priority English CWR, highlighted spots of CRW diversity in sites including as far as the Lizard Peninsula, the Dorset coast and Cambridgeshire. They suggested sites for the establishment of network of genetic reserves and also identified individual *in situ* and *ex situ* priorities for each English CWR. Based on their findings, they have recommended implementations for an effective long-term conservation of English CWR.

Fielder et al. (2015) have pointed that the conservation of 148 CWR identified as a priority in the context of food security (derived in consultation with Natural England) is currently incomplete and point out areas for further study such as ensuring comprehensive conservation of all other valuable plant-derived supplies and their wild relatives. Fielder et al. (2015) further point out that the methodology described in their paper is applicable to all types of CWR (not only those with a role in improving food security) and can be used to achieve comprehensive coverage of all wild relatives.

12.4.1 CWR Inventory

As per Fielder et al. (2015), the English national inventory of priority CWR has 148 taxa (126 species and 22 subspecies), which represents 10% of the taxa listed in the English CWR checklist (Full inventory at the Plant Genetic Resources Diversity Gateway can be seen at: http://pgrdiversity.bioversity-international.org.). Of the 148 priorities CWR, 76% are related to food crops whilst the remaining 24% are related only to forage or fodder crops. Thirteen plant families, with Poaceae, Brassicaceae, and Fabaceae contain the most genera (16, 7, and 7). Fielder et al. (2015) have pointed that genera with the highest taxon richness are *Trifolium* L. (clovers, 18 taxa), *Vicia* L. (vetches, 12 taxa), and *Chenopodium* L. (goosefoots, 11 taxa).

12.5   National Laws and International Obligations

The UK's habitats are subject to various protective measures and a number of UK laws, international conventions and European Directives apply to them including The Convention on the Conservation of European Wildlife and Natural Habitats 1979, The Wildlife and Countryside Act 1981, The EC Habitats Directive 1992 and the Convention on Biological Diversity 1993 to name a few. The UK has commitments to meet international environmental goals, such as those in the Convention on Biological Diversity's Aichi Targets and the United Nation's Sustainable Development Goals. However, the findings of the report (Hayhow et al., 2016) suggest that UK is not on course to meet the Aichi 2020 targets, and that much more action needs to be taken towards the 2030 Agenda for Sustainable Development if UK is to meet the Sustainable Development Goals.

12.6   Conclusion

Basic taxonomy is vital to what we have in the environment. Increasingly over the years, it has been noticed that taxonomy and taxonomic expertise is slowly eroding and often projects which are funded for a limited period of time, very few manage to get the projects extended or funded twice by various funding bodies and then it becomes arduous to maintain the sustainability which is the need of the hour if we can safeguard our nature for future generations. The biological recording has a long and rich history in UK. In UK and the OTs, although, there are numerous organizations (see some in Appendix 1) who are recording, researching and are involved in conservation; we must not forget the majority of biological recording is by volunteers. The question worth asking is how many of the organizations work as independent silos and how many of them come together and share their data? No doubt, all these bodies have a common bridge that joins them together;

i.e. nature and the future is all about open data; won't it help immensely if all these organizations are joined up under a common platform? For example at the recent plenary session on 7th September 2017, as pointed by one of the speakers at the Linnean Society, *Bellis perennis* when searched in BSBI (http://bsbi.org/maps?taxonid=2cd4p9h.xbs) and NBN (https://species. nbnatlas.org/species/NBNSYS0000004445), retrieved a different number of records. This example highlighted that new solutions must be found to fund the aspirations of public, society members, various recording bodies and the government.

Although it is seen that increasingly, Apps attract more novice readers, there are new monitoring initiatives, eDNA, citizen science surveys, open data, field and identification guides, molecular techniques, the biggest threat to all these is lack of funding, not new methods!

Fox (2003: 204) has pointed out that the potential effects of global warming will heighten problems for many species. Vegetation zones and habitats might shift, many species might be unable to follow and colonize new areas. Eutrophication, drainage, intensification of agricultural methods, grazing both over in uplands and under in lowlands, changes in land use and non-native species have severely impacted the wildlife of the British Isles, this is nowhere more acute than in the most populous areas of lowland England. This is exemplified by *Drosera anglica* Huds, a carnivorous plant commonly known as Greater Sundew, once widely spread across lowland England, now, due to various reasons such as peat digging, eutrophication and modern technological advances in land drainage has resulted in irreversible habitat loss, fragmentation of populations resulting in *D. anglica* now being considered Endangered in England (BSBI, 2017).

Major challenges and questions to answer in the coming years would be effects of climate change and we all know that is a threat globally. UK is now Europe's one of the least wooded nations (Hayhow et al., 2016).

As pointed by Hayhow et al. (2016), factors such as climate change, hydrological change, urbanization, decreasing forest management (e.g., cessation of traditional management practices, coppicing) and decreasing management of other habitats (e.g., abandonment of traditional management, including grazing, burning and cutting, which is crucial for the maintenance of habitats such as heathland and grassland) which have resulted in changes to the UK's wildlife over recent decades, policy-driven agricultural change was by far the most significant driver of declines.

Another important factor is pollution (CBD, 2017, see https://www.cbd. int/countries/profile/default.shtml?country=gb#facts).

Hayhow et al. (2016) have further pointed out how well-planned conservation projects could help the future of wildlife. Kew recently provided a synthesis of current knowledge of the world's plant in State of the World Plants (2017, see https://stateoftheworldsplants.com/) and points that more than one-fifth of the World's plant species are now threatened with extinction.

An integrated approach between governments, non-governmental organizations, businesses, communities and individuals to safeguard nature will always be required and volunteers will continue to play an important role in conservation efforts helping to monitor and protect the UK's wildlife.

Keywords

• crop wild relatives   • flora

References

Anderson, R., (2005). An annotated list of the non-marine molluscs of Britain and Ireland. *J. Conchology.*, *38*(6), 607–637.

Asher, J., Warren, M., Fox, R., Harding, P., Jeffcoate, G., & Jeffcoate, S., (2001). *The Millennium Atlas of Butterflies in Britain and Ireland*. Oxford University Press.

Ball, S., (2015). *Britain's Hoverflies: A Field Guide*. Princeton, NJ, Princeton University Press. https://muse.jhu.edu/books/9781400866021/.

Bee, L., Oxford, G., & Smith, H., (2017). *Britain's Spiders, a Field Guide*. Princeton University Press.

Blockeel, T. L., & Long, D. G., (1998). *A Check-List and Census Catalogue of British and Irish Bryophytes*. British Bryological Society.

Brodie, J., Wilbraham, J., Pottas, J., & Guiry, M. D., (2016). A revised check-list of the seaweeds of Britain. *J. Marine Biol. Association of the United Kingdom, 96*(05), 1005–1029. doi: 10.1017/S0025315415001484.

BSBI. *A Vascular Plant Red List for England*, (2017), available from:http://bsbi.org/england.

Carreck, N., (2016). *Field Guide to the Bees of Great Britain and Ireland*, By Stephen Falk. Illustrated by Richard Lewington.

Chinery, M., (1977). *A Field Guide to the Insects of Britain and Northern Europe*. London, Collins.

Druce, G. C., (1932). *The Comital Flora of the British Isles: (Flora Comitalis Britannicae: Fl. Com. Brit.): Being the Distribution of British (Including a Number of Non-Indigenous) Plants Throughout the 152 Vice-Counties of Great Britain, Ireland, and the Channel Islands, with the Place of Growth, Elevation, World-Distribution, Grade, Chief Synonyms, and First Names by Which the Plants Were Recorded as British*. Buncle.

Emmet, A. M., Heath, J., et al., (1990). The butterflies of great Britain and Ireland. *The Moths and Butterflies of Great Britain and Ireland*, Part 1 (Hesperiidae to Nymphalidae). Harley Books, Colchester, UK, *vol. 7*.

Fielder, H., Brotherton, P., Hosking, J., Hopkins, J. J., Ford-Lloyd, B., & Maxted, N., (2015). Enhancing the conservation of crop wild relatives in England. *PLoS ONE, 10*(6), e0130804. https://doi.org/10.1371/journal.pone.0130804.

Fox, R., (2003). *Butterflies and Moths*, (Hawskworth, D., ed.). The changing wildlife of great Britain and Ireland. Taylor and Francis Limited. London.

Hawksworth, D. L., James, P. W., & Coppins, B. J., (1980). Checklist of British lichen-forming, lichenicolous and allied fungi. *The Lichenologist, 12*(1), 1–115.

Hawsworth, D. L., (2003). *The Changing Wildlife of Great Britain and Ireland*. Taylor and Francis Limited. London.

Hayhow, D. B., Burns, F., Eaton, M. A., Al Fulaij, N., August, T. A., Babey, L., Bacon, L., Bingham, C., Boswell, J., Boughey, K. L., Brereton, T., Brookman, E., Brooks, D. R., Bullock, D. J., Burke, O., Collis, M., Corbet, L., Cornish, N., De Massimi, S., Densham, J., Dunn, E., Elliott, S., Gent, T., Godber, J., Hamilton, S., Havery, S., Hawkins, S., Henney, J., Holmes, K., Hutchinson, N., Isaac, N. J. B., Johns, D., Macadam, C. R., Mathews, F., Nicolet, P., Noble, D. G., Outhwaite, C. L., Powney, G. D., Richardson, P., Roy, D. B., Sims, D., Smart, S., Stevenson, K., Stroud, R. A., Walker, K. J., Webb, J. R., Webb, T. J., Wynde, R., & Gregory, R. D., (2016). State of Nature 2016. *The State of Nature Partnership*. Online, https://ww2.rspb.org.uk/globalassets/downloads/documents/conservation-projects/state-of-nature/state-of-nature-uk-report-.pdf, accessed 08/09/2017.

Macadam, C., & Bennett, C., (2010). *A Pictorial Guide to British Ephemeroptera*. FSC publications.

Merrett, P., & Murphy, J. A., (2000). A revised checklist of British spiders. *Bull. British Arachnological Soc., 11*(9), 345–358.

Prakash, R. O., (2016). Wallich and his contribution to the Indian natural history. *Rheedea, 26*(1), 13–20.

Ratcliffe, P. R., (1987). *Distribution and Current Status of Sika Deer, Cervus Nippon, in Great Britain*. Mammal review, DOI: 10. 1111/j. 1365–2907. 1987. tb00047. x.

Roy, H., Brown, P., Frost, R., & Poland, R., (2011). *Ladybirds (Coccinellidae) of Britain and Ireland*. Natural Environment Research Council (NERC).

Royal Entomological Society of London, (1954). *Handbooks for the Identification of British Insects*.

Stace, C. A., Ellis, R. G., Kent, D. H., & McCosh, D. J., (2003). *Vice-County Census Catalogue of the Vascular Plants of Great Britain, the Isle of Man and the Channel Islands*, London: BSBI.

Stubbs, A. E., & Drake, C. M., (2014). *British Soldier flies and Their Allies: An Illustrated Guide to Their Identification and Ecology*. BENHS, Reading.

Stubbs, A. E., (1996). *British Hoverflies: An Illustrated Identification Guide*. Second (revised and enlarged) supplement. London, British entomological & natural history society.

Thomas, J., & Lewington, R., (2014). *The Butterflies of Britain & Ireland*. British Wildlife Publishing.

Tomlinson, D., & Still, R., (2002). *Britain's Butterflies*. WildGuides, Old Basing, UK.

Wilbraham, J., & Brodie, J., (2013). Seaweed Collections Online: Mobilizing data from national and regional museums. *SPNHC, 246* 6–246.

Yesilyurt, J. C., Thomas, A. L., Cesar, E. A., Broom, Y. S., Bhatia, R., & Miller, R., (2016). A tangible embrace with the invisible: How a curator can achieve collections goals in partnership with volunteers and the public. *JoNSC, 4, NatSCA.*, pp. 12–21.

Online Sources (all accessed between 1 July 2017 to 13 September 2017):

http://news.bbc.co.uk/1/hi/world/europe/country_profiles/6233450.stm.

http://sounds.bl.uk/Environment/British-wildlife-recordings/022M-W1CDR0001382–0200V0.

http://www.atlapedia.com/online/countries/unitedki.htm.

http://www.bats.org.uk/data/files/Amazing_Bats_2015.pdf.

http://www.bennevisweather.co.uk/index.asp.

https://www.cbd.int/countries/profile/default.shtml?country=gb#facts.

http://diatoms.myspecies.info/.

https://www.eurotunnel.com/uk/build/.

https://www.gov.uk/government/policies/uk-overseas-territories.

http://www.independent.co.uk/travel/uk/the-complete-guide-to-the-scottish-islands-633851.html.

http://www.justice.gov.uk/downloads/about/moj/our-responsibilities/Background_Briefing_on_the_Crown_Dependencies2.pdf.

https://www.metoffice.gov.uk/public/weather/climate.

http://www.nationalparks.gov.uk/.

http://www.nationsencyclopedia.com/Europe/United-Kingdom-FLORA-AND-FAUNA.html#ixzz4rvhlz1Sj.

http://www.nhm.ac.uk/our-science/data/uk-species/checklists/NBNSYS0000000020/index.html.

http://www.nhm.ac.uk/our-science/data/uk-species/checklists/NHMSYS0000527885/index.html.

http://naturalhistory.museumwales.ac.uk/diatoms/.

https://www.nationaltrust.org.uk/wasdale/features/enjoying-englands-highest-mountain--scafell-pike-in-wasdale.

https://www.ons.gov.uk/peoplepopulationandcommunity/populationandmigration/populationestimates.

https://scottishwildlifetrust.org.uk/our-work/our-projects/saving-scotlands-red-squirrels/.

https://stateoftheworldsplants.com/.

Appendices

Appendix 1 A List of Some of the Organizations Involved in the UK and Its Overseas Territories (OTs) in Recording, Researching, and Conservation of Nature (Adapted from Hayhow et al., 2016)

- A Focus on Nature, afocusonnature.org
- A Rocha, arocha.org.uk
- Amphibian and Reptile Conservation (ARC), arc-trust.org
- Association of Local Environmental Records Centers (ALERC), alerc.org.uk
- Bat Conservation Trust (BCT), bats.org.uk
- Biological Records Centre (BRC), brc.ac.uk
- Botanical Society of Britain and Ireland, bsbi.org
- British Bryological Society (BBS), britishbryologicalsociety.org.uk
- British Dragonfly Society (BDS), british-dragonflies.org.uk
- British Lichen Society, britishlichensociety.org.uk
- British Pteridological Society (BPS), ebps.org.uk
- British Trust for Ornithology (BTO), bto.org
- Buglife, buglife.org.uk
- Bumblebee Conservation Trust, bumblebeeconservation.org
- Butterfly Conservation, butterfly-conservation.org
- Centre for Ecology & Hydrology (CEH), ceh.ac.uk
- Chartered Institute of Ecology and Environmental Management (CIEEM), cieem.net
- Conchological Society of Great Britain and Ireland, conchsoc.org
- Durrell Wildlife Conservation Trust (Durrell), durrell.org
- Earthwatch, eu.earthwatch.org
- Freshwater Habitats Trust, freshwaterhabitats.org.uk
- Friends of the Earth, foe.co.uk
- Froglife, froglife.org
- Fungus Conservation Trust, abfg.org
- iSpot, ispotnature.org
- Jersey Government Department of the Environment, gov.je/Government/Departments/PlanningEnvironment
- Mammal Society, mammal.org.uk
- Manx BirdLife, manxbirdlife.im

- Marine Biological Association (MBA), mba.ac.uk
- MARINELife, marine-life.org.uk
- Marine Conservation Society, mcsuk.org
- Marine Ecosystems Research Programme, marine-ecosystems.org.uk
- National Biodiversity Network (NBN), data.nbn.org.uk
- National Forum for Biological Recording, nfbr.org.uk
- National Trust, nationaltrust.org.uk
- Natural History Museum, nhm.ac.uk
- ORCA, orcaweb.org.uk
- People's Trust for Endangered Species (PTES), ptes.org
- Plantlife, plantlife.org.uk
- PREDICTS, predicts.org.uk
- Rothamsted Research, rothamsted.ac.uk
- Royal Society for the Protection of Birds (RSPB), rspb.org.uk
- Shark Trust, sharktrust.org
- States of Guernsey, gov.gg
- Sir Alister Hardy Foundation for Ocean Science (SAHFOS), sahfos.ac.uk
- University of Sheffield, sheffield.ac.uk
- Vincent Wildlife Trust, vwt.org.uk
- Whale and Dolphin Conservation (WDC), uk.whales.org
- Wildfowl & Wetlands Trust (WWT), wwt.org.uk
- Wildlife Trusts, wildlifetrusts.org
- Woodland Trust, woodlandtrust.org.uk
- WWF, wwf.org.uk
- Zoological Society of London (ZSL), zsl.org

Appendix 2   Sources of Nonnative Invasive Species

1   European Alien Species Information Network – Species Mapper (http://alien.jrc.ec.europa.eu/SpeciesMapper)
2   The NBN Atlas (http://www.gov.scot/topics/marine)
3   National Biodiversity Data Centre (http://maps.biodiversityireland.ie/)
4   Centre for Environmental Data and Recording (CEDaR) (http://www2.habitas.org.uk/records/maps)
5   iRecord (Botanical Records Centre) (https://www.brc.ac.uk/irecord/)
6   British Society of Britain & Ireland (BSBI) Plant Atlas (Botanical Records Centre) (http://www.brc.ac.uk/plantatlas/)

Index

A

Printed and bound by CPI Group (UK) Ltd, Croydon, CR0 4YY

23/10/2024

01777705-0014